SCIENCE, TECHNOLOGY, AND NATIONAL POLICY

Science, Technology, and National Policy

EDITED BY

Thomas J. Kuehn

Alan L. Porter

CORNELL UNIVERSITY PRESS
ITHACA AND LONDON

Copyright © 1981 by Cornell University

All rights reserved. Except for brief quotations in a review, this book, or parts thereof, must not be reproduced in any form without permission in writing from the publisher. For information address Cornell University Press, 124 Roberts Place, Ithaca, New York 14850.

First published 1981 by Cornell University Press.
Published in the United Kingdom by Cornell University Press Ltd.,
Ely House, 37 Dover Street, London W1X 4HQ.

International Standard Book Number 0-8014-1343-5 (cloth)
International Standard Book Number 0-8014-9876-7 (paper)
Library of Congress Catalog Card Number 80-66900
Printed in the United States of America
*Librarians: Library of Congress Cataloging information appears
on the last page of the book.*

We dedicate this book to Edward Wenk, Jr.,
and Dael Wolfle in recognition of their
contributions to science, technology, and
public affairs, as well as to the
education of the editors.

Contents

Preface ... 9
Introduction: The Agenda for Science and Technology Policy ... 11

Part I. The Social Context of Science and Technology

A. The Society ... 33
   1. Technology, Evolution, and Purpose ... 35
      Harvey Brooks
   2. How Technology Will Shape the Future ... 57
      Emmanuel G. Mesthene
   3. The Artificial Environment: Disneyland Is Better ... 80
      Theodore Roszak

B. The Polity ... 94
   4. The Spectrum from Truth to Power ... 95
      Don K. Price
   5. The Scientist and the Politician ... 132
      Roger Revelle

C. The Economy ... 147
   6. The National Climate for Technological Innovation ... 148
      Donald Schon
   7. Trends in U.S. Technology: A Political Economist's View ... 161
      Michael Boretsky

D. World Affairs ... 189
   8. Military Research and Development: A Postwar History ... 190
      Herbert F. York and G. Allen Greb
   9. The Transfer of Technology to Underdeveloped Countries ... 215
      Gunnar Myrdal
   10. The International Functional Implications of Future Technology ... 226
      Eugene B. Skolnikoff

Contents

Part II. Science, Technology, and American Government

E. The Federal Executive — 249
  11. Organization for Science and Technology in the Executive Branch — 251
    American Association for the Advancement of Science
  12. The Scientific Advisory System: Some Observations — 262
    Martin L. Perl
  13. Office of Management and Budget: Skeptical View of Scientific Advice — 274
    John Walsh and Barbara Culliton

F. The Congress — 296
  14. Congress and the Science Budget — 297
    Herbert Roback
  15. Scientists on the Hill — 315
    Barry M. Casper
  16. The Rhetoric and Reality of Congressional Technology Assessment — 327
    Barry M. Casper

G. The Courts — 346
  17. Towards a New Technological Ethic: The Role of Legal Liability — 347
    Laurence H. Tribe
  18. Risk and Responsibility — 356
    David L. Bazelon

H. State and Local Government — 366
  19. Science Policy in American State Government — 367
    Harvey M. Sapolsky
  20. Federal Technology Policy: Innovation and Problem Solving in State and Local Governments — 397
    J. David Roessner

I. Citizen Participation — 415
  21. Participatory Technology — 416
    James D. Carroll
  22. Information and the New Movements for Citizen Participation — 434
    Hazel Henderson

J. Public Choice — 449
  23. Technology Assessment and Social Control — 451
    Michael S. Baram
  24. The Politics of Technology Assessment — 474
    Mark R. Berg
  25. Science and the Formation of Policy in a Democracy — 496
    Duncan MacRae, Jr.

Annotated Bibliography — 515
Index — 525

Preface

This book explores the relationships among science, technology, society, and policy makers. The readings we selected focus primarily on technology policy rather than on science policy, although the line between the two is not always definable. More specifically, the book looks at technological change in relation to society and to American government. Thus, the readings are oriented toward understanding the practical application of science and technology in order to expand the realm of human possibility. Part I provides some insight into the underlying social, political, economic, and international concerns in science and technology policy. Part II is an inquiry into the ways that various government institutions—the federal executive, Congress, courts, and state and local governments—deal with science and technology, as well as aspects of citizen participation in the process. To cover this territory, we can only sample the rich and diverse literature with no pretense of being comprehensive. Our intent is to provide a broad perspective, to illustrate a wide spectrum of important concerns in science and technology policy, and to stimulate critical thinking. We urge the interested reader to pursue additional literature, suggested in the annotated bibliography, which contains a rich reservoir of ideas that could not be captured herein.

Who do we perceive to be our audience? First, we are addressing upper-division or graduate-level students. Second, we hope that this compilation of classic and timely readings is a convenient source for those engaged in technology-intensive policy processes, including professionals, decision makers, and other interested parties. We hope to engage these students and practitioners in a dialogue with those who speak to the issues from different vantage points. For engineers and scientists, we hope to highlight the potent impact of their work on society and the processes by which society strives to manage it. For political scientists, economists, and other social scientists, we hope to point out some ways in which technology affects policymaking and

Preface

institutions. We feel that these readings could serve as a textbook for a course in science and technology policy. They could also be used to complement a wide range of courses in science, technology, and society; engineering; political science; public administration; and public policy.

This book has its own history. It began in 1973 in conjunction with the Program in Social Management of Technology, directed by Edward Wenk, Jr., at the University of Washington for the College of Engineering and the Graduate School of Public Affairs. The editors and Steven Flajser (currently with the Senate Committee on Commerce, Science, and Transportation) sought out appropriate readings. We benefited from the guidance of Ed Wenk, and also of colleagues Dael Wolfle, Kai N. Lee, Brewster Denny, and Fremont Lyden, who reviewed early efforts and suggested appropriate directions. Since that time, we have made several more literature searches and have had to make some agonizing choices from among the abundant literature in the field. Our choices were often based on a desire to represent a particular point of view or approach. Cornell University Press editors James Twiggs and Daniel Snodderly have fostered the process since 1975.

We would like to thank our editors and colleagues associated with the Program in Social Management of Technology, just noted, as well as Melvin Kranzberg of the Georgia Institute of Technology for his guidance, and our anonymous reviewers for their helpful suggestions. Above all, we thank the authors for their insights and their publishers for allowing us to include the selections herein.

THOMAS J. KUEHN
ALAN L. PORTER

Washington, D.C.

Introduction: The Agenda for Science and Technology Policy

> Perfection of means and confusion of goals seem, in my opinion, to characterize our age.
> —Albert Einstein
>
> The true purpose of knowledge lies in the consequences of directed action.
> —John Dewey

For the last century, we have witnessed unprecedented technological and economic achievements and, paradoxically, the great devastations of world wars and poverty. While the engines of material well-being and warfare have been perfected, society's institutions have not been successful at reconciling the conflicting uses of technology for the purposes of peace and war. It may be that scientific progress and technological innovation are not enough. If we are to manage our technological and social actions wisely, we also need to improve the integration of scientific and technical knowledge with public policy. This book of readings is concerned with the use of technology to meet social goals—which we view as the primary purpose of technology policymaking.

In the realm of national policymaking, the present rift between technology's means and goals may be the result of a narrow view of technology. A preoccupation with the material products of industry may have overwhelmed the human attributes of technology. Perhaps the definition of technology must be broadened to embody both the problems and the solutions—to focus on both the ends and the means. Bernard Gendron defines technology as "any systematized practical knowledge based on experimentation and/or scientific theory, which enhances the capacity of a society to produce goods and services and which is embodied in productive skills, organization, or machinery." In the same vein, Emmanuel Mesthene has defined technology as the

Introduction

organization of knowledge for the achievement of practical purposes because "it is in this broader meaning that we can best see the extent and variety of the effects of technology on our institutions and values." Both these definitions emphasize human organization, institutions, and values and encompass technological means and ends. The ability to put ends and means together, to organize and make collective decisions about technological change, is the essence of technology and public choice.

The application of technology to satisfy human needs is a lofty and deceptively simple statement. The acquisition, control, invocation, or use of scientific knowledge and technology for any public purpose is a political act that is bound to conflict with many self-interested views of the world. This conflict between technology and politics is characterized by Don K. Price, writing in *The Scientific Estate,* as a spectrum between truth and power:

> The four broad functions in government and public affairs—the scientific, the professional, the administrative and the political—are by no means sharply distinguished from one another even in theory, but fall along a gradation or spectrum within our political system. At one end of the spectrum, pure science is concerned with knowledge and truth; at the other end, pure politics is concerned with power and action. But neither ever exists in its pure form.

The conflicts between science and politics are as old as the concepts of truth and power themselves, but have new meaning today because of the degree of complexity and the consequences of technological choices that face us.

When the stakes are high, political consensus is difficult to achieve. Technology means different things to different people. Technology is seen as an instrument of profit to industry—as the powerhouse behind yesterday's industrial revolution and today's multinational corporation. It is seen as an instrument of power to nation-states waging war, or waging the threat of war, to maintain or to extend national wealth or sovereignty. More recently, technology is seen by many people as an instrument of ecological self-destruction, laying waste to the landscape and diminishing humanity. Less-developed nations of the world see technology as both the hope of economic development (or even survival) and an intimidating assault on traditional values.

How, then, can technology be used more effectively to improve the human condition? This question has no easy answer. The new field of science, technology, and public policy is presently characterized by more questions than answers, more supposition than theory. But a

Introduction

systematic inquiry may help the reader place many of the selections that follow in some perspective. Assuming that there is indeed much room for improvement in technology policymaking, where are the failings of existing technology policy? There are at least four avenues that need to be explored in searching for the shortfalls in the social management of technology:

1. *Institutional malfunctions*—How well do social institutions employ technology for social good?

2. *Inadequate knowledge, research resources, and methods*—Are present theory and knowledge about technological problems and decision processes inadequate? Are present research methods appropriate for the assessment of technology?

3. *Insufficient political power*—Are scientists and technologists as an interest group adequately represented in the policymaking process? What is the political motivation for rational policymaking and the use of available scientific and technological information?

4. *Inadequate public education*—How well prepared are individual citizens to make decisions about technological change or participate in technology policymaking?

Underlying the problem of how to convert technological knowledge into effective social choice and action are still other questions: How is scientific and technological knowledge used in the political process? Can the policy analyst be objective and politically neutral? Is information an attribute of power, and if so, how can society prevent the abuse of this power? How do political and social institutions deal with science and technology? Do American institutions embody both the problem and the solution to complex technological issues?

These and other questions are discussed in the remainder of this introduction to help focus on the central issues in the readings that follow.

The Social Context of Science and Technology

The technological content of public policy has greatly increased in the past two decades. A whirlwind tour of recent history reminds us that many of the economic, cultural, and political developments since World War II have their origin in twentieth-century science and engineering. The war itself stimulated the growth of research and development under government sponsorship to forge weapons and provide the mobility that characterizes global conflict. The dramatic advances in aircraft, atomic bombs, nuclear submarines, and intercontinental missiles illustrate the potency of organized science and

Introduction

technology. These advances gave rise to the postwar military-industrial complex and a new level of expectations regarding science and technology.

The notion that "high technology" could be harnessed for explicit peaceful purposes has gained momentum in recent years with "atoms for peace," the "moon race," and the "war on poverty." Each of these developments generated progress and new wealth, but also new problems, social disruptions, and an awareness of the less desirable effects of growth on society.

The environmental and the energy crises illustrate paradoxical aspects of science and technology. Technology is the precursor of industrial and social development, while the careless use of technology has created a raft of new problems and unintended consequences. The mortgage of past achievements and progress is now past due in the form of today's environmental and energy problems. The American response to these problems is equally paradoxical. It illustrates the love-hate relationship between a society and its technology. The alienation and social dissonance characteristic of the environmental and nuclear energy protests of recent times have made an indelible impression on public policy. Yet these same "crises" have spawned new technologies for environmental measurement and protection, and development of alternative energy sources such as solar energy. Technology has helped create a more complex society that has adopted the habit of turning to technology for solutions to its problems. Yet the validity and permanence of those solutions are limited by the political and social institutions that must design, implement, and manage technological change.

The impact of technology on society has become an articulated theme in social science and philosophy. An understanding of the pertinent relationships between technology and society is essential to get to the roots of most technological issues. Many studies of innovation have cited the unanticipated effects of invention, showing that people and institutions have rarely anticipated or been prepared for all of the secondary effects of a new development. The automobile did much more than speed travel; it appears to have fundamentally affected the family, the city, business, and commerce. Electronic miniaturization fostered by the space program wrought not only a new age of information processing for science and business, but also a boom in pocket calculators and home computers.

Our institutions have never been prepared to deal systematically with rapid technological change. They are characterized more by bureaucratic inertia than as swift purveyors or capable managers of

Introduction

technological progress. Today issues such as energy, peace, and environmental quality seem more complex and our institutions harder pressed to cope with them.

Some say that meaningful public participation in policy issues has been replaced by technocratic and elitist politics. On the other hand, scientists and engineers bemoan the lack of accurate technical information in political decision making. Both of these viewpoints challenge the adequacy of democratic institutions to deal with technological issues. How can present institutions be improved? What new institutions or methods are needed?

Still others see technology as a constructive force that creates new opportunities rather than new problems. Technology leads to changes in values and society as it creates new choices and options. If this is so, how can society exploit these opportunities? How can we harvest the benefits while avoiding the undesirable consequences?

These questions permeate the political process in which budget allocations for technology R & D, the access of technologists to political power, and the influence of technical information on public decisions and popular opinion come into play. These political interrelationships present opportunity for conflict and cooperation. The politician's expertness in social values and needs can complement the knowledge of scientists and engineers. In what ways are the worlds of politics and technology conducive to mutual advising to provide solutions to problems?

Conflicts inevitably arise when knowledge contradicts popular opinion and special-interest politics. Other kinds of conflicts arise when the American government wants a "technological fix" for a problem that is essentially political, social, or organizational in nature. Often, too, technological solutions may not be neatly bounded by political jurisdictions—another source of conflict. Since the problems of energy, transportation, water resources, and communication transcend even national borders, greater cooperation and social integration may be required to manage these technological resources. In some cases, greater centralization of decision making may be required to provide uniform standards and solutions to universal problems. Conversely, some problems may require decentralized decision making so that local governments can deal directly with the problems.

Science and technology, however, have had little impact on decentralized decision making. State and local governments have not made effective use of the resources of science and technology. One can only wonder whether the federal government has preempted state and local initiative in solving technical problems such as energy and trans-

Introduction

portation. Or is scientific and technological expertise an unwanted intrusion on local sovereignty and parochial interests? What technological problems are best solved by centralized, decentralized, or hierarchical decision-making processes? Are new institutions needed to maintain the separation of powers while coordinating solutions to national problems in the federal system? Are regional centers or new cooperative institutions (for example, university-industry-government) for research and development required to promote innovation and respond to public technology needs?

The press of technological issues inevitably affects our governments and institutions. Technology may force institutions together by tightening interdependencies among government, private industry, and universities to solve complex problems, even though our political values and ideology favor separation of powers. Ironically, the desire for separation of the private and public sector, federal and state governments, and executive and legislative powers, and so forth, contradicts the basic trend of technological progress toward greater social integration. Can industrial societies develop democratic institutions for governing technology? Are there internal contradictions between political and technological ideology, and can these be reconciled? Should the resources of science and technology be more accessible to all segments of a pluralistic society?

The fundamental dilemma of technology policy may be reduced to this uncomfortable relationship between knowledge and power. In a pluralistic system of government, the means and ends of technological power can easily be confused and abused. Control of technology and technical information is as important today as the control of resources and capital that has been the basis historically for real political power. The ideal flow of "objective" scientific and technological information into the political process can be manipulated by special interests to promote private ends. Does this pose a grave challenge to democracy?

Turning from the political to the economic domains, technology has made a fundamental contribution to the American economy. Throughout the American industrial revolution, the relationship between technology, industry, and government has been both intimate and antagonistic. The role of government in supporting industrial innovation and in promoting a strong scientific and technological capability has become an important tenet of national policy, especially since World War II. The military-industrial complex has thrived on government spending for research and development of weapons and spacecraft. These developments have stimulated profitable civil appli-

Introduction

cations in computers, materials, instrumentation, jet aircraft, and weather and communication satellites. More recently, the burdens of economic regulation, archaic patent policies, competition from heavily subsidized foreign cartels, austere research and development budgets, and foreign trade restrictions may have been eroding the American technological base.

Attitudes toward federal and private roles in innovation vary widely according to political and economic outlook, ranging from laissez-faire to social welfare models. The lack of philosophical consensus in this area has resulted in a hodgepodge of conflicting purposes, regulations, incentives, and disincentives to innovation. Technology policies inevitably intersect with major economic issues such as economic regulation, productivity, and public investment in innovation.

Technology is a crucial, yet inadequately understood element in the economic system. For instance, the failure to provide technological solutions to energy production, conversion, and conservation may add heavily to inflation or recession in ways that are not easily explained by present economic models and policies. Excessive dependence on foreign oil leads to inflation and trade deficits which could be offset by energy innovations and capital investments to improve the efficiency of fuel utilization or conservation. In addition, exporting advanced energy-conversion or renewable-resource technology, such as solar, could offset the trade deficits produced by importing foreign oil. There are many economic questions related to technology policy that need to be answered: Are declines of innovation at the heart of decreasing rates of productivity? How do economic regulations affect the rate of technological innovation? Are unsatiated demands for technological consumer items outstripping the supply of energy and material resources? Can technology policies be designed to improve the use of science and technology to achieve economic goals such as conservation and reduced inflation?

Turning from economic to international affairs, technology policy intersects with economic, cultural, and political issues—especially those concerning war and peace. Technology exaggerates the gap between rich and poor nations and, as such, adds to international conflict, competition, and tensions. The development and proliferation of nuclear arsenals, the arms race between the superpowers, and increased inventories of conventional weapons can only further destabilize a world of scarce resources, rising expectations, and population growth. At the same time, advances in health and sanitation, medicine, and agriculture are vital to the survival of every nation, rich or poor. The term *appropriate technology*, or *intermediate technology*, has

Introduction

been coined to describe the need for technology that meets the transitional needs of economic and social change in underdeveloped countries. The questions of technology policy are especially acute in world affairs since much of the world's population lives on a thin margin between progress and poverty, life and death, chaos and civilization. Can the chain of poverty, population growth, ignorance, and despair be broken? Can world economic and political stability be improved by exporting appropriate technology?

The technology-intensive issues of environmental pollution, management of the oceans, health, and population control are not confined to individual nations. As such, international institutions and perhaps multinational corporations will also be severely tested in the years ahead. As the issues gain political urgency, tension may increase. Yet another dilemma for technology policy is the conflict between the need for international cooperation and the increased requirement for national sovereignty in order to control scarce technology and natural resources. Can this conflict be managed by international institutions? Will technology issues such as energy and environment become highly politicized like national security issues and become more central concerns of nationalistic governments?

Military technology development has had a dramatic impact on domestic and international affairs. Technological determinism is dramatically illustrated by the cycle of weapons and antiweapons development. U.S. defense policy is vulnerable to new technological developments as antiballistic missiles, cruise missiles, and neutron bombs dictate changes in policy. The SALT II negotiations, for instance, suffered from the uncertainties raised by new weapon capabilities. Negotiations took careful note of prototypes such as antisatellite weapons and MX and cruise missiles. In addition to the weapons of destruction, sensor technology became an important issue because it affects each nation's ability to detect cheating and verify compliance to the treaty. Spy satellites and electromagnetic intelligence networks played a vital role in the negotiation of the SALT treaty. Antisatellite weapons presented a threat to the ability to maintain adequate intelligence. The technological dynamics of weapon, antiweapon, and intelligence systems may play a very real role in shaping the future course of world armament or disarmament. Do the anticipation and fear of future military technology deter or incite conflict? How do advances in technology affect the rate and course of the arms race? How can technological dynamics be changed to slow or arrest the arms race? Are there "peace technologies" that can be developed to curb the arms race? Does advanced technology such as telecommunications

Introduction

and microromputers shift the advantage to defensive weapons? Will superior development of defensive weapons stem the tide of the strategic arms race and improve the possibilities for peace or disarmament?

Technology is important in every sphere of national and world affairs. The articles in Part I of this book explore the social, political, economic, and international dimensions of technology policy. From these perspectives, it is clear that technology is neither a blessing nor a curse; it can be used for good *and* evil. As such, technology is too important to take for granted or leave in the hands of politicians, bureaucrats, industrialists, or the scientists and technologists themselves. Society must make a deliberate effort to develop or improve institutions for the application of science and technology for social benefit. The alternative to wise management of technological resources is to leave humanity in the grips of technological determinism, with goals deriving from means. In a complex modern world, technological determinism may lead down the road of self-destruction, since there are no assurances that the negative and unanticipated consequences of technological change will be contained. Technology policy must provide the social link between ends and means.

Science, Technology, and American Government

The second part of this book examines the interactions of governmental units as they deal with technological issues raised in the previous chapters. Each branch of government has a unique vantage point on technology policy based on its constitutional or statutory roles. Most of the papers presented focus on the problem of bringing scientific and technological information to bear in the policymaking process.

By the time technological issues have reached the upper echelons of government decision making—the federal executive branch, Congress, or the courts—they have emerged as full-fledged social, political, and economic issues. Issues such as nuclear energy and toxic waste disposal often involve conflicts between one or more sectors or values in the society. The degree of technological content of these issues may vary widely, but usually a large measure of scientific or technological or social uncertainty is involved. One should not underestimate the influence on technology of other levels of decision making where perhaps the vast majority of technical decisions are made within individual firms and government agencies. Profound decisions

Introduction

affecting technology and society are often "routinely" made within the patent courts, regulatory agencies, and executive departments.

The process of technology policymaking in government agencies and industrial firms is not well understood. However, scientific and technological agencies have developed appropriate techniques and values for making policy choices. The goals and resources of an individual organization help structure the policy process to the extent required for the organization to function. Straightforward analytic techniques may work within the more orderly confines of a single government agency or firm, but they are poorly adapted for national policymaking. Problems that reach the status of national issues rarely lend themselves to neat analytical solutions since they may involve fundamental conflicts of value rather than differences in fact. Here the policy tools are inadequate, scientific and technological "advice" seems unheeded, and scientific "objectivity" is often at odds with other considerations. How does scientific "fact" influence public policy decision? How are controversies over different interpretations of the "facts" to be resolved? How do the federal executive, Congress, the courts, state and local governments, and public institutions affect technology policy?

Literature on science, technology, and public policy provides valuable insights into the pervasive influence of science and technology on the many departments of the federal government. Yet this rich literary mosaic is incomplete and depends more on personal insight than exhaustive social scientific inquiry. The technology policymaking process within such agencies as the National Bureau of Standards and the Department of Energy deserves careful and thorough study since these agencies deal routinely with complex technological decisions. These federal agencies seem to acquire and use technology adequately as long as their problems and projects do not cross jurisdictional borders. Within these narrow technological confines, however, very few societal problems can be solved comprehensively and holistically. Policies generated are limited by parochial, short-term, and special-interest objectives. It is little wonder that American government is preoccupied with the technological fix; it may be the easiest policy to pursue.

The critical problems for the federal executive are diverse. How can priorities be set among technological programs? What means must be provided to resolve differences of scientific and technological opinion? What new institutions are required to assess the costs and benefits of technological developments and impacts? How can policy analysis and policy design be improved to provide more comprehen-

Introduction

sive and holistic solutions including both social and technological innovations? What new institutions are required to cut across the narrow jurisdictions of federal agencies to plan, implement, and evaluate comprehensive technology policies? Such questions, of course, make up a tall order that will take decades to understand, let alone achieve.

Congress has an altogether different style than the federal executive in setting policy for technology. To a certain degree, the executive branch is hierarchial, fanning out from the president. In contrast, the policy business of Congress is directed by a collection of committees and subcommittees, competing for jurisdiction and political advantage. In spite of these structural differences, the two branches of government are closely intertwined, especially through the budget and oversight processes within Congress. The agencies, which form the bulk of the executive pyramid, have strong ties to the congressional committees in whose jurisdiction they lie. Of course, the executive is responsible for implementing and carrying out the laws established by Congress.

One of the most serious challenges for Congress in dealing with technology is the difficulty in acquiring independent information and advice. The executive agencies have extensive expertise to pursue technological programs and to provide a vigorous defense of their activities and decisions. The Congress is thought to be at a disadvantage in this respect. In an attempt to remedy the situation, Congress in the early 1970s created two staff agencies—the Congressional Budget Office and the Office of Technology Assessment. The former provides a central, overall budget coordination function previously lacking in Congress. The latter provides information to Congress on the implications of technological change.

The people responsible for handling science and technology in Congress change with each new session as new members of Congress replace old stalwarts. But the basic process changes very slowly. Congressional committee chairpersons work closely with agency heads to guide budget authorizations and appropriations and to oversee agency functionings. The fate of technological projects and programs seems to depend heavily on the economic climate and personal politics. The role of scientific and engineering advice is strictly circumscribed by political necessity and strategy.

The relationship between science and politics is a difficult one in the legislative process. Scientists and technologists do not wish to subjugate values such as "objectivity" to the inconsistencies and imperfections of political values. It is not likely that the reconciliation of science and politics can be forged in the halls of Congress between the persis-

Introduction

tent pressure of special interest (including that of the scientific communities) and the scramble for position and political posturing. There are obvious limits to how and what technological information will be used in a political arena concerned not with a search for "objective" truth but with satisfying political demands. To protect the amorphous "public interest," Congress must be knowledgeable enough about science and technology to detect the large and small lie.

The institution of Congress has inherent shortcomings in dealing with science and technology. The timetables for congressional elections and for research and development and innovation have nothing in common. The time required to accomplish a comprehensive technology policy analysis or technology assessment does not conform well with the legislative calendar. However, by increasing emphasis on and extending the time frame for technology policy analysis and assessments, Congress would grant more autonomy to the analyst. This raises another dilemma for technology policy. Who would oversee the policy analysis, and under what criteria would independent technology assessments be conducted? What is to guarantee the objectivity, impartiality, and honesty of the technology assessor?

The relationship between Congress and technology is perhaps analogous to that between the larger society and technology. Technology policy in Congress may be the proving ground for innovative relations and institutions dealing with the interface of science, technology, and society. How can society take full advantage of science and technology in public affairs? How do scientific and political values conflict, and can they be reconciled? What are the role and responsibility of scientists and engineers to educate and inform the general public on important technological issues? How should educators prepare scientists and engineers to assume that role?

Turning to the judicial system, the courts play two important roles in managing technology. First, the legal system can be used to adjudicate disputes between private parties, many of which may have scientific and technological dimensions. Second, the courts can counteract actions of the executive agencies that are determined to be outside the law.

Common law is one means to control the consequences of technological development. The present trend in contract, tort, and nuisance law is to hold technologists liable for damages sustained—even if these damages were unintentional or unforeseen. The effects of lawsuits may go beyond direct damages; by dramatizing injustices and reflecting public values a lawsuit can also cause the public to exert pressure for more responsible technology. However, the use of com-

Introduction

mon law to guide technological development poses concerns. How can negative impacts be borne by the parties responsible without stifling innovation? Furthermore, how can the law "manage" technology when it can only respond to complaints, not anticipate them?

Recent significant changes in judicial control over the actions of the executive branch have many implications for technology policy. Court action must be initiated by someone, and that someone must have sufficient stake or "standing" in the controversy. Liberalization of standing requirements recently has increased the role of the courts. Filing suit against the government has also been eased as legislatures restrict the government's sovereign immunity. In addition, judicial review of agency decisions, particularly under the National Environmental Policy Act of 1969, has begun to address substantive concerns about environmental quality. While it is difficult for judges to cope with many complex issues in science and technology, the courts are playing a significant role in holding executive agencies and their experts accountable for explaining the basis for their actions. Even though the courts cannot initiate policy actions, they appear to have a growing role in the formulation, or at least interpretation, of technology policy.

The challenge for the courts in improving their prowess and wisdom in dealing with technology is threefold. First, the courts must deal with more and more cases involving scientific and technological issues, such as product liability, nuclear risks, environmental protection, electronic invasion of privacy, medical malpractice, and even the legality of scientific experimentation with human subjects, dangerous substances, recombinant DNA, and test-tube babies. The courts will thus be required to develop an ever-increasing sophistication about science and technology. Is a special court system required to handle highly technological issues? Are special courts of appeal, such as the U.S. Patent Court or Maritime Court, required to adjudicate technological disputes?

Second, the court system has yet to take full advantage of technological innovations that can be applied to improve the efficiency of the courts themselves. For instance, the courts should look into the more effective use of information science, management science, and computer and telecommunications technology. Can legal research be automated? Can technology be used to transfer information on criminal and civil cases between court systems? Can improved information systems provide feedback on social problems and issues that are raised in the court system?

Third, improved information technology may make the courts a

Introduction

more effective social institution. While the courts may react only to existing law and policy, adjudication of these technological issues may provide an important source of social and institutional feedback about the impacts of technological change. The problems that are identified, the human costs, and the environmental impacts of technology policies may be measurable by the litigation that is brought before the courts. This information may be a valuable social asset if used to help initiate needed reforms.

The American federal system of government has preserved the autonomy and authority of states and localities. Many of the issues confronting state executive and legislative branches—for instance, highway development, mass transportation, environmental protection, and public health and safety—are highly technical. Yet the role of scientists and engineers in influencing state policy remains small by comparison with that at the federal level. Many of the historical precedents that fostered the growth of science and technology advisory mechanisms in the federal government have no counterparts in the states. Much of the impetus for federally sponsored research and development stems from nonlocalized issues such as defense and health. Nevertheless, the example of the federal government has stimulated the interest of both scientists and politicians at the state level, and the lure of federal R & D money has prompted many states to try to get their share.

Local governments have various regional consortiums, planning groups, and special metropolitan districts for managing water, mass transportation, air quality, and the like. Many of these agencies have clear technological responsibilities and therefore require considerable expertise. County, city, and other local governments could also draw on the nation's technological resources to help meet their needs and to help solve increasingly complex public problems. It is at the local level of government that the weaknesses of technology policy are most apparent.

State and local governments have been experimenting with science and technology advisory functions, but the results have been mixed, and few have lived up to expectations. Many of today's technological problems, however, cannot be solved solely from the top down. The nature of these problems, and the complexity of state and local political, economic, and social structures, often prevent federal "formula" solutions from working.

There are at least four weaknesses in scientific and technological resources available to the states and localities: (1) underutilization of indigenous resources to be found in business, industry, and universi-

Introduction

ties; (2) scarcity of managerial and technical talent to implement both federal and state programs; (3) inadequate research and institutional structures required to identify problems and design acceptable solutions; and (4) insufficient incentives for long-range capital investments in public technology.

The need for technology policy analysis may be far greater than the demand in state and local government. On the other hand, the conventional definition of technology may be too narrow to allow a good match between available technology and state and local needs. If we use the broader definition, discussed above, of technology as technique and organization, the range of technological solutions is extended to include social as well as mechanical innovations. The process of policy analysis admits a more realistic set of "solutions" and thus liberates the decision maker from the technological fix. For instance, traffic congestion may be reduced by synchronized traffic lights, automated traffic control, mass transportation, or, alternatively, better planning, staggered work hours, or car pooling. Two key questions regarding the role of state and local technology policy are: What are the respective roles of federal, state, and local governments in making and implementing technology policy? Are new institutions required to foster technology innovation or the improved use of indigenous scientific and technological resources to solve public problems?

Every citizen experiences the consequences of technological policies from a unique vantage point. These experiences have resulted in reactions ranging from unequivocal support to profound alienation. It is, therefore, important to consider the role of direct citizen participation in making technology policy. In order to participate in technology decision making, though, nonprofessionals must understand the issues. Even with such understanding, present channels for citizen participation are a difficult course. Lawsuits are expensive, direct participation is time consuming, and conflict is inescapable. Yet, participation offers promise of technological developments that are more responsive and more attuned to human needs.

Public participation may take many forms—from radical participation in specific scientific and technological decisions to diffuse public support for efforts to control technological development. Participation may take one of several forms including litigation, public technology assessment, and organized lobbying.

The mechanism for public participation in technology policy is part of a larger challenge to democratic institutions. The creation and dissemination of knowledge can be a political act, no matter how

Introduction

objective the information may be. Public participation advocates call for more direct influence of individual citizens at all levels of private and public decision making in addition to the normal legislative representation. Of course, there are a number of problems with direct citizen participation, including (*a*) the synthesis of expert knowledge and individual value, and (*b*) preserving a balanced representation of public opinion while increasing the voice of individual citizens.

Several different mechanisms may be required to provide legitimate citizen participation in different levels of decision making. Citizens may organize into collective consumer, environmental, and other types of interest groups to exercise their right of petition within the established legislative process and to inform the public of their point of view. Citizen advisory groups may be a good source of social feedback and counsel to individual decision makers. Surveys and questionnaires using statistically valid samples of the population can also be used to formulate and evaluate the effectiveness of technology policies. Whatever the mechanism, individual citizens' reactions to technological change are a first-hand source of information.

This human resource can be put to better use by both the policy analyst and the decision maker. What innovative means can be used to tap the "conventional wisdom" of individuals who have intimate or first-hand knowledge of technology policy problems? What are the best means to bring public opinion into technology policy making and assessment processes? Are new institutions feasible that would permit direct election of public representatives to oversee the conduct of technology policy assessments or decisions? How would this function be differentiated from the legislative process? What are the best means to give individual citizens a greater voice in technology policy decisions?

Public Choice

One consistent theme present in many of the articles presented later in this book is that science, technology, and politics are worlds apart. In order to harness the full potential of scientific discovery and technological innovation and direct them more efficiently toward social gains, we need to bridge these worlds. Public choices about technology are as difficult and complex as the interrelationships between science and democracy.

Traditionally, experts have addressed technical means, leaving the citizen, and even the politician, unenlightened to judge the technical considerations. One alternative approach to technology and public

Introduction

choice returns to the notion of the well-informed citizen by providing for the education of the citizenry in technological matters. Another approach is to develop representative science by infusing a sense of social responsibility in scientists and technologists. Bridging the worlds of science, technology, and society will likely require both of these approaches, but it cannot happen as a matter of course. The new field of science, technology, and public policy is needed to build the bridges and manage the traffic. Technology policy analysis is a promising way to draw technical competence into the policy realm.

Technology policy analysis and assessment provide the factual base required for making public policy decisions about complex technological and scientific issues. Scientific data must first be summarized, evaluated, and interpreted in order to identify public policy alternatives and inform public decision making. Of this process, Jurgen Schmandt writes in a *Science* editorial:

> Policy analysis establishes an important link between the worlds of science and public policy. Obviously, it has other important components, having to do with consideration of social, economic, and political factors. However, the treatment of scientific evidence is of critical importance for the quality of the total analytical effort. Policy analysis belongs to the realm of science to the extent that it makes use of the analytical tools of various scientific disciplines. But, given its location and function in the policy process, the standards and methods of other scientific activities do not necessarily apply to it.[1]

Technology policy analysis is a difficult task in that it must accomplish what society has failed thus far to do—integrate scientific, technical, and social information for use in public policy. This process is noted by M. Granger Morgan, also writing in *Science:*

> Good science and good policy analysis are not the same thing and do not serve the same ends. Many traditional scientists find policy analysis alien. Good science has as its objective the discovery of physical truth. Opinions, preferences, and values play a limited role in the exercise of good science. Except as it contributes to the design of future experimental and theoretical research, good science does not engage in speculation. It waits for full understanding.... Good policy analysis recognizes that physical truth may be poorly or incompletely known. Its objective is to evaluate, order, and structure incomplete knowledge so as to allow decisions to be made with as complete an understanding as possible of the current state of knowledge, its limitations, and its implications. Like good science, good policy analysis does not draw hard conclusions unless they are warranted by unambiguous data or well-founded theoretical insight.

[1]Jurgen Schmandt, *Science*, vol. 201, Sept. 8, 1978, p. 869.

Introduction

Unlike good science, good policy analysis must deal with opinions, preferences, and values, but it does so in ways that are open and explicit and that allow different people, with different opinions and values, to use the same analysis as an aid in making their own decisions.... Examples of good policy analysis are much harder to find than examples of good science. There are too many problems and too few skilled and qualified practitioners. This is particularly true for many problems involving science and technology, where good policy analysis requires a thorough understanding of the technical issues involved and an ability to sort out good science from bad.[2]

Of course, all of this will require the collective effort of many scientists, engineers, and social scientists to develop the theoretical basis, the knowledge, and the experience to practice good technology policy analysis. To accomplish this, the present institutional arrangements for education, research, application, and utilization of technology policy analysis must be greatly strengthened. The difficulty of this challenge is eclipsed only by the larger problem of technology and public choice. Even when the realms of science, technology, and public policy have been reconciled, the social choice regarding the means and ends of technological progress will continue to pose great ethical dilemmas.

Returning to Albert Einstein's thought, "Perfection of means and confusion of goals seem, in my opinion, to characterize our age," we have certainly witnessed a century of technological and economic achievements that have created boundless expectations in science and technology. Even much of the criticism levied at technology represents a reaffirmation of technological progress. The complaints are often that technology has not done enough to improve the human condition or that it has not been done well enough to avoid the negative consequences of technological excesses. However, we ask how many of these faults are the results of institutional and ethical failures rather than technological ones. As Einstein's observation suggests, we may have perfected the means for technological change and neglected the social institutions required to determine the appropriate goals. Do science and technology create new conditions, ways of thought, and definitions of political and economic power that weaken the foundations on which present institutions are built?

Institutional questions pervade many of the papers presented later in this book. There are many different kinds of institutional changes

---

[2]Granger Morgan, *Science*, vol. 201, Sept. 15, 1978, p. 971. Copyright 1978 by the American Association for the Advancement of Science.

Introduction

that may be required to improve the ability of society to manage technological change. Perhaps the most important step is improving and strengthening institutions for research and education in science, technology, and public policy. With improved knowledge, theory, and experience in technology policy, society will be in a better position to institute and utilize technology policy analysis. Some of the challenges in the field of science and technology policy to help develop these capabilities are summarized below:

1. Research into the *causes and effects of technological change,* including the development of a transdisciplinary theory of technological change and social action;

2. Research into the *determinants and outcomes of technological policy* in order to develop improved policy designs, implementation, and evaluation;

3. Improved methods and greater use of *public opinion measurement and public participation* in policy research to set goals, criteria and objectives for technology policy assessments;

4. *Professional standards* for technology policy analysis, including consideration of arrangements for the organization, funding, and quality control;

5. Improved methods for *policy implementation* including adequate provisions for training, organization, and infrastructure development;

6. Increased attention to and capability for *evaluation* of policies;

7. Research and education in the *history and ethics* of science, technology, and public policy issues, and techniques to include these perspectives in the technology policy analysis and decision process.

These are but a few of the many challenges of science, technology, and public policy. The student and practitioner of technology policy will be working within or around several different institutional settings. This book will provide a brief excursion into many of these settings and examine some of the issues of concern. Like this introductory essay, the book raises more questions than answers, but it is hoped that these will stimulate debate and perhaps lead to eventual solutions. The field of technology policy is wide open for such thought and innovation. The stakes grow higher every day as society must learn to manage technological complexity and change.

The outcomes of technology policies may be nothing short of war or peace, economic decay or prosperity, human freedom or enslavement. But technology itself will play a minor role in determining the outcomes. Science and technology can help only in making things possible and offering society new choices. In so doing, technology has helped fulfill long-standing human needs for clothing, food, shelter,

Introduction

and leisure, but at a price of an irrevocable dependence on this same technology, at least in the Western world. In the final analysis, the outcomes will depend on the morality and the ethics of the technological choices that are made. However, the means of technological change are better known today than the ethical ends of human progress. Then we are left with two final questions: Can society use science and technology wisely, and to what ends shall they be applied?

PART I

The Social Context of Science and Technology

A. The Society

The interactions between technological development and society pose concerns that extend beyond the scope of this book. This section cannot possibly address that whole spectrum of issues. Instead, it presents three perspectives on the heavily technological nature of our society.

In Chapter 1, Harvey Brooks provides a broad perspective on the evolution and purpose of technology. He notes that "a technology must include the managerial and social supporting systems necessary to apply it on a significant scale." Recent trends in technological development such as centralization and the substitution of labor, energy, and materials may be the products of cultural forces as much as any inherent technological logic. Within the last decade, however, increasing attention has been given to collective channeling of technology to realize social goals. In spite of the material progress, it is still uncertain whether the actual application of technological means is entirely compatible with other social goals such as democracy, personal liberty, and environmental quality. The next decade will be critical in relation to the application of technology to economic and social development.

Emmanuel Mesthene views technology less as a source of problems than as a creator of opportunities. In Chapter 2, he agrees that technology leads to changes in social organization and values, but these are not necessarily destructive changes; indeed, they are responses to new options. Technology does induce societal goals—"space vehicles spawn moon programs." And it requires new organizations to exploit the new opportunities.

In Chapter 3, Theodore Roszak uses powerful imagery ("The Coca-colonization of the world") to assert that technology has led us into a plastic culture. He warns us that technology is not a servant to man's needs; it is the active force altering our social values. He marvels at our unquestioning acceptance of technology, our casual contempt for the environment, and our self-assuredness about it all. But technology is more; it is the dominant source of political power. The high technical content of current policy issues reduces the opportunity for citizen participation. We are left with technocratic policies, utterly beholden to technical expertise for our society's survival and prosperity.

Brooks, Mesthene, and Roszak seem to agree on the importance of knowledge and expertise in a technological society characterized by rapid change. Government is more dependent on information and analysis for decision making. And decision making is more critical as a result of the pace of change, increased complexity of interconnecting systems, and the heightened role of the public sector at the expense of market forces.

1. Technology, Evolution, and Purpose
Harvey Brooks

It has been traditional to define technology in terms of its physical embodiments, as novel physical objects created by man to fulfill certain human purposes. In my opinion, this is too limited a view and one that is becoming increasingly obsolete. Nevertheless, this narrow conception of technology is perhaps one source of the current malaise with respect to technology and our technological society.

Hannay and McGinn have said that the "basic function of technology is the expansion of the realm of practical human possibility."[1] This is a good definition, but it seems to me that it already implies a good deal more than physical artifacts. Technology must be sociotechnical rather than technical, and a technology must include the managerial and social supporting systems necessary to apply it on a significant scale. Most highly original inventions have usually involved social as well as technical innovation. The Edison electric light, the Xerox copier, the computer, the Polaroid camera, the automobile, the television system—all involved concepts of complete technological systems that included supporting organizations and markets. In the case of TV, very different managerial and financing systems grew up in different countries with the same hardware building blocks. To be sure, not all innovations are of the sociotechnical system type. Many are simply components that fit into existing systems, such as magnetic bubbles or integrated circuits, and their introduction requires little social or market innovation. Ultimately, however, even such purely "hardware" type innovations can have major social impact through the changes they make possible in the systems of which they are a part. Other innovations may arise largely out of basic scientific discoveries such as lasers and many new materials.

Reprinted by permission of the American Academy of Arts and Sciences from *Daedalus*, Winter 1980.

Harvey Brooks

In the past, the social aspect was largely restricted to the market, to envisioning a social need not previously imagined to exist that could be so marketed as to create a self-sustaining technological system. In this sense, many of the most important inventions have been systems inventions in which the organization of the market has been part of the system. Today, managerial innovations are becoming an increasingly important aspect of technology. We see this particularly in the case of computers and communications systems, where not only the software, but also the organization that goes with the system, are inseparable from the physical embodiment of the technology, and are often the most expensive and innovative parts of it. Some enterprises such as McDonald's or the supermarket are based almost entirely on managerial innovation, with simple inventions in hardware being added gradually and later as incidental improvements to the original managerial concept.[2]

It seems to me that the defining feature of technology is that it is "public knowledge," in the sense proposed by J. M. Ziman with respect to science.[3] But rather than knowledge of how and why things are as they are, it is knowledge of how to fulfill certain human purposes in a specifiable and reproducible way. The characteristic aspect of public knowledge is its communicability and reproducibility; it is something that can be reconstructed in principle through specifiable algorithms.[4] To an extent, it is independent of culture and can be reproduced in any culture. Yet, not all cultures are equally receptive to it, and the process of assimilation may require great effort and patience. Furthermore, different cultures will tend to generate or select different technologies, so that the actual technological system that is chosen will be culture-dependent. Technology, therefore, does not consist of artifacts but of the public knowledge that underlies the artifacts and the way they can be used in society. Management, insofar as it can be described by fully specifiable rules, is thus a technology, and indeed every large bureaucratic organization can be considered an embodiment of technology just as much as a piece of machinery. Thus it has been suggested that the greatest innovation in the Apollo program was not the hardware, but the managerial system. This system made possible the degree of reliability and technical discipline necessary to bring the project to a successful conclusion, through the coordination of the activities of hundreds of contractors and subcontractors.[5] In contrast, the report of the Kemeny Commission suggests that it was the lack of similar managerial innovation in the nuclear industry that led to the mishaps of Three Mile Island and Brown's Ferry.[6]

Technology, Evolution, and Purpose

By identifying technology with knowledge rather than artifacts, we do not intend to imply that technology is the same as science, or even that it is based on science, which is conceptual knowledge involving mental models applicable in a large number of concrete situations. Technological knowledge can be highly scientific and abstract, but it can also be highly concrete and empirical. Usually it consists of a mix, with the science component being much larger in the technologies of more recent vintage. It is also important to observe that the process of creating technology for the first time is quite different from the process of specifying it so that it can be reproduced by others. The creation process is culture-dependent, but its reproducibility makes it technology rather than art.

The assertion that technology is culture-*in*dependent is, of course, controversial, but it is certainly true that most cultures behave as though this were so. The attitude toward Western technology in Islamic cultures is particularly interesting in this respect. Even a country as conservative as Saudi Arabia believes it can adopt modern technology while retaining all the essentials of Islamic culture and beliefs. Japan is often cited as an example of a different culture that has assimilated modern technology and become a creator of it without giving up its own cultural norms. That this is questionable does not alter the fact that the transferable elements that specifically define technology are not part of culture. The social effects of technology are secondary consequences of living with the embodiments of technology, not a part of the technology itself. This is true even if the social consequences turn out to be inevitable for practical purposes. We know, for example, that people of widely differing political, cultural, and religious persuasions can both create and apply the same technologies.

The Innovation Process

The process by which technology is conceived, developed, codified, and deployed on a large scale is called innovation. In fact, it is its reproducible and transferable nature that makes it possible for technology to be diffused widely, often with surprising speed. We know today that innovation is the major source of economic growth in industrial societies, almost certainly more important than physical factors such as labor and capital.[7] Indeed, growth probably occurs primarily as a result of the embodiment of new knowledge both in physical capital and in human labor and organizations. In this sense, innovation consists in the creation of sociotechnical systems, and it is these systems that are the source of economic growth.

Because of the relationship between the creation of technology and economic growth, industrial societies have become increasingly preoccupied with diagnosis of the innovative process in order to stimulate and nurture it as a source both of domestic growth and comparative advantage in international trade.[8] In addition, it has become an ever larger component of military power; indeed, the capacity of the United States to innovate rapidly and efficiently in weapons systems is frequently put forward as the key source of continuing U.S. military advantage over the Soviets, offsetting their greater rate of investment in military hardware during the past decade.[9] Since the early 1970s, increasing concern has been expressed in the United States about an alleged decline in our rate of innovation, especially in comparison with the dynamic economies of Germany and Japan.[10] The U.S. trade deficit and declining rates of growth of productivity in the domestic economy are attributed in part to declining innovative capacity. However, there is little agreement as to whether the declining innovation rate is a problem in itself or a symptom of other changes in U.S. society, such as a low savings and investment rate and an increasing shift of both innovative effort and investment toward environmental improvement, energy savings, and social services. The product of these improvements is not measured in customary productivity indices, but it may be of equal or greater value from the standpoint of overall social welfare.[11]

Innovation is not the same thing as either R & D or invention, although these are both important parts of the innovation process. The process includes the evolution of a whole technological system, from research through invention, to design for manufacturing, and marketing or operational application. As indicated earlier, innovation includes organizational change and the creation of social support systems to make possible the deployment and use of artifacts on a large scale. Thus the system of service stations and repair shops, of highways and highway maintenance, of credit and insurance, of traffic controls and law enforcement, all comprise the technological system of automobile transportation. Their creation constitutes a part of the innovation process in which the automobile is the central artifact.[12] In this sense, we see that innovations are not conceived and created all at once, but, rather, that they evolve in close interaction with society. Very often the driving force may be the central artifact—in our example, the automobile—and this pulls along with it a host of ancillary technologies, from gasoline pumps to oil refineries to radar speed

Technology, Evolution, and Purpose

measuring devices. One can, in fact, imagine a variety of alternative organizational and supporting systems embracing the same basic technological elements. For example, the automobile system could have developed entirely on a rental rather than an ownership basis, much like the telecommunication system.

Most innovations are directed not at the final consumer, but at the development of capital goods and intermediate products that are inputs to the manufacturing process of the distribution system. These innovations affect the consumer only indirectly through their influence on the costs of production and distribution. They may also interface with the public through their effect on the environment or public health.

The fact that most innovation is directed at capital and intermediate inputs is the reason that labor productivity, or factor productivity more generally, can be used as measures of the rate of technological change. Innovation directed at the final consumer, by contrast, often does not show up in Gross National Product measures. For example, home appliances that have enormously increased the productivity of work in the household, or power tools that have increased the productivity of do-it-yourself work, hardly show up at all in conventional measures of GNP or productivity. Only the tools and appliances themselves, not the continuing stream of services they provide, are valued in the GNP. Similarly, new products or processes that reduce pollution or save energy have become an increasingly prominent target for innovation, yet are measured only minimally in the GNP. An auto tire that has a much longer lifetime or that results in substantial fuel savings over the life of a car will be counted in the GNP only to the extent that it is more expensive than the tires it replaces. This extra initial cost, however, will generally be more than offset by lower sales of replacement tires and by fuel savings over the lifetime of the car. Thus, the net effect of such an innovation could be a reduction of measured GNP, even though society is obviously materially enriched by its substitution for older products.[13]

In the past, a very dynamic area of innovation has been that of materials with new properties, ranging from high-purity silicon fabricated into integrated circuits to high-strength alloy steels. Yet, these material innovations result almost exclusively in intermediate goods that are sold to manufacturers either to improve production or to be incorporated into capital goods or consumer goods. New materials have often made goods cheaper and more durable, as in the replacement of vacuum tube radios and TV by solid state electronic devices that are both cheaper and longer lasting. Here, again, we see an important package of innovations that does not show up directly in measures of economic output or personal consumption. They may affect GNP, but only indirectly, by reducing the cost of final goods, and thus generating increased purchases through the mechanism of price elasticity of demand, or simply because the goods are more attractive to consumers.[14]

It is often useful to think of biological evolution as a metaphor for technological evolution and innovation, although technological evolution takes place a million times faster than biological evolution.[15] In this metaphor the part of genetic inheritance is played by the inherent logic or technological development, whereas the part of natural selection by the environment is played by the social mechanisms of decision, including the market. Just as other species form part of the environment that exercises selection pressures on a given species, so do competing technologies form part of the selective environment that determines the evolution of a given technology. Technologies have ecological relationships with one another, and occupy ecological niches in the overall technological system, as do species in the biological world. The biological metaphor is most apt in the case of markets, where selection is exercised in millions of decentralized and uncoordinated decisions. It is less clearly applicable when society exercises collective choices, as, for example, in the case of political decisions to go forward with or cancel large technological programs such as the SST or the breeder reactor. Social decisions to regulate environmental impact, or occupational health and safety, similarly have no clear counterpart in natural biological evolution, except in the sense that they are culture-dependent and thus part of the social context of technological evolution.

The biological analog of conscious political choice in technological evolution is, perhaps, best viewed as analogous to the artificial selection used in creating domesticated species. Here, man has intervened in natural selection to produce new biological results that are the product of conscious collective choice. Just as man has learned how to direct natural evolution in parts of the ecosystem through artificial selection, so have societies gradually learned how to take over the direction of technological evolution from the market through collective regulation or government investment. In this sense, the art of technology assessment becomes analogous to the art of plant or animal breeding. The possibilities for the channeling of technological evolution are still constrained by the internal logic of technology at a particular stage of its development, just as the possibilities for creating new properties of plants and animals are constrained by the varieties

Technology, Evolution, and Purpose

that exist in the present generation and by the laws of genetics. One can, perhaps, carry the analogy still further by pointing out that the modern phenomenon of organized innovation in large firms and laboratories may bear some resemblance to a kind of genetic engineering in the biological field.

The biological metaphor is useful because it illuminates the debate between those who see technology as proceeding inexorably by an inner logic and those who see technology and innovation as being largely driven by social forces or class interests. Just as the genetic variations on which natural selection acts are determined by internal genetic events, so evolution from one generation of technology to the next is determined by logic internal to the technological system. But just as the number of genetic variations is very large compared with the number that are propagated in the next generation as a result of natural selection, so is the number of technical possibilities very large compared with those that actually survive in the development process, and even more so in the market or society. Thus the influence of society and culture on the inner logic of technology is similar to the influence of the environment on genetic inheritance between successive generations. In each case the inheritance mechanism produces a large redundancy of possibilities, while the environment selects those that survive to the next generation. In technological evolution, what survives provides the knowledge base that generates the full range of possibilities for the next generation of technology. The genealogy of ideas in the evolution of technology is similar to the genealogy of genetic variations in biological evolution.

Some of the critics of recent technological trends argue from perspectives somewhat similar to those who are critical of human intervention in natural evolution, as in the use of artificial monocultures in modern intensive agriculture. Just as ecologists deplore man's upsetting of the "balance of nature" through his agricultural and building or industrial practices, so some critics of recent technology argue that consumer sovereignty no longer controls its development; technology, rather, is forced into the market by high-powered sales techniques and by the influence of concentrated corporate power on the political process. These critics would agree that the biggest revolution in the last half century is the development of systems for the organized generation of new technology, the creation of institutions to manufacture new technology in almost the same way that in the nineteenth century we created institutions to manufacture and market goods and services on a large scale. But it is exactly this revolution that they condemn as taking technology out of the context of "natural" evolution in the marketplace.[16] Only in the days of the lone inventor was the evolution of technology truly "organic," allowing the social selection process time and trial-and-error to determine the true "fitness" of a new technology. By contrast, today's "artificial selection" or forced development of technology is creating "monsters" that are vulnerable and nonresilient.

It is interesting to note, in fact, that this criticism emanates in different ways from both the Right and the Left. The Left sees the generation of new technology as having been taken over by concentrated corporate power in alliance with government, especially the military; the Right sees the evolution of technology as being increasingly distorted by political intervention and the creation of perverse incentives by government. Neither, however, usually condemns the same end results of technological evolution. Critics of the Left focus their condemnation on large-scale technologies, especially military development and large-scale energy generation technologies such as nuclear power; critics of the Right, on the diversion of technological innovation to meet what they see as unreasonable environmental requirements or health and safety regulations. Nevertheless, some critics on the Right have asserted that nuclear power has run into trouble just because it was forced prematurely into the market through government pressures on industry and special economic incentives that distorted the form and rate of evolution of the technology.[17] Industry, they say, was led down the primrose path by government. In two other respects the critics would probably agree: that government regulation has given a comparative advantage to larger firms, and that it has tended to erode competitiveness in the economy.[18]

Recent Trends in Technological Development

What can we say about recent trends in the character of technology? To what extent are these trends inherent in technological logic; to what extent are they a product of social forces; and to what extent are they merely transient phenomena as compared with irreversible tendencies? Just as biological evolution has often led to dead ends, is it possible that technological evolution is leading us into cultural dead ends, as the critics of the New Left maintain? Is technology the cause for, or the solution of the *problematique* of our newly global society, or is it some combination of both?[19]

Following are analyses of a number of trends or alleged trends and attempts to diagnose their future course.

Technology, Evolution, and Purpose

Scale

Economics of scale have been a major driving force in the evolution of technology in the twentieth century. These economies apply both to the size of individual embodiments of technology—supertankers, nuclear power plants, wide-bodied aircraft, energy and communications networks—and to the size of the market for consumer technologies—automobiles, TV and radio, pocket computers. The scale of markets, of course, implies an accompanying large-scale sociotechnical system for marketing and service, so that it equally requires large organizations and control systems. There is some evidence that we may have come to the end of the road as far as the scale of individual technological embodiments are concerned. In the past decade, for example, no additional economies of scale have been realized with electric generating plants, and there is even some indication that the reliability of such plants is lower than for smaller plants. Political difficulties in the siting of power plants have increased rapidly as the scale of individual plants has increased, in part because of the concentration of environmental impacts. Even though the cost of pollution control tends to be a smaller fraction of plant cost as size increases, the environmental impact is more concentrated, and hence visible, and tends to provide a more obvious target for opposition.[20] Supertankers have become so large that they can enter fewer and fewer ports, and will be subject to ever tighter navigation restrictions. Ecological damage from the shipwreck of one supertanker is more obvious and concentrated, even though such tankers account for only a small fraction of the oil entering the marine environment.

Many environmental problems are associated not with new technologies *per se* but with the scale on which they are applied and diffused. Individually, automobiles are less polluting and place smaller demands on natural resources than did the horse-and-buggy. Electric generating plants pollute the air less than wood stoves relative to the amount of energy produced. Traffic and congestion are problems of scale, not of the basic building blocks of technology. Some of these problems of scale arise because a particular technology tends to become less adaptive to its environment as its scale of application increases: designs become standardized in order to realize economies of scale, making it more difficult to adapt designs as environmental problems resulting from this same scale become manifest. Technological evolution, which is highly plastic and responsive in its early phases, tends to rigidify as a result of its market success.

Yet, scale is a complex concept, and not all trends are in the direction of centralization. When examined carefully, the concepts of scale and centralization become rather ambiguous. The automobile, in its individual embodiment, is a decentralized technology that permits much more personal control over mobility than does public transit. Indeed, when the value of time is taken into account, the automobile appears to be the cheapest form of transportation in most circumstances. Other small-scale technologies that facilitate personal control and choice are home appliances, television, phonographs, personal computers, telephone services, and credit cards. From the standpoint of the user, these seem to be "appropriate technologies" in most of the senses usually put forward by the advocates of such technologies. Yet, they are seldom cited favorably, apparently because all require enormous production, distribution, and service networks that are associated with centralization and bureaucracy. The booming do-it-yourself home tool and shop business, cable TV, citizen's band radio, self-powered camping vehicles—all are technologies that appear to increase individual control, without depending very much on large centralized infrastructures. Individual solar energy installations are frequently cited as an example of appropriate technology that frees the individual from dependence on impersonal energy networks; yet, it is hard to see how solar energy could be deployed on a significant scale without standardization and mass production, and without large distribution and service networks differing little from those associated with, say, automobiles or home appliances.[21] The advocates of decentralized energy systems often envision a do-it-yourself construction and repair industry, but, in this case, massive education and information dissemination would be necessary, even assuming most families would want to build their own. Clearly, this is a prime example of Spreng and Weinberg's trade-off with both time and information.[22]

Centralization and Decentralization

The automobile and cable TV may be examples of two decentralized technologies that have displaced or will displace centralized large-scale technologies. The automobile has already largely displaced the passenger railroad and urban mass transportation in most of the United States. It has done this by making possible decentralized living patterns (associated also with the availability of home appliances) and by introducing, through competition, diseconomies of small scale into public transportation. Public transport has the characteristic that both the quality and the cheapness of the service improve rapidly with increasing patronage. But patronage of rail and public transit is de-

Technology, Evolution, and Purpose

clining everywhere in the industrialized world, and we see a descending spiral of poorer service and further loss of patronage.

It could well be that we will see a similar phenomenon occurring between broadcast and cable TV and, even conceivably, between solar energy and centralized electric grids. Cable TV could well draw off the most affluent patronage from broadcast TV, thus setting off a spiral of declining advertising revenue and even poorer programming, which could result in the demise of "poor man's" home entertainment, just as the automobile threatens the demise of "poor man's" public transportation. Similarly, the advent of decentralized power sources, available at first largely to the more affluent, could so degrade the economics of energy distribution networks as to make them uneconomical or inordinately expensive without public subsidy, such as is now necessary to sustain public transit.

These examples show that the trend toward centralization in modern technology is less clear and less certain than is sometimes asserted. Many modern technologies are, in fact, decentralizing rather than centralizing, much as in the case of TV in comparison to the movie theater of a generation ago.

Standardization

One of the ways to achieve economies of scale has been through standardization of products, especially in the basic technology. This standardization is now appearing on a world scale through dissemination by enterprises that transcend national boundaries. One hears, for instance, increasing talk of a "world automobile."[23] We have seen that the widespread use of small-scale decentralized technologies requires standardization in production and service aspects if the technology is to become inexpensive enough to be accessible to a large fraction of the population. Thus, even for decentralized "appropriate" technologies, it is the form of centralization that is changed rather than the need for centralization in some aspect of the system.

The advent of sophisticated and cheap information technology may also be greatly decreasing the significance of standardization for achieving economies of scale in production and distribution. Assembly systems can now be programmed to turn out many individualized versions of a technology on the same production line, and this is probably only the beginning of a trend toward substituting cheap information for standardization. Indeed, following the argument of Spreng and Weinberg, the old-fashioned mass-production factory was a device for economizing on both time and information in the production process; but as information becomes cheaper, it is no longer so necessary to economize on it in order to obtain an inexpensive product accessible to all. Production can be much more individualized, even crafted. Similarly, in the distribution process, the ability to store information has made it possible to reduce inventories and to offer wide choices to consumers. In the serving of products, the replacement of mechanical by electronic functions should have two effects: to make products more durable and to make diagnostics and repair much more automatic, hence, cheaper. Thus one impact of information technology may be to change the economic balance between cheap capital-intensive production and expensive labor-intensive repair that has been the underlying driving force of our "throw-away" society. This change will come about because repair will be cheaper and materials and energy that go into the original production more expensive.

Labor Versus Energy and Materials

An important characteristic of technological progress in the last one hundred years has been that it is labor-saving (or time-saving) rather than materials- and energy-saving. Although outputs per unit of labor, material, and energy input have increased, the output per unit of labor has increased much more rapidly than the other two. This is a consequence of the fact that the cost of labor has risen much faster than the cost of materials and energy, and it is thus logical that a major part of technological innovation should be directed at saving on the most expensive input. A man-hour of labor today buys many times more raw materials and energy than it did in 1900.[24] Indeed, until the early 1970s, the trend in real costs of energy and raw materials had been almost uniformly downward for more than a century.

While there are few experts who would agree that we have reached "limits to growth" in terms of basic materials and energy, or that we will "run out" of the resources necessary to support continued economic growth within the foreseeable future, there *is* general agreement that resources will become more expensive, and that we are entering a period of transition from a world economy based on the extraction of cheap resources from localized, high-grade locations, to an economy based on the extraction of more abundant but much lower grade, and hence more expensive to extract, resources that are more widely distributed.[25] This transition will involve, as well, extensive substitution for relatively rare and expensive resources—as it were, substituting information (and ingenuity) for unusually, but only temporarily, accessible resources. What this implies, of course, is that

Technology, Evolution, and Purpose

much more innovative effort in the future will be directed at saving resources and energy, or substituting more abundant, but more difficult to convert and use, resources for relatively rare, but easier to convert and use, resources.[26] Natural gas and oil are prime examples of resources that are remarkably easy to extract and use; they will be replaced by coal, uranium, and solar energy, all of which are more abundant and widely distributed, but much harder and trickier to convert to usable form. But in the intermediate term there is general agreement that the most cost-effective policy is to extend the life of our highest-quality resources through investments in more efficient end use.[27]

A question difficult to answer is whether the shift of innovative effort toward materials and energy conservation will result in slower growth of labor productivity. There is evidence of a trend in this direction in the United States, but there is no consensus that this has much to do with a shift of innovative effort. Some attribute it to emphasis on meeting environmental restrictions, but the question still appears open at present.

There is little question, however, that the entrepreneurial opportunities in technology for energy and materials efficiency are very large. This is especially true because the new technologies associated with the word "microelectronics" have very high potential for substituting ingenuity (information) for materials and energy consumption. A major part of the recent improvement in fuel efficiency of automobiles has been due to the use of electronic techniques in controlling combustion. The possibilities of using sophisticated controls on the direction and flow of energy in industrial processes and in commercial and residential use are great and have barely been started. Of course, microelectronics can also replace human labor; to the extent that higher energy costs are reflected in increased wage inflation, there may be equal pressures to use technology to replace labor. But it may also be that here is an opportunity to channel consciously an emerging technology into applications that are preferentially resource-saving rather than labor-saving, and this may require more than normal market forces to make it come about. At all events, this appears to be a prospective sea change in the direction of technological progress in the future.

Consumer Sovereignty Versus Complexity

To the degree that the products of innovation go directly to the consumer, rather than being sold as intermediate goods or capital equipment to producers, they have been characterized by an increasing mismatch between the complexity of the product and the ability of the purchaser to assess its qualities and performance. Thus the traditional model of the market in which producers and consumers bargain rationally, based on complete information regarding the properties of the products in question, breaks down. To be sure, various countervailing mechanisms have arisen—consumer research organizations, consumer protection legislation and agencies such as the Consumer Products Safety Commission, stricter product liability interpretations in the courts, and voluntary industry standards. All of these mechanisms are intended to substitute for the lack of knowledge and time for the consumer to evaluate the host of products he buys. Where the balance now lies, therefore, can be endlessly debated without any clear conclusion.

In the nineteenth century *caveat emptor* was the rule, and the purchaser of a defective or hazardous product had little recourse other than not to purchase the product a second time. But products were also much simpler and fewer in number and variety. The question today is whether the countervailing mechanisms have kept up with the increased complexity of products. This question applies not only to issues of safety, but also to matters of durability, maintainability, energy consumption, and environmental effects, none of which can really be assessed by the consumer before he buys. In the courts, the burden of proof has shifted strongly toward the producer to prove the safety and performance of his product. For high-technology products, such as prescription drugs, an elaborate assessment mechanism has been created to protect the consumer, despite the fact that the physician is an intermediary between the producer and developer of the drug and the patient. This is because the technical sophistication of the product is beyond the capacity of even a professional who is not a specialist on pharmaceuticals to evaluate.

Of course, the inability of the consumer to properly evaluate high-technology products has led to the centralization of decision-making with respect to the safety of such products. Organizations such as the Food and Drug Administration, the Consumer Product Safety Commission, the Automobile Safety Administration, and the Federal Trade Commission intervene between producer and consumer. Even for producer goods, such as manufacturing machinery, the Occupational Health and Safety Administration has intervened to protect the worker. Products such as passenger aircraft, nuclear reactors, public transportation equipment, and residential and commercial buildings are all subject to evaluation by public intermediaries on behalf of the safety and health of the consumer.

Technology, Evolution, and Purpose

In summary, while the increasing complexity of products has led to the erosion of consumer sovereignty as a mechanism for quality control and product assessment in the market, substitute intermediary mechanisms have evolved for making such decisions in a centralized way on behalf of the consumer. These mechanisms, however, constitute another source of centralization and bureaucratic structures that have been generated by modern technological trends.

Environmental Pollution and Externalities

There is a widespread public perception that technology has increased the pollution of the environment, and multiplied the hazards to the health and safety of the public that arise from the deployment of technology. This perception may be only partially correct. The rising public and scientific concern is not yet matched by statistical indicators that show life is more hazardous on the average than it used to be. Quite the contrary is the case.[28] With the exception of the rise in the incidence of lung cancer, which is generally acknowledged as attributable mainly, if not exclusively, to smoking, there are no trends in the incidence of specific forms of cancer that suggest the appearance of a major new threat to human health. Certain other forms of cancer, such as melanoma, have shown increases in incidence, but are statistically minor in the overall picture. Some diseases thought to be environmentally caused—notably cardiovascular diseases—have shown improving trends in the United States within the last few years.

However, all this does prove that technology may not present a latent hazard that has not yet caught up with us. New substances are constantly being introduced into the environment, and new technology deployed in such a way that a major commitment is often made in a period that is short compared with the induction period of cancer produced by low-level exposures to environmental contaminants. Our knowledge of the biological mechanisms of cancer induction is too rudimentary for us to rule out such future delayed effects from current industrial activities. There have been numerous examples in which occupational exposure to chemicals, or exposure of patients to new drugs, has resulted in serious delayed effects. The cases of asbestos and vinyl chloride monomer come to mind, as well as several examples of mercury poisoning. Fortunately, the population affected has been sufficiently small so that these episodes have not affected national health statistics. But the potential is there. In the case of saccharin, for example, more than seventy million people routinely use saccharin, and the rate of saccharin use among children under ten has risen dramatically within a few years.[29] With such large popula-

49

tions exposed, even very minute cancer risks could result in a large number of people being affected in decades to come. Hence, the concern and debate over whether saccharin should be banned. Fatigue failure of man-made structures is something that can occur after many years of use; it represents another example of a delayed effect that could appear long after an apparently safe technology had been deployed on a large scale. Fortunately, we know a good deal more about the physics of metal fatigue than about the biology of cancer induction, but the possibility of hazard is there. The recent problem with the DC-10 is a sobering reminder.

Alterations of natural ecosystems can also have human effects that are long delayed and may not be felt until future generations.[30] The intergenerational question has been raised especially forcefully in connection with radioactive waste management from the nuclear power industry. The effect of fluorocarbons from spray cans on stratospheric ozone is a worry only many decades in the future, but it will result from current activities and take many decades to disappear even after the use of fluorocarbons is banned.[31] The increase of carbon dioxide in the atmosphere as a result of burning fossil fuels and the conversion of forests to agriculture is another essentially irreversible, but long-in-the future, effect.[32] Thus the absence of contemporaneous indicators of trouble is not cause for compacency.

Another source of concern is that, whereas earlier environmental threats could be apprehended by the senses (for example, smoke), or had an immediate effect (disease due to contaminated water), today's hazards of technology can often be identified only with the aid of very sophisticated science. They usually have to be measured with sensitive scientific instruments, and their human effects lie at the end of a long, complex, and frequently uncertain causal chain, as in the case of fluorocarbons or sulfate aerosols from coal-fired electric power plants.

Innovation for What?

Until World War II the process of technological change was largely directed by the market. Only in the military field was it clearly dirven by other forces, but until the postwar period the magnitude of military R & D and the rate of military innovation were modest compared with civilian fields. In 1938 agricultural research constituted 40 percent of all government-supported research in the United States. By the mid-sixties it had dropped to 1.6 percent. In 1938 federally supported research constituted less than one-sixth of national R & D expenditures. In the mid-sixties it was nearly 70 percent, and it has fallen off to about 45 percent today.[33] During the postwar period technological development has also been increasingly driven by new technological opportunities, as perceived in the political process—civilian nuclear energy and the space program are among the clearest examples.

Within the last decade increasing attention has been given to the collective channeling of technology to realize certain social goals outside the purview of the market. The biomedical research program of the National Institutes of Health and rising expenditures on energy and environmental research are here the clearest examples. Environmental regulation has also played an important technology-forcing role in the private sector, as illustrated by the dramatic growth of research expenditures by the automobile industry in connection with emission controls and fuel conservation. More precisely stated, regulation has created a market for technologies that would not have been demanded by consumers because their benefits represent a public good—for example, pollution controls. Nobody has made an accurate estimate of what fraction of the national R & D effort, private and public, is now directed at meeting goals established by government through regulation or by citizens through collective action in the courts.

We do know that in the last two decades the percentage of technical articles published in *Science* that deal with risk assessment has increased eightfold, and this is certainly symptomatic of a broad trend in scientific and technological activity and interest.[34] Despite all this, however, the overwhelming proportion of nonmilitary research is still directed at meeting traditional market demands, directly through innovations in consumer products, and indirectly through advances in manufacturing or service technology or in the materials and components used in consumer products and capital goods.

Robert Morison's paper raises the issue of whether this laissez faire (or *laissez innover*) approach to technology is appropriate under modern circumstances.[35] Much of today's R & D, it is argued, is aimed at extending or elaborating the amenities available in already affluent societies. These may represent improvements in the quality of life that are marginal at best, and that may be increasingly offset by unforeseen side effects a little further in the future. The question raised is whether technology can and should be directed in a more concentrated way at meeting fundamental human needs of the future.

In the eyes of many, humanity is at a crossroads, on the threshold of a precarious future that could bring catastrophe and collapse or material abundance for all. This "transition crisis" arises because many of man's activities are now of sufficient magnitude to make significant inroads on the natural world we have inherited from geological ages. For the first time we are depleting some of the mineral concentrations laid down over geological history. Several natural geochemical cycles of nitrogen and of sulfur are now being appreciably accelerated by man's activities, and we have the capacity to accelerate many others, such as the hydrological cycle, in the near future. We are now changing the trace substance composition of the global atmosphere sufficiently to alter climate drastically early in the next century. Agriculture now accounts for about 5 percent of all photosynthetic fixation of carbon, and the burning of fossil fuels similarly corresponds to 5 percent of the global carbon fixation rate.[36]

What are the fundamental human needs to which advances in science and technology could contribute? They are the standard list: food, energy supply, health, a cleaner environment, materials supply, transportation, shelter, personal security, and a social system that facilitates rapid adaptive change while containing the possibility of violent conflict. The achievement of a minimum level for all people in each of these areas is for the first time within reach from a technical and scientific viewpoint, although science and technology cannot provide a solution by themselves. They only generate the conditions in which society can develop a solution.

Should or can science and technology be channeled toward providing a minimum standard in each of these areas for every living person? I start from the premise that such a standard is technically attainable; not everybody would agree. The existence of absolute poverty in the world can no longer be justified as a result of the scarcity of resources or the lack of technical capacity to alleviate it. This is also an important respect in which we are at a crossroads: ours is the first generation in which the technical possibility of a materially secure future can be envisioned concretely without implausible extrapolation from existing knowledge.[37]

Socially and managerially we may not be able to manage the transition. It might, for example, entail a sacrifice of other values that are too cherished in advanced industrial societies to be given up willingly. A central argument of antinuclear activists, for example, is that safe and secure management of a global nuclear power industry would inevitably entail infringement of civil liberties.

Continued economic growth is most frequently attacked on the grounds that it destroys environmental values, which are cherished more for their own sake than for their biological or physical suste-

Technology, Evolution, and Purpose

nance of human existence.[38] Legal philosophers debate seriously whether objects of nature should have "standing" in the courts in the same sense as young children or other helpless people who cannot defend themselves.[39] Whether the plight of the poor in the world can be alleviated without some sacrifice of these values is at least open to question. Those who defend them most earnestly argue that the problem of poverty should be solved by redistribution rather than growth, but it is hard, for me at least, to believe that the proponents of this view are really serious.

Improvement of the material condition of mankind inevitably involves *some* redistribution, of course, even if it results mainly from economic and technological growth. Such items as land and wilderness are in finite supply, and thus tend to become increasingly shared in one way or another.[40] In affluent societies the most privileged groups can no longer command the personal services of others to the degree possible in poor societies. One suspects this may be one of the unconscious reasons the condemnation of economic growth is confined mainly to an affluent minority in such societies. The redistribution that has taken place is not so much in income or wealth (though rich countries generally have much more equal income distributions than poor ones), but in what wealth or income can command. The poor can buy TV sets and automobiles, but the rich can no longer buy personal services or privacy.

What is implied by the foregoing, then, is that, although science and technology have provided us with the means to overcome scarcity for everybody, it is less than self-evident that the actual application of these means is compatible with such other goals as democracy, personal liberty, an aesthetically satisfying environment, the preservation of pristine nature, or individual privacy and dignity. These, of course, are culturally derived values, and the culture is certainly part of the environment that selects future technology. Also, the personality types that emerge in various technological systems are different, and may in fact be contradictory with the character types we idealize and regard as "civilized" in the best sense.

Perhaps the greatest fear of some critics of technological societies is not that progress will destroy or eliminate these values and character types, but that, if they do, we will not even miss them. At all events, it seems to me that this is where the central debate on "innovation for what?" lies. Is the society of material abundance for all, which is within our reach from a technical point of view, a society that we want? The problem is, those who have not attained to the state of the affluent countries seldom consider that there is a real choice. They want material progress; they are willing to mitigate the social costs only to the extent that this does not interfere with progress. If the materially most advanced societies decide that they have had enough of material progress, and look to other values, they may not be able to arrest the process of change at the present level. The rest of the world may simply sweep over them in its demand for what the advanced societies have already achieved.

The other problem is, of course, that innovative capacity is largely concentrated in the advanced societies and is mostly concentrated on the problems and aspirations of those societies, whereas the objectives of innovation are increasingly related to the poor societies. Some objectives, of course, are to some extent common, energy supply perhaps being the best example. But even here needs may be different and technologies not directly transferable.

We have learned that innovative capacity is the hardest thing in the world to transfer. Technology *can* be transferred, because, as we have seen, the very definition of technology implies transferability in some degree, even though the receptivity of a different culture may be in doubt. But innovative *capacity* may be much more culture-dependent. One of the characteristics of all innovation (as opposed to research or development *per se*) is that it can only be carried to closure in very close relationship to the final user of the technology. Even in developed societies many innovations fail because in the end they don't quite "fit." This problem is compounded when we try to innovate in a society at one cultural and material level for the benefit of another society at quite a different level. Indeed, several participants at the conference that led to this volume have severely condemned the whole notion of innovation *for* someone else. Yet, if the world cannot make use of the innovative capacity that already exists, but has to wait for the development of indigenous innovative capacity in an almost autonomous or self-sufficient fashion, it may be too late for the transition. There are some promising developments, although it is too early to say how they will finally turn out. During the past eight years the international agricultural research system, with its network of independent but cooperating institutes, has grown to be extraordinarily effective, especially in innovating close to the ultimate user. An increasing number of developing countries show signs of having crossed a sort of take-off threshold of economic development, although the revolution in Iran illustrates how precarious this transition can be. The next decade will be critical in relation to a host of developments in the application of technology to economic and social

Technology, Evolution, and Purpose

development. It is unclear whether crises like that in Iran will be a common feature of rapidly developing societies, especially those with non-European cultural traditions, or whether rapid economic development can follow a relatively smooth path. I remain an optimist because of my belief that man will not be foolish enough to reject what is within his grasp, and that we will succeed in reconciling further material progress with the other values we hold dear.

Notes

1. *Daedalus,* Winter, 1980.
2. Theodore Levitt, "Management and the 'Post-Industrial' Society," *The Public Interest,* 44 (Summer 1976), 69–103.
3. J. M. Ziman, *Public Knowledge: The Social Dimension of Science* (New York: Cambridge University Press, 1968).
4. H. Brooks, "A Framework for Science and Technology Policy," Arleigh House IEEE Workshop on Goals and Policy, 2, no. 5 (November 1972), 584–88.
5. L. A. Sayles and M. K. Chandler, *Managing Large Systems* (New York: Harper & Row, 1971).
6. J. G. Kemeny (ch.), *The Need for Change: the Legacy of TMI,* Report of the President's Commission on the Accident of Three Mile Island, October 1979, USGPO.
7. National Science Foundation, Science and Technology Annual Report to the Congress, August 1978, USGPO, 1978.
8. The White House. Fact Sheet. The President's Industrial Innovation Initiatives, October 31, 1979.
9. R. Perry, F. A. Long, and J. Reppy, "U.S. Programs Military R & D: An Introductory Overview," and "R & D American Style," prepared for Cornell/Rockefeller Workshop on Decision-Making for U.S. Programs of Military R & D, March 1–2, 1979 (to be published).
10. M. Katz (ch.), *Technology, Trade, and the U.S. Economy,* report of NAE Workshop, Woods Hole, August 22–31, 1976, Office of Foreign Secretary, NAS, 1978.
11. H. Brooks, "Science and the Future of Economic Growth," *Journal of the Electrochemical Society,* 121, no. 2 (February 1974), 35c–42c.
12. NAS, *Technology: Processes of Assessment and Choice,* Committee on Science and Astronautics, U.S. House of Representatives, July, 1969, USGPO, pp. 16–17.
13. W. Heller, "On Economic Growth," in S. H. Schurr, ed., *Energy, Economic Growth and the Environment,* published for Resources for the Future, Inc. (Baltimore: Johns Hopkins Press, 1972), pp. 3–29.
14. H. Brooks, "Applied Research: Definitions, Concepts, Themes," in *Applied Science and Technological Progress,* Committee on Science and Astronautics, U.S. House of Representatives, June 1967, USGPO, pp. 21–55.
15. H. Brooks, "Technology Assessment as a Process," *International Social Science Journal,* UNESCO, 25, no. 3 (1973), 247–56.
16. G. Wald, and discussants D. Bell, H. Brooks, and J. B. Fisk, "Man and

the Machine: Prospects for the Human Enterprise," in M. Greenberger, ed., *Computers, Communications, and the Public Interest* (Baltimore: Johns Hopkins Press, 1971), pp. 255–93.

17. I. C. Bupp and J. C. Derian, *Light Water* (New York: Basic Books, 1978).

18. Douglas H. Ginsburg and W. J. Abernathy, eds., *Government, Technology, and the Future of the Automobile* (New York: McGraw-Hill, 1980).

19. H. Brooks, "Technology: Hope or Catastrophe?" *Technology in Society*, 1 (1979), 3–17.

20. H. Brooks, "The Energy Problem," *Oak Ridge Bicentennial Lectures, Technology and Society*, sponsored by Oak Ridge National Laboratory, 1977, pp. 101–19, ORNL/PPA-77/3, USGPO, no. 060–000–0081–6.

21. H. Brooks, "Critique of the Concept of Appropriate Technology," *Appropriate Technology and Social Values—A Critical Appraisal*, ed. F. A. Long and A. Oleson, published for the American Academy of Arts and Sciences (Cambridge, Mass.: Ballinger, 1980).

22. *Daedalus*, Winter, 1980.

23. C. Kenneth Orski, "Urban Transportation: The Role of Major Actors." Address delivered at the Aspen Conference on Future Urban Transportation, sponsored by the American Planning Association, June 3–7, 1979.

24. D. B. Rice, (ch.), *Government and the Nation's Resources*, report of the National Commission on Supplies and Shortages, Washington, USGPO, December 1976.

25. H. Brooks, "Notes on the Energy Problematique," prepared for IFIAS Seminar "From Vision to Action," held in Mexico City under auspices of El Colegio de Mexico, October 22–26, 1979 (to be published).

26. H. Brooks, "Resources and the Quality of Life in 2000," *Annual Review of Materials Science*, 8 (1978), 1–19.

27. H. Brooks and J. M. Hollander, "United States Energy Alternatives to 2010 and Beyond: The CONAES Study," *Annual Review of Energy*, 4 (1979), 1–70.

28. R. W. Kates, "Assessing the Assessors: The Act and Ideology of Risk Assessment," *Ambio*, 6, no. 5 (1977), 247–52.

29. NAS/NRC, *Saccharin: Technical Assessment of Risk and Benefits*, part 1[F], Committee for a Study on Saccharin and Food Safety Policy, Assembly of Life Sciences, NAS-NRC, 1979.

30. H. Brooks, "Environmental Decision-Making: Analysis and Values," in L. H. Tribe, C. Schelling, and J. Voss, eds., *When Values Conflict* (Cambridge, Mass.: Ballinger, 1976), pp. 115–35.

31. J. W. Tukey, et. al., *Halocarbons: Environmental Effects of Chlorofluoromethane Release*, NAS-NRS, September 10, 1976.

32. J. Charney (ch.), *Carbon Dioxide and Climate: A Scientific Assessment*, Summary report of an Ad Hoc Study Group on Carbon Dioxide and Climate, Woods Hole, July 23–27, 1979, National Academy of Sciences (to be published).

33. V. Bush, et al., *Science, the Endless Frontier*, a report to the president on a program for postwar scientific research, July 1945, republished by the National Science Foundation, July 1960, cf. table 1, p. 86.

34. R. W. Kates, "Assessing the Assessors."

35. *Daedalus*, Winter, 1980.

36. W. Strumm, ed., *Global Chemical Cycles and Their Alterations by Man* (Berlin: Dahlem Konferenzen, 1977).
37. For a striking presentation of this perspective, see C. Marchetti, *On 10^{12}: A Check on Earth Carrying Capacity for Man*, Research Report RR-78-7, May 1978, International Institute for Applied Systems Analysis. Laxenburg, Austria.
38. L. Tribe, "Ways Not to Think About Plastic Trees," in L. H. Tribe, C. Schelling, and J. Voss, eds., *When Values Conflict* (Cambridge, Mass.: Ballinger, 1976), pp. 115–35.
39. C. D. Stone, "Should Trees Have Standing?—Toward Legal Rights for Natural Objects," *Southern California Law Review* 45 (1972), 490.
40. H. Brooks, "Environmental Decision-Making."

2. How Technology Will Shape the Future

EMMANUEL G. MESTHENE

There are two ways, at least, to approach an understanding of how technology will affect the future. One, which I do not adopt here, is to try to predict the most likely technological developments of the future along with their most likely social effects (*1*). The other way is to identify some respects in which technology entails change and to suggest the kinds or patterns of change that, by its nature, it brings about in society. It is along the latter lines that I speculate in what follows, restricting myself largely to the contemporary American scene.

New Technology Means Change

It is widely and ritually repeated these days that a technological world is a world of change. To the extent that this statement is meaningful at all, it would seem to be true only of a world characterized by a more or less continuous development of new technologies. There is no inherent impetus toward change in tools as such, no matter how

Copyright © 1968 by the American Association for the Advancement of Science. Reprinted from *Science*, Vol. 161, July 12, 1968, pp. 135–143, by permission.

Emmanuel G. Mesthene

many or how sophisticated they may be. When new tools emerge and displace older ones, however, there is a strong presumption that there will be changes in nature and in society.

I see no such necessity in the technology-culture or technology-society relationship as we associate with the Marxist tradition, according to which changes in the technology of production are inevitably and univocally determinative of culture and social structure. But I do see, in David Hume's words, a rather "constant conjunction" between technological change and social change as well as a number of good reasons why there should be one, after we discount for the differential time lags that characterize particular cases of social change consequent on the introduction of new technologies.

The traditional Marxist position has been thought of as asserting a strict or hard determinism. By contrast, I would defend a position that William James once called a "soft" determinism, although he used the phrase in a different context. (One may also call it a *probabilistic* determinism and thus avoid the trap of strict causation.) I would hold that the development and adoption of new technologies make for changes in social organization and values by virtue of creating new possibilities for human action and thus altering the mix of options available to men. They may not do so necessarily, but I suggest they do so frequently and with a very high probability.

Technology Creates New Possibilities

One of the most obvious characteristics of new technology is that it brings about or inhibits changes in *physical* nature, including changes in the patterns of physical objects or processes. By virtue of enhancing our ability to measure and predict, moreover, technology can, more specifically, lead to controlled or directed change. Thus, the plow changes the texture of the soil in a specifiable way; the wheel speeds up the mobility (change in relative position) of people or objects; and the smokebox (or icebox) inhibits some processes of decay. It would be equally accurate to say that these technologies respectively *make possible* changes in soil texture, speed of transport, and so forth.

In these terms, we can define any new (nontrivial) technological change as one which (i) makes possible a new way of inducing a physical change; or (ii) creates a wholly new physical possibility that simply did not exist before. A better mousetrap or faster airplane are examples of new ways and the Salk vaccine or the moon-rocket are instances of new possibilities. Either kind of technological change will

How Technology Will Shape the Future

extend the range of what man *can* do, which is what technology is all about.

There is nothing in the nature or fact of a new tool, of course, that requires its use. As Lynn White has observed, "a new device merely opens a door; it does not compel one to enter" (*2*). I would add, however, that a newly opened door does *invite* one to enter (*3*). A house in which a number of new doors have been installed is different from what it was before and the behavior of its inhabitants is very likely to change as a result. Possibility as such does not imply actuality (as a strict determinism would have to hold), but there is a high probability of realization of new possibilities that have been deliberately created by technological development, and therefore of change consequent on that realization.

Technology Alters the Mix of Choices

A correlative way in which new technology makes for change is by removing some options previously available. This consequence of technology is derivative, indirect, and more difficult to anticipate than the generation of new options. It is derivative in that old options are removed only after technology has created new ones. It is indirect, analogously, because the removal of options is not the result of the new technology, but of the act of choosing the new options that the technology has created (*4*). It is more difficult to anticipate, finally, to the degree that the positive consequences *for* which a technology is developed and applied are seen as part of the process of decision to develop and apply, whereas other (often negative) consequences of the development are usually seen only later if at all.

Examples abound. Widespread introduction of modern plumbing can contribute to convenience and to public health, but it also destroys the kind of society that we associate with the village pump. Exploitation of industrial technology removes many of the options and values peculiar to an agricultural society. The automobile and airplane provide mobility, but often at the expense of stabilities and constancies that mobility can disturb.

Opportunity costs are involved in exploiting any opportunity, in other words, and therefore also the opportunities newly created by technology. Insofar as the new options are chosen and the new possibilities are exploited, older possibilities are displaced and older options are precluded or prior choices are reversed. The presumption, albeit not the necessity, that most of the new options will be chosen is therefore at the same time a presumption that the choice will be made to pay the new costs. Thus, whereas technology begins by simply adding to the options available to man, it ends by altering the spectrum of his options and the mix or hierarchy of his social choices.

Social Change

The first-order effect of technology is thus to multiply and diversify material possibilities and thereby offer new and altered opportunities to man. Different societies committed to different values can react differently (positively or negatively, or simply differently) to the same new possibilities, of course. This is part of the explanation, I believe, for the phenomenon currently being referred to as the "technological gap" between Western Europe and the United States. Moreover, as with all opportunities when badly handled, the ones created by technology can turn into new opportunities to make mistakes. None of this alters the fact that technology creates opportunity.

Since new possibilities and new opportunities generally require new organizations of human effort to realize and exploit them, technology generally has second-order effects that take the form of social change. There have been instances in which changes in technology and in the material culture of a society have not been accompanied by social change, but such cases are rare and exceptional (5). More generally:

> ... over the millennia cultures in different environments have changed tremendously, and these changes are basically traceable to new adaptations required by changing technology and productive arrangements. Despite occasional cultural barriers, the useful arts have spread extremely widely, and the instances in which they have not been accepted because of pre-existing cultural patterns are insignificant (6).

While social change does not necessarily follow upon technological change, it almost always does in fact, thus encouraging the presumption that it generally will. The role of the heavy plow in the organization of rural society and that of the stirrup in the rise of feudalism provide fascinating medieval examples of a nearly direct technology-society relationship (7). The classic case in our era, of course—which it was Karl Marx's contribution to see so clearly, however badly he clouded his perception and blinded many of his disciples by tying it at once to a rigid determinism and to a form of Hegelian absolutism—is the Industrial Revolution, whose social effects continue to proliferate.

When social change does result from the introduction of a new technology, it must, at least in some of its aspects, be of a sort conducive to exploitation of the new opportunities or possibilities created by that technology. Otherwise it makes no sense even to speak of the social effects of technological change. Social consequences need not be and surely are not uniquely and univocally determined by the character of innovation, but they cannot be entirely independent of that character and still be accounted consequences. (Therein lies the distinction, ultimately, between a hard and soft determinism.) What the advent of nuclear weapons altered was the military organization of the country, not the structure of its communications industry, and the launching of satellites affects international relations much more directly than it does the institutions of organized sport. (A change in international relations may affect international competition in sports, of course, but while everything may be connected with everything else in the last analysis, it is not so in the first.)

There is a congruence between technology and its social effects that serves as intellectual ground for all inquiry into the technology-society relationship. This congruence has two aspects. First, the subset of social changes that can result from a given technological innovation is smaller than the set of all possible social changes and the changes that do in fact result are a still smaller subset of those that can result—that is, they are a sub-subset. In relation to any given innovation, the spectrum of all possible social changes can be divided into those that cannot follow as consequences, those that can (are made possible by the new technology), and those that do (the actual consequences).

It is the congruence of technology and its social consequences in this sense that provides the theoretical warrant for the currently fashionable art of "futurology." The more responsible practitioners of this art insist that they do not predict unique future events but rather identify and assess the likelihood of possible future events or situations. The effort is warranted by the twin facts that technology constantly alters the mix of possibilities and that any given technological change may have several consequences.

The second aspect of the congruence between technology and its social consequences is a certain "one-wayness" about the relationship—that is, the determinative element in it, however "soft." It is after all only technology that creates new *physical* possibilities (though it is not technology *alone* that does so, since science, knowledge, social organization, and other factors are also necessary to the process). To be sure, what technologies will be developed at any particular time is dependent on the social institutions and values that prevail at that particular time. I do not depreciate the interaction between technology and society, especially in our society which is learning to create scientific knowledge to order and develop new technologies for already established purposes. Nevertheless, once a new technology is created, it is the impetus for the social and institutional changes that follow it. This is especially so since a social decision to develop a particular technology is made in the principal expectation of its predicted first-order effects, whereas evaluation of the technology after it is developed and in operation usually takes account also of its less-foreseeable second- and even third-order effects.

The "one-wayness" of the technology-society relationship that I am seeking to identify may be evoked by allusion to the game of dice. The initiative for throwing the dice lies with the player, but the "social" consequences that follow the throw are initiated *by the dice* and depend on how the dice fall. Similarly, the initiative for development of technology in any given instance lies with people, acting individually or as a public, deliberately or in response to such pressures as wars or revolutions. But the material initiative remains with the technology and the social adaptation to it remains its consequence. Where the analogy is weak, the point is strengthened. For the rules of the game remain the same no matter how the dice fall, but technology has the effect of adding new faces to the dice, thus inducing changes in society's rules so that it can take advantage of the new combinations that are created thereby. That is why new technology generally means change in society as well as in nature.

Technology and Values

New technology also means a high probability of change in individual and social values, because it alters the conditions of choice. It is often customary to distinguish rather sharply between individual and social values and, in another dimension, between tastes or preferences, which are usually taken to be relatively short-term, trivial, and localized, and values, which are seen as higher-level, relatively long-term, and extensive in scope. However useful for some purposes, these distinctions have no standing in logic, as Kenneth Arrow points out:

> One might want to reverse the term "values" for a specially elevated or noble set of choices. Perhaps choices in general might be referred to as "tastes." We do not ordinarily think of the preference for additional bread over additional beer as being a value worthy of philosophic inquiry. I believe, though, that the distinction cannot be made logically.... (8)

How Technology Will Shape the Future

The *logical* equivalence of preferences and values, whether individual or social, derives from the fact that all of them are rooted in choice behavior. If values be taken in the contemporary American sociologist's sense of broad dominant commitments that account for the cohesion of a society and the maintenance of its identity through time, their relation to choice can be seen both in their genesis (historically in the society, not psychologically in the individual) and in their exemplification (where they function as criteria for choice).

Since values in this sense are rather high-level abstractions, it is unlikely that technological change can be seen to influence them directly. We need, rather, to explore what difference technology makes for the choice behaviors that the values are abstractions from.

What we choose, whether individually or as a society (in whatever sense a society may be said to choose, by public action or by resultant of private actions) is limited, at any given time, by the options available. (Preferences and values are in this respect different from aspirations or ideals in that the latter can attach to imaginative constructs. To confuse the two is to confuse morality and fantasy.) When we say that technology makes possible what was not possible before, we say that we now have more options to choose from than we did before. Our old value clusters, whose hierarchical ordering was determined in the sense of being delimited by antecedent conditions of material possibility, are thus now subject to change because technology has altered the material conditions.

By making available new options, new technology can, and generally will, lead to a restructuring of the hierarchy of values, either by providing the means for bringing previously unattainable ideals within the realm of choice and therefore of realizable values, or by altering the relative ease with which different values can be implemented—that is, by changing the costs associated with realizing them. Thus, the economic affluence that technological advance can bring may enhance the values we associate with leisure at the relative expense of the value of work and achievement, and the development of pain-killing and pleasure-producing drugs can make the value of material comfort relatively easier of achievement than the values we associate with maintaining a stiff upper lip during pain or adversity.

One may argue further that technological change leads to value change of a particular sort in exact analogy to the subset of possible social changes that a new technology may augur (as distinct from both the wider set that includes the impossible and the narrower, actual subset). There are two reasons for this. First, certain attitudes and values are more conducive than others to most effective exploitation of the potentialities of new tools or technologies. Choice behavior must be somehow attuned to the new options that technology creates, so that they will in fact be chosen. Thus, to transfer or adapt industrial technologies to underdeveloped nations is only part of the problem of economic development; the more important part consists in altering value predispositions and attitudes so that the technologies can flourish. In more advanced societies, such as ours, people who hold values well adapted to exploitation of major new technologies will tend to grow rich and occupy elite positions in society, thus serving to reinforce those same values in the society at large.

Second, whereas technological choices will be made according to the values prevailing in society at any given time, those choices will, as previously noted, be based on the foreseeable consequences of the new technology. The essence of technology as creative of new possibility, however, means that there is an irreducible element of uncertainty—of unforeseeable consequence—in any innovation (9). Techniques of the class of systems analysis are designed to anticipate as much of this uncertainty as possible, but it is in the nature of the case that they can never be more than partially successful, partly because a new technology will enter into interaction with a growing number and variety of ongoing processes as societies become more complex, and partly—at least in democratic societies—because the unforeseeable consequences of technological innovation may take the form of negative *political* reaction by certain groups in the society.

Since there is an irreducible element of uncertainty that attends every case of technological innovation, therefore, there is need for two evaluations: one before and one after the innovation. The first is an evaluation of prospects (or ends-in-view, as John Dewey called them). The second is an evaluation of results (of outcomes actually attained) (10). The uncertainty inherent in technological innovation means there will usually be a difference between the results of these two evaluations. To that extent, new technology will lead to value change.

Contemporary Patterns of Change

Our own age is characterized by a deliberate fostering of technological change and, in general, by the growing social role of knowledge. "Every society now lives by innovation and growth; and it is theoretical knowledge that has become the matrix of innovation" (11, p. 29; 12).

In a modern industrialized society, particularly, there are a number of pressures that conspire toward this result. First, economic pressures argue for the greater efficiency implicit in a new technology. The principal example of this is the continuing process of capital modernization in industry. Second, there are political pressures that seek the greater absolute effectiveness of a new technology, as in our latest weapons, for example. Third, we turn more and more to the promise of new technology for help in dealing with our social problems. Fourth, there is the spur to action inherent in the mere availability of a technology: space vehicles spawn moon programs. Finally, political and industrial interests engaged in developing a new technology have the vested interest and powerful means needed to urge its adoption and widespread use irrespective of social utility.

If this social drive to develop ever more new technology is taken in conjunction with the very high probability that new technology will result in physical, social, and value changes, we have the conditions for a world whose defining characteristic is change, the kind of world I once described as Heraclitean, after the pre-Socratic philosopher Heraclitus, who saw change as the essence of being (13).

When change becomes that pervasive in the world, it must color the ways in which we understand, organize, and evaluate the world. The sheer fact of change will have an impact on our sensibilities and ideas, our institutions and practices, our politics and values. Most of these have to date developed on the assumption that stability is more characteristic of the world than change—that is, that change is but a temporary perturbation of stability or a transition to a new (and presumed better or higher) stable state. What happens to them when that fundamental metaphysical assumption is undermined? The answer is implicit in a number of intellectual, social, and political trends in present-day American society.

Intellectual Trends

I have already noted the growing social role of knowledge (14). Our society values the production and inculcation of knowledge more than ever before, as is evidenced by sharply rising research, development, and education expenditures over the last 20 years. There is an increasing devotion, too, to the systematic use of information in public and private decision-making, as is exemplified by the President's Council of Economic Advisers, by various scientific advisory groups in and out of government, by the growing number of research and analysis organizations, by increasing appeal to such techniques as program planning and budgeting, and by the recent concern with assembling and analyzing a set of "social indicators" to help gauge the social health of the nation (*15*).

A changing society must put a relatively strong accent on knowledge in order to offset the unfamiliarity and uncertainty that change implies. Traditional ways (beliefs, institutions, procedures, attitudes) may be adequate for dealing with the existent and known. But new technology can be generated and assimilated only if there is technical knowledge about its operation and capabilities, and economic, sociological, and political knowledge about the society into which it will be introduced.

This argues, in turn, for the importance of the social sciences. It is by now reasonably well established that policy-making in many areas can be effective only if it takes account of the findings and potentialities of the natural sciences and of their associated technologies. Starting with economics, we are gradually coming to a similar recognition of the importance of the social sciences to public policy. Research and education in the social sciences are being increasingly supported by public funds, as the natural sciences have been by the military services and the National Science Foundation for the last quarter of a century. Also, both policy-makers and social scientists are seeking new mechanisms of cooperation and are exploring the modifications these will require in their respective assumptions and procedures. This trend toward more applied social science is likely to be noticeable in any highly innovative society.

The scientific mores of such a society will also be influenced by the interest in applying technology that defines it. Inquiry is likely to be motivated by and focused on problems of the society rather than centering mainly around the unsolved puzzles of the scientific disciplines themselves. This does not mean, although it can, if vigilance against political interference is relaxed, (i) that the resulting science will be any less pure than that proceeding from disinterested curiosity, or (ii) that there cannot therefore be any science motivated by curiosity, or (iii) that the advancement of scientific knowledge may not be dependent on there always being some. The research into the atomic structure of matter that is undertaken in the interest of developing new materials for supersonic flight is no less basic or pure than the same research undertaken in pursuit of a new and intriguing particle, even though the research strategy may be different in the two cases. Even more to the point, social research into voting behavior is not *ipso facto* less basic or pure because it is paid for by an aspiring candidate rather than by a foundation grant.

There is a serious question, in any event, about just how pure is pure in scientific research. One need not subscribe to such an out-and-out Marxism as Hessen's postulation of exclusively social and economic origins for Newton's research interests, for example, in order to recognize "the demonstrable fact that the thematics of science in seventeenth century England were in large part determined by the social structure of the time" (16). Nor should we ignore the fashions in science, such as the strong emphasis on physics in recent years that was triggered by the military interest in physics-based technologies, or the very similar present-day passion for computers and computer science. An innovative society is one in which there is a strong interest in bringing the best available knowledge to bear on ameliorating society's problems and on taking advantage of its opportunities. It is not surprising that scientific objectives and choices in that society should be in large measure determined by what those problems and opportunities are, which does not mean, however, that scientific objectives are identical with or must remain tied to social objectives.

Another way in which a society of change influences its patterns of inquiry is by putting a premium on the formulation of new questions and, in general, on the synthetic aspects of knowing. Such a society is by description one that probes at scientific and intellectual frontiers, and a scientific frontier, according to the biologist C. H. Waddington, is where "we encounter problems about which we cannot yet ask sensible questions" (17). When change is prevalent, in other words, we are frequently in the position of not knowing just what we need to know. A goodly portion of the society's intellectual effort must then be devoted to formulating new research questions or reformulating old ones in the light of changed circumstances and needs so that inquiry can remain pertinent to the social problems that knowledge can alleviate.

Three consequences follow. First, there is a need to reexamine the knowledge already available for its meaning in the context of the new questions. This is the synthetic aspect of knowing. Second, the need to formulate new questions coupled with the problem-orientation, as distinct from the discipline-orientation, discussed above requires that answers be sought from the intersection of several disciplines. This is the impetus for current emphases on the importance of interdisciplinary or cross-disciplinary inquiry as a supplement to the academic research aimed at expanding knowledge and training scientists. Third, there is a need for further institutionalization of the function of transferring scientific knowledge to social use. This process, which began in the late 19th century with the creation of large central research laboratories in the chemical, electrical, and communications industries, now sees universities spawning problem- or area-oriented institutes, which surely augur eventual organizational change, and new policy-oriented research organizations arising at the borderlines of industry, government, and universities, and in a new no-man's land between the public and private sectors of our society.

A fundamental intellectual implication of a world of change is the greater theoretical utility of the concept of process over that of structure in sociological and cultural analysis. Equilibrium theories of various sorts imply ascription of greater reality to stable sociocultural patterns than to social change. But as the anthropologist Evon Vogt argues, "change is basic in social and cultural systems ... Leach [E. R.] is fundamentally correct when he states that 'every real society is a process in time.' Our problem becomes one of describing, conceptualizing, and explaining a set of ongoing processes ... ," but none of the current approaches is satisfactory "in providing a set of conceptual tools for the description and analysis of the *changing* social and cultural *systems* that we observe" (*18*).

There is no denial of structure: "Once the processes are understood, the structures manifested at given time-points will emerge with even greater clarity," and Vogt goes on to distinguish between short-run "recurrent processes" and long-range and cumulative "directional processes." The former are the repetitive "structural dynamics" of a society. The latter "involve alterations in the structures of social and cultural systems" (*18*, pp. 20-22). It is clear that the latter, for Vogt, are more revelatory of the essence of culture and society as changing (*19*).

Social Trends

Heraclitus' philosophy of universal change was a generalization from his observation of physical nature, as is evident from his appeal to the four elements of ancient physics (fire, earth, water, and air) in support of it (*20*, p. 37, fragments 28 to 34). Yet he offered it as a metaphysical generalization. Change, flux, is the essential characteristic of all of existence, not of matter only. We should expect to find it central also, therefore, to societies, institutions, values, population patterns, and personal careers.

We do. Among the effects of technological change that we are beginning to understand fairly well even now are those (i) on our principal institutions: industry, government, universities; (ii) on our

How Technology Will Shape the Future

production processes and occupational patterns; and (iii) on our social and individual environment: our values, educational requirements, group affiliations, physical locations, and personal identities. All of these are in movement. Most are also in process; that is, there is direction or pattern to the changes they are undergoing, and the direction is moreover recognizable as a consequence of the growing social role of knowledge induced by proliferation of new technology.

It used to be that industry, government, and universities operated almost independently of one another. They no longer do, because technical knowledge is increasingly necessary to the successful operation of industry and government, and because universities, as the principal sources and repositories of knowledge, find that they are adding a dimension of social service to their traditional roles of research and teaching. This conclusion is supported (i) by the growing importance of research, development, and systematic planning in industry; (ii) by the proliferation and growth of knowledge-based industries; (iii) by the changing role of the executive, who increasingly performs sifting, rearranging, and decision operations on ideas that are generated and come to him from below; (iv) by the entry of technical experts into policy-making at all levels of government; (v) by the increasing dependence of effective government on availability of information and analysis of data; (vi) by the importance of education and training to successful entry into the society and to maintenance of economic growth; and (vii) by the growth, not only of problem-oriented activities on university campuses, but also of the social role (as consultants, advisory boards, and so forth) of university faculties.

The economic affluence that is generated by modern industrial technology accelerates such institutional mixing-up by blurring the heretofore relatively clear distinction between the private and public sectors of society. Some societies, like the Scandinavian, can put a strong emphasis on the acquisition of public or social goods even in the absence of a highly productive economy. In our society, affluence is a precondition of such an emphasis. Thus, as we dispose increasingly of resources not required for production of traditional consumer goods and services, they tend to be devoted to providing such public goods as education, urban improvement, clean air, and so forth.

What is more, goods and services once considered private more and more move into the public sector, as in scientific research and graduate education or the delivery of medical care. As we thus "socialize" an increasing number of goods once considered private, however, we tend also to farm out their procurement to private institutions, through such devices as government grants and contracts. As a result,

> ... our national policy assumes that a great deal of our new enterprise is likely to follow from technological developments financed by the government and directed in response to government policy; and many of our most dynamic industries are largely or entirely dependent on doing business with the government through a subordinate relationship that has little resemblance to the traditional market economy (*21*).

Another observer says:

> Increasingly it will be recognized that the mature corporation, as it develops, becomes part of the larger administrative complex associated with the state. In time the line between the two will disappear (*22*).

This fluidity of institutions and social sectors is not unreminiscent of the more literal fluidity that Heraclitus immortalized: "You cannot step twice into the same river, for other waters are continually flowing on" (*20*, p. 29, fragment 21). But also like the waters of a river, the institutional changes of a technologically active age are not aimless; they have direction, as noted, toward an enhancement of the use of knowledge in society:

> Perhaps it is not too much to say that if the business firm was the key institution of the past hundred years, because of its role in organizing production for the mass creation of products, the university will become the central institution of the next hundred years because of its role as the new (*sic*) source of innovation and knowledge (*11*, p. 30).

One should recall, in this connection, that the university is the portal through which more and more people enter into productive roles in society and that it increasingly provides the training necessary for leadership in business and government as well as in education.

The considerable debate of the last few years about the implications of technological change for employment and the character of work has not been in vain. Positions originally so extreme as to be untenable have been tempered in the process. Few serious students of the subject believe any longer that the progress of mechanization and automation in industry must lead to an irreversible increase in the level of *involuntary* unemployment in the society, whether in the form of unavailability of employment, or of a shortening work week, or of lengthening vacations, or of an extension of the period of formal schooling. These developments may occur, either voluntarily, because people choose to take some of their increased productivity in the form of leisure, or as a result of inadequate education, poor social management, or failure to ameliorate our race problem. But reduction of the overall level of employment is not a necessary consequence of new industrial technology.

Too much is beginning to be known about what the effects of technology on work and employment in fact are, on the other hand, for them to be adequately dealt with as merely transitional disruptions consequent on industrialization. A number of economists are therefore beginning to move away from explanation in terms of transition (which is typical of traditional equilibrium theory) to multi-level "steady state" models, or to dynamic theories of one sort or another, as more adequate to capturing the reality of constant change in the economy (*23*).

The fact is that technological development has provided substitutes for human muscle power and mechanical skills for most of history. Developments in electronic computers are providing mechanical substitutes for at least some human mental operations. No technology as yet promises to duplicate human creativity, especially in the artistic sense, if only because we do not yet understand the conditions and functioning of creativity. (This is not to deny that computers can be useful aids to creative activity.) Nor are there in the offing mechanical equivalents for the initiatives inherent in human emotions, although emotions can of course be affected and modified by drugs or electrical means. For the foreseeable future, therefore, one may hazard the prediction that distinctively human work will be less and less of the "muscle and elementary mental" kind, and more and more of the "intellectual, artistic, and emotional" kind. (This need *not* mean that only highly inventive or artistic people will be employable in the future. There is much sympathy needed in the world, for example, and the provision of it is neither mechanizable nor requisite of genius. It is illustrative of what I think of as an "emotional" service.)

While advancing technology may not displace people by reducing employment in the aggregate, therefore, it unquestionably displaces some jobs by rendering them more efficiently performed by machines than by people. There devolves on the society, as a result, a major responsibility for inventing and adopting mechanisms and procedures of occupational innovation. These may range from financial and organizational innovations for diverting resources to neglected public needs to social policies which no longer treat human labor as a market commodity. Whatever the form of solution, however, the problem is more than a "transitional" one. It represents a qualitative and permanent alteration in the nature of human society consequent on perception of the ubiquity of change.

This perception and the anticipatory attitude that it implies have some additional consequences, which are not less important but are as yet less well understood even than those for institutional change and occupational patterns. For example, lifetime constancy of trade or profession has been a basis of personal identity and of the sense of individuality. Other bases for this same sense have been identification over time with a particular social group or set of groups as well as with physical or geographical location.

All of these are now subject to Heraclitean flux. The incidence of life-long careers will inevitably lessen, as employing institutions and job contents both change. More than one career per lifetime is likely to be the norm henceforth. Group identities will shift as a result: every occupational change will involve the individual with new professional colleagues, and will often mean a sundering from old friends and cultivation of new ones. Increasing geographical mobility (already so characteristic of advanced industrial society) will not only reinforce these impermanencies, but also shake the sense of identity traditionally associated with ownership and residence upon a piece of land. Even the family will lose influence as a bastion of personality, as its loss of economic *raison d'être* is supplemented by a weakening of its educational and socializing functions and even of its prestige as the unit of reproduction (*24*).

I have alluded elsewhere to the implications for education of a world seen as essentially changing (*13*). Education has traditionally had the function of preparing youth to assume full membership in society (i) by imparting a sense for the history and accumulated knowledge of the race, (ii) by imbuing the young with a sense of the culture, mores, practices, and values of the group, and (iii) by teaching a skill or set of skills necessary to a productive social role. Philosophies of education have accordingly been elaborated on the assumption of stability of values and mores, and on the up-to-now demonstrable principle that one good set of skills well learned could serve a man through a productive lifetime.

This principle is undermined by contemporary and foreseeable occupational trends, and the burden of my general argument similarly disputes the assumption of unalterable cultural stabilities. There are significant implications for the enterprise of education. They include (i) a decline in the importance of manual skills, (ii) a consequent rising emphasis on general techniques of analysis and evaluation of alternatives, (iii) training in occupational flexibility, (iv) development of management skills, and (v) instruction in the potentialities and use of modern intellectual tools. The major problem of contemporary education at the primary and secondary levels is that the educational establishment is by and large unprepared, unequipped, and poorly organized to provide education consonant with these realities.

Above all, perhaps, higher education especially will need to attend more deliberately and systematically than it has in recent decades to developing the reflective, synthetic, speculative, and even the contemplative capacities of men, for understanding may be at a relatively greater premium henceforth than particular knowledge. When we can no longer lean on the world's stabilities, we must be able to rely on new abilities to cope with change and be comfortable with it.

There is an analogous implication for social values and for the human enterprise of valuing. There is concern expressed in many quarters these days about the threat of technology to values. Some writers go so far as to assert an incompatibility between technology and values and to warn that technological progress is tantamount to dehumanization and the destruction of all value (25).

There is no question, as noted earlier, that technological change alters the mix of choices available to man and that choices made ipso facto preclude other choices that might have been made. Some values are destroyed in this process, which can thus involve punishing traumata of adjustment that it would be immoral to ignore. It is also unquestionably the case that some of the choices made are constrained by the very technology that makes them available. In such cases, the loss of value can be tragic, and justly regretted and inveighed against. It is in the hope of anticipating such developments that we are currently investigating means to assess and control technological development in the public interest.

On the other hand, I find no justification for the contention that technological progress must of necessity mean a progressive destruction of value. Such fears seem rather to be based partly on psychological resistance to change and partly on a currently fashionable literary mythology that interprets as a loss the fact that the average man today does not share the values that were characteristic of some tiny elites centuries ago. To the extent that it is more than that, this contention is based, I think, on a fundamental misunderstanding of the nature of value.

The values of a society change more slowly, to be sure, than the realities of human experience; their persistence is inherent in their emergence as values in the first place and in their function as criteria, which means that their adequacy will tend to be judged later rather than earlier. But values do change, as a glance at any history will show. They change more quickly, moreover, the more quickly or extensively a society develops and introduces new technology. Since technological change is so prominent a characteristic of our own society, we tend to note inadequacies in our received values more quickly than might have been the case in other times. When that perception is coupled with the conviction of some that technology and value are inherently inimical to each other, the opinion is reinforced that the advance of technology must mean the decline of value and of the amenities of distinctively human civilization.

While particular values may vary with particular times and particular societies, however, the activity of valuing and the social function of values do not change. That is the source of the stability so necessary to human moral experience. It is not to be found, nor should it be sought, exclusively in the familiar values of the past. As the world and society are seen increasingly as processes in constant change under the impact of new technology, value analysis will have to concentrate on process, too: on the process of valuation in the individual and on the process of value formation and value change in the society. The emphasis will have to shift, in other words, from values to valu*ing*. For it is not particular familiar values as such that are valuable, but the human ability to extract values from experience and to use and cherish them. And that value is not threatened by technology; it is only challenged by it to remain adequate to human experience by guiding us in the reformulation of our ends to fit our new means and opportunities.

Political Trends

There are a number of respects in which technological change and the intellectual and social changes it brings with it are likely to alter the conditions and patterns of government. I construe government in this connection in the broadest possible sense of the term, as governance (with a small "g") of a *polity*. Better yet, I take the word as equivalent to govern*ing*, since the participle helps to banish both visions of statism and connotations of public officialdom. What I seek to encompass by the term, in other words, is the social decision-making function in general, whether exemplified by small or large or public or private groups. I include in decision-making, moreover, both the values and criteria that govern it and the institutions,

How Technology Will Shape the Future

mechanisms, procedures, and information by means of which it operates.

One notes that, as in other social sectors and institutions, the changes that technology purports for government are of a determinate sort—they have direction: they enhance the role of government in society and they enhance the role of knowledge in government.

The importance of decision-making will tend to grow relative to other social functions (relative to production, for example, in an affluent society): (i) partly because the frequency with which new possibilities are created in a technologically active age will provide many opportunities for new choices, (ii) partly because continuing alteration of the spectrum of available choice alternatives will shorten the useful life of decisions previously made, (iii) partly because decisions in areas previously thought to be unrelated are increasingly found to impinge on and alter each other, and (iv) partly because the economic affluence consequent on new technology will increase the scope of deliberate public decision-making at the expense, relatively, of the largely automatic and private charting of society's course by market forces. It is characteristic of our time that the market is increasingly distrusted as a goal-setting mechanism for society, although there is of course no question of its effectiveness as a signaling and controlling device for the formulation of economic policy.

Some of the ways in which knowledge increasingly enters the fabric of government have been amply noted, both above and in what is by now becoming a fairly voluminous literature on various aspects of the relation of science and public policy (26). There are other ways, in addition, in which knowledge (information, technology, science) is bound to have fundamental impacts on the structures and processes of decision-making that we as yet know little about.

The newest information-handling equipments and techniques find their way quickly into the agencies of federal and local government and into the operations of industrial organizations, because there are many jobs that they can perform more efficiently than the traditional rows of clerks. But it is notorious that adopting new means in order to better accomplish old ends very often results in the substitution of new ends for old ones (27). Computers and associated intellectual tools can thus, for example, make our public decisions more informed, efficient, and rational, and less subject to lethargy, partisanship, and ignorance. Yet that possibility seems to imply a degree of expertise and sophistication of policy-making and implementing procedures that may leave the public forever ill-informed, blur the lines between executive and legislature (and private bureaucracies) as all increasingly rely on the same experts and sources of information, and chase the idea of federalism into the history books close on the heels of the public-private separation.

There is in general the problem of what happens to traditional relationships between citizens and government, to such prerogatives of the individual as personal privacy, electoral consent, and access to the independent social criticism of the press, and to the ethics of and public controls over a new elite of information keepers, when economic, military, and social policies become increasingly technical, long-range, machine-processed, information-based, and expert-dominated.

An exciting possibility that is however so dimly seen as perhaps to be illusory is that knowledge can widen the area of political consensus. There is no question here of a naive rationalism such as we associate with the 18th-century enlightenment. No amount of reason will ever triumph wholly over irrationality, certainly, nor will vested interest fully yield to love of wisdom. Yet there are some political disputes and disagreements, surely, that derive from ignorance of information bearing on an issue or from lack of the means to analyze fully the probable consequence of alternative courses of action. Is it too much to expect that better knowledge may bring about greater political consensus in such cases as these? Is the democratic tenet that an informed public contributes to the commonweal pure political myth? The sociologist S. M. Lipset suggests not:

> Insofar as most organized participants in the political struggle accept the authority of experts in economics, miliary affairs, interpretations of the behavior of foreign nations and the like, it becomes increasingly difficult to challenge the views of opponents in moralistic "either/or" terms. Where there is some consensus among the scientific experts on specific issues, these tend to be removed as possible sources of intense controversy (*28*).

Robert E. Lane of Yale has made the point more generally:

> If we employ the term "ideology" to mean a comprehensive, passionately believed, self-activating view of society, usually organized as a social movement... it makes sense to think of a domain of knowlege distinguishable from a domain of ideology, despite the extent to which they may overlap. Since knowledge and ideology serve somewhat as functional equivalents in orienting a person toward the problems he must face and the policies he must select, the growth of the domain of knowledge causes it to impinge on the domain of ideology (*12*, p. 660).

How Technology Will Shape the Future

Harvey Brooks, finally, draws a similar conclusion from consideration of the extent to which scientific criteria and techniques have found their way into the management of political affairs. He finds an

> increasing relegation of questions which used to be matters of political debate to professional cadres of technicians and experts which function almost independently of the democratic political process.... The progress which is achieved, while slower, seems more solid, more irreversible, more capable of enlisting a wide consensus (29).

I raise this point as fundamental to the technology-polity relationship, not by way of hazarding a prediction. I ignore neither the possibility that *value* conflicts in political debate may become sharper still as factual differences are muted by better knowledge, nor the fact that decline of political ideiology does not *ipso facto* mean a decline of political disagreement, nor the fear of some that the hippie movement, literary anti-intellectualism, and people's fears of genuine dangers implicit in continued technological advance may in fact augur an imminent retreat from rationality and an interlude—perhaps a long interlude—either of political know-nothingism reminiscent of Joseph McCarthy or of social concentration on contemplative or religious values.

Yet, if the technology-values dualism is unwarranted, as I argued above, it is equally plausible to find no more warrant in principle in a sharp separation between knowledge and political action. Like all dualisms, this one too may have had its origins in the analytic abhorrence of uncertainty. (One is reminded in this connection of the radical dualism that Descartes arrived at as a result of his determination to base his philosophy on the only certain and self-evident principle he could discover.) There certainly is painfully much in political history and political experience to render uncertain a positive correlation between knowledge and political consensus. The correlation is not necessarily absent therefore, and to find it and lead society to act on it may be the greatest challenge yet to political inquiry and political action.

To the extent that technological change expands and alters the spectrum of what man can do, it multiplies the choices that society will have to make. These choices will increasingly have to be deliberate social choices, moreover, rather than market reflections of innumerable individual consumer choices, and will therefore have to be made by political means. Since it is unlikely—despite futurists and technological forecasters—that we will soon be able to predict future opportunities (and their attendant opportunity costs) with any significant degree of reliability in detail, it becomes important to investigate the conditions of a political system (I use the term in the wide sense I assigned to "government" above) with the flexibility and value presuppositions necessary to evaluate alternatives and make choices among them as they continue to emerge.

This prescription is analogous, for governance of a changing society, to those advanced above for educational policy and for our approach to the analysis of values. In all three cases, the emphasis shifts from allegiance to the known, stable, formulated, and familiar, to a *posture* of expectation of change and readiness to deal with it. It is this kind of shift, occurring across many elements of society, that is the hallmark of a truly Heraclitean age. It is what Vogt seeks to formalize in stressing processual as against structural analysis of culture and society. The mechanisms, values, attitudes, and procedures called for by a social posture of readiness will be different in kind from those characteristic of a society that sees itself as mature, "arrived," and in stable equilibrium. The most fundamental *political* task of a technological world, in other words, is that of systematizing and institutionalizing the social expectation of the changes that technology will continue to bring about.

I see that task as a precondition of profiting from our accumulating knowledge of the effects of technological change. To understand those effects is an intellectual problem, but to do something about them and profit from the opportunities that technology offers is a political one. We need above all, in other words, to gauge the effects of technology on the *polity,* so that we can derive some social value from our knowledge. This, I suppose, is the 20th-century form of the perennial ideal of wedding wisdom and government (*30*).

Notes

1. Herman Kahn and his associates at the Hudson Institute have made a major effort in this direction in *Toward the Year 2000: A Framework for Speculation* (New York: Macmillan, 1967).

2. L. White, Jr., *Medieval Technology and Social Change* (New York: Oxford Galaxy Book, 1966), p. 28.

3. Melvin Kranzberg has made this same point in *Virginia Quarterly Review* 10, no. 4 (1964): 591.

4. I deal extensively (albeit in a different context) with the making of new possibilities and the preclusion of options by making choices in my *How Language Makes Us Know* (The Hague: Nijhoff, 1964), ch. 3.

5. One such case is described by E. Z. Vogt, *Modern Homesteaders: The Life of a Twentieth-Century Frontier Community* (Cambridge, Mass.: Harvard University Press, 1955). The Mennonite sects in the Midwest are another example.

How Technology Will Shape the Future

6. J. H. Steward, *Theory of Culture Change: The Methodology of Multi-linear Evolution* (Urbana: University of Illinois Press, 1955), p. 37. Steward is generally critical of such fellow anthropologists as Leslie White and Gordon Childe for adopting strong positions of technological determinism. Yet even Steward says, "White's . . . 'law' that technological development expressed in terms of man's control over energy underlies certain cultural achievements and social changes [has] long been accepted" (p. 18).

7. See L. White, Jr. (*2*, pp. 44 ff. and 28 ff.). Note especially White's contention that analysis of the influence of the heavy plow has survived all the severe criticisms leveled against it.

8. K. J. Arrow, "Public and Private Values," in S. Hook, ed., *Human Values and Economic Policy* (New York: New York University Press, 1967), p. 4.

9. R. L. Heilbroner points up this unforeseeable element—he calls it the "indirect effect" of technology—in *The Limits of American Capitalism* (New York: Harper and Row, 1966), p. 67.

10. J. Dewey, "Theory of Valuation," in *International Encyclopedia of Unified Science*, 2, no. 4 (Chicago: University of Chicago Press, 1939). The model of the ends-means continuum that Dewey developed in this work should prove useful in dealing conceptually with the value changes implicit in new technology.

11. D. Bell, *The Public Interest*, no. 6 (Winter, 1967). See also R. E. Lane (*12*) for evidence and a discussion of some of the political implications of this development.

12. R. E. Lane, *American Sociological Review*, 31, no. 5 (October 1966), 652.

13. E. G. Mesthene, *Technology and Culture*, 6, no. 2 (Spring 1965), 226. D. A. Schon also has recently recalled Heraclitus for a similar descriptive purpose and has stressed how thoroughgoing a revolution of attitudes is implied by recognition of the pervasive character of change [*Technology and Change* (New York: Delacorte Press, 1967), p. xi ff.].

14. In addition to Bell and Lane (*11*), see also L. K. Caldwell, *Public Administration Review*, 27, no. 3 (June 1967); R. L. Heilbroner, (*9*); and A. F. Westin, *Columbia Law Review*, 66, no. 6 (June 1966), 1010.

15. See "Social Goals and Indicators for American Society," *Annals of the American Academy of Political and Social Sciences*, 1 (May 1967); and 2 (September 1967).

16. R. K. Merton, *Social Theory and Social Structure* (Glencoe, Ill.: Free Press, 1949), p. 348. Hessen's analysis is in "The Social and Economic Roots of Newton's Mechanics," *Science at the Crossroads* series (London: Kniga, no date). The paper was read at the Second International Congress of the History of Science and Technology (June 29–July 3, 1931).

17. Quoted in *Graduate Faculties Newsletter* (Columbia University, March 1966).

18. E. Z. Vogt, *American Anthropologist*, 62 (1960), 1, 19–20.

19. The structure-process dualism also has its familiar philosophical face, of course, which a fuller treatment than this paper allows should not ignore. Such a discussion would recall at least the metaphysical positions that we associate with Aristotle, Hegel, Bergson, and Dewey.

20. P. Wheelwright, *Heraclitus* (New York: Atheneum, 1964). In his commentary on fragments 28 to 34, Wheelwright makes clear that the element of

fire which looms so large in Heraclitus remains a physical actuality for him, however much he may also have stressed its symbolic character (pp. 38-39).

21. D. K. Price, *The Scientific Estate* (Cambridge, Mass.: Harvard University Press, 1965), p. 15.

22. J. K. Galbraith, *The New Industrial State* (Boston: Houghton Mifflin, 1967), p. 393.

23. I am indebted to conversations with John R. Meyer and to a personal communication from Robert M. Solow for clarification of this point.

24. These points are made and discussed by R. S. Morison, "Where Is Biology Taking Us?' in E. and E. Hutchings, eds., *Scientific Progress and Human Values* (New York: Elsevier, 1967), p. 121 ff.

25. Examples of such apocalyptic literature are: J. Ellul, *The Technological Society* (New York: Knopf, 1964); D. Michael, *Cybernation: The Silent Conquest* (Santa Barbara, Calif.: Center for the Study of Democratic Institutions, 1962); and J. W. Krutch, *New York Times Magazine*, July 30, 1967.

26. See, for example, the section "Science, Politics and Government" in L. K. Caldwell, *Science, Technology and Public Policy: A Selected and Annotated Biliography* (Bloomington: Indiana University Press, 1968).

27. For a more extended discussion, see E. G. Mesthene, *Public Administration Review*, 27, no. 2 (June 1967).

28. S. M. Lipset, *Daedalus*, 93 (1964), 273.

29. H. Brooks, "Scientific Concepts and Cultural Change," in G. Holton, ed., *Science and Culture* (Boston: Houghton Mifflin, 1965), p. 71.

30. I thank my colleagues for comments during the preparation of the article, and especially Irene Taviss of the Harvard Program on Technology and Society.

3. The Artificial Environment: Disneyland Is Better

THEODORE ROSZAK

The Technological Coalescence: Its Use and Abuse

By far the most highly (and deservedly) celebrated aspect of urban-industrial expansion is the worldwide technological coalescence it has brought about. The physical isolation of cultures has been abolished and all people everywhere are now drawn into a common fate. No age before ours has possessed such a wealth of materials with which to draw the self-portrait of the human race. Not that we needed modern transport and communication to inspire the vision of human universality. The fellowship of mankind and the ethical energy it demands (need one recall?) have been before us as an ideal at least since the age

From *Where the Wasteland Ends* by Theodore Roszak. Copyright © 1972 by Theodore Roszak. Reprinted by permission of Doubleday & Company, Inc. and by permission of Faber and Faber Ltd.

The Artificial Environment: Disneyland Is Better

of Buddha and the rhapsodic prophets. What technology has brought us is the superior means of shared experience and, under the threat of thermonuclear annihilation, the necessity of universal caring. It has brought the apparatus that enables us, and the crisis that forces us to turn the old vision into an historical project.

But in the very process of achieving this new world unity, we—and by "we" I here mean peculiarly the white western middle classes—have suffered a gross distortion of the sensibilities. It is as if in amassing the sheer physical energy the technological coalescence demanded of us, we have burned away our deeper awareness of the opportunity and responsibility that are uniquely ours. In the chapters that follow I hope to suggest how this degeneration of consciousness has come about in our culture.

Our technology ought to be the means of a universal cultural sharing, and we who "discovered" the rest of the world should be the most eager to share, learn, and integrate. Instead, from the outset, we have mistaken the invention and possession of this means as the self-evident sign of cultural superiority, and have at last made the technology itself (and the science on which it is based) a culture in its own right—the *one* culture to be uniformly imitated or imposed everywhere. Our technology, freighted with all the prestige the western hegemony can lend it, communicates nothing so efficiently as itself: its attendant values and ideologies, its obsession with power-knowledge, above all the underdeveloped worldview from which it derives. Though some may celebrate such ironies, what is it but the death of dialogue when the medium blocks out every message and asserts only itself? In the communist nations of Europe and Asia alone during this century, we have seen more than one third of the human race scrap its finest traditions wholesale in order to adopt a Marxist-Leninist ideology of development which is but a comic caricature of nineteenth-century bourgeois scientism.

There have been the few—the artists, the philosophers, the scholars—who have risen to the level of the age and have used the technological coalescence as a true forum where the cultures of past and present may meet as self-respecting equals. And perhaps the young people one now sees back-packing their way across the globe have tried to enter the world more authentically than the legions of tourists who bounce from one vacation paradise to another. But at this point the currents of change do not run in the direction of dialogue and sharing. The international unities that matter significantly remain those of trade, warfare, and technics: the unities of power. The world is being bound together by the affluent societies in ingenious networks of investment, military alliance, and commerce which, in themselves, can only end by propagating an oppressive urban-industrial uniformity over the earth. Yet there is no lack of "forward-looking" opinion makers who accept that uniformity as the highest expression of a world culture. They mistake the homogenized architecture of airports, hotels, and conference centers—which is as much as many jet-set intellectuals ever see of the world—for an authentic sharing and synthesis of sensibilities.

Either that, or they identify as a glorious multiplication of "options" the glossy cosmopolitan eclecticism now available to the educated and citified. But even as urban-industrial populations have the varied cultural fare of the world laid before them in their theater, television, cinema, art galleries, restaurants, supermarkets, book shops, and newsstands, urban-industrialism expands aggressively to blanket the globe, crowding out every alternative that does not agree to reduce itself to the status of museum piece. Thus, while urban-industrial society grows intellectually fat on a smorgasbord of cultural tidbits, the world as a whole is flattened beneath the weight of its appetite and becomes steadily poorer in real *lived* variety. Herman Kahn gazes into the future and sees a "mosaic society" containing innumerable alcoves and enclaves of minority taste and custom. So too Alvin Toffler sees our society as one that enjoys "an explosive extension of freedom," "a crazy-quilt pattern of evanescent lifestyles."[1] Yes, but both the mosaic and the crazy quilt have a design—and the design is that of an urban-industrial global monopoly.

Within that cultural monopoly and on its terms, we may have a variety of nostalgic fads and eccentric novelties, hobbies and entertainments; but nothing for real and for keeps. And outside the monopoly, nothing, nothing, nothing at all. In another two generations, there will be no primitive or tribal societies left anywhere on earth—and they are not all giving up their traditional ways because they freely choose to. In another three generations, no self-determining rural life. In another four generations, no wildlife or wilderness on land or sea outside protected areas and zoos. Today there are few societies where official policy works to preserve wilderness and the old ways of life as serious alternatives to the urban-industrial pattern; at best, they are being embalmed and tarted up as tourist attractions. Of course there are those who think the accessible counterfeit is far superior to any reality one must take pains to approach and know. After all, the whole force of urban-industrialism upon our tastes is to convince us that artificiality is not only inevitable, but better—perhaps finally to shut the real and original out of our awareness entirely. It was about such corrupted tastes that Hans Christian Andersen wrote his story of the nightingale, one of the supreme parables of our time. How many still read and understand it?

The Artificial Environment: Disneyland Is Better

As things stand now, we are apt to have a worldwide artificial environment long before an authentic world culture or an ethical community of mankind has had the chance to become more than the failed aspiration of a few sensitive, inquring minds. The Oneness of humanity properly means communion of vision and concern, and from this a new integration of consciousness. Instead, we have settled for the artificial environment under the debased leadership of the technocratic elites we shall discuss in the chapter that follows; we have begun with a mere material sameness and physical integration, and now use that environment as the procrustean bed to whose size all values and sensibilities must be tailored.

And how rapidly it is happening. The Coca-colonization of the world within little more than a century's time, the emblems and political imperatives of urban-industrialism carried to every remote corner of the earth. How can any people be so sure about so much?

Is the alternative to the urban-industrial dominance some manner of anti-technological, anti-urban fanaticism? I think not, and, as a born and bred urbanite who craves the life of the city, I would hardly be the person to champion such an extreme position. The fact is that much of my resistance to the artificial environment stems from its poisonous effect upon the qualities I value in an authentic city life. Megalopolis is as much the destroyer of cities as of wilderness and rural society—a point we will return to where I hope to suggest a more modest and life-enhancing place for the city and the machine in our world.

Disneyland Is Better

Until somewhere within the last few generations, the prevailing popular attitude toward untamed nature in western society was one of belligerent distrust grounded in fear. The face of nature was the face of wolf and tiger, plague and famine, insurmountable obstruction and inscrutable mystery. Nature was Goliath, oppressor and enemy against whom we held no more likely weapon than the slingshot of our monkey cunning. The odds were not in our favor.

But then, quite suddenly, the war against nature came to an end for the urban-industrial millions; it was over and decisively won. The turning point is marked by no single event, but by the stupendous accumulation of discoveries and new technical possibilities that have followed the advent of atomic energy. True, journalists may still speak of outer space as hostile territory needing to be "conquered"; a few recalcitrant diseases hold out against us; now and then we suffer a natural disaster or two. But who (besides a few worrywart ecologists) doubts that even these remaining obstacles to human dominance will

Theodore Roszak

in time be brought under control? And meanwhile we know how to make good the damage, rebuild to our profit, and carry on to still greater heights. Clearly the old adversary's teeth have been drawn. In the stronghold of our artificial environment, what have we any longer to fear from the non-human? All the mountains have been scaled, all the sea-depths explored. The course of great rivers and the behavior of our genes can now be redesigned to our specifications.

So nature—or at least what most of us still know of her through the mediation of technical manipulation and scientific authority—no longer wears the look of a serious opponent. Instead, she has come to seem, in her humiliation, rather pathetic and (once we puzzle out her tricky mechanisms) highly incompetent, needing to be taken in hand and much improved. There seems to be nothing she sets before us that we could not dismantle if we put our mind to it; nothing she does that we cannot find a way of doing bigger, better, faster. Even her grandeur survives only by our sufferance. Not long ago, there were those in the Pentagon who talked seriously of using the moon as a missile testing target, perhaps of knocking off a few sizable chunks. What a show of military prowess that would be!

This is the monumental novelty the artificial environment has introduced into the mindscape of our time: that the non-human world should have become for us an object of such casual contempt. As if our vision, closed in by these glass and concrete urban contours, could no longer look before and after and discern the context of our being. Of course it is only proper that people, like all living things, should strive to make a livable place for themselves in the world. Who can imagine it otherwise? But the state of soul in which we undertake the project—*that* is what makes the difference.

A personal anecdote: a sign of the times. Several years ago, on a cross-country trip, my family and I were foolish enough to visit Yellowstone National Park during the summer crush. We expected it to be crowded but hardly—as it was—more congested than downtown San Francisco. The traffic for forty miles leading into the park and all the way through was packed bumper to bumper. There was not an inch of solitude or even minimal privacy to be found anywhere during the two days we stayed before giving up and leaving. Never once were we out of earshot of chattering throngs and transistor radios or beyond the odor of automobile exhaust. We spent a major part of each day waiting in line to buy food or get water. And at night, the coming and going of traffic and crowds made sleep nearly impossible. Ah, wilderness...

All this was bad enough to convince me that, if any true respect for wildlife survived in me, my duty hereafter was to stay away from it

The Artificial Environment: Disneyland Is Better

and spare it my contribution to such desecration. But the worst came on the day we decided to visit Old Faithful. At the site we discovered what might have been a miniature football stadium: a large circle of bleachers packed thick with spectators devouring hot dogs and swilling soft drinks while vendors passed among them hawking souvenirs.

The center of attraction was the geyser which—as all the world must know—erupts promptly once each hour. To prove the point, a large clock stood nearby with a sign indicating exactly when the next eruption would happen. Loudspeakers were located around the area, and through them a park guide kept up a steady monologue giving every conceivable statistic about the geyser's geology and performance—as if to certify its status as a veritable marvel of nature. But most of what he said was drowned in the roar of the crowd.

Beside us on the bleacher seat we occupied was a family of five, the father and three children equipped with ornate and expensive cameras. They were extremely impatient for the show to begin. All were guzzling Coke and timing the geyser, somewhat skeptically, on their wrist watches. Finally the great event came. The guide made one loud, urgent appeal for attention, the crowd stilled momentarily, and Old Faithful went off on schedule. For all the carnival atmosphere, it was an impressive phenomenon to witness. There was even a scattering of applause and cheers as the water gushed up in several ascending spurts.

But many of the spectators, like the father and three children beside me, were far too preoccupied to ogle the spectacle. For no sooner did the geyser begin to blow than their eyes vanished behind their cameras, as if for sure the best way to have this experience was through a lens and on film. The children, eager to catch Old Faithful at its very zenith, kept asking their father throughout the eruption, "Now, Dad? Now?" But the father, apparently expecting the geyser to split the heavens, restrained them. "No, not yet . . . wait till she really gets up there."

But Old Faithful only rises so high, and then the show is over. Before this family of photographers had snapped their pictures, the veritable marvel of nature had bubbled back beneath the earth. The father's face emerged from behind his camera, frowning with disbelief. "Is that all?" he asked, with the belligerent voice of a man who knows when he is not getting his money's worth. "I thought it was supposed to go higher," his wife said. The children were no less crushed with disappointment. As they got up to leave, one of the boys sourly remarked, "Disneyland is better." And the whole family agreed.

Theodore Roszak

The Scientization of Culture

Since Charlie Chaplin's *Modern Times,* people as the victims of their own technical genius have been a standard subject of popular comedy. But the ironies of progress are far more than laughingstock for casual satire. Even crusading reformers like Ralph Nader or Paul Ehrlich who gamely challenge each new criminality or breakdown of the system, often fail to do justice to the depth of the problem. While each failure of the technology may allow us to pillory a few profiteering culprits for incompetence or selfishness, the plain fact is that few people in our society surrender one jot of their allegiance to the urban-industrial system as a whole. They cannot afford to. The thousand devices and organizational structures on which our daily survival depends are far more than an accumulation of technical appendages that can be scaled down by simple subtraction. They are an interlocking whole from which nothing can easily be dropped. How many of us could tolerate the condition of our lives if but a single "convenience" were taken away . . . the telephone . . . automobile . . . air conditioner . . . refrigerator . . . computerized checking account and credit card . . . ? Each element is wedded to the total pattern of our existence; remove any one and chaos seems to impend. As a society, we are addicted to the increase of environmental artificiality; the agonies of even partial withdrawal are more than most of us dare contemplate.

From this fact springs the great paradox of the technological mystique: its remarkable ability to grow strong by virtue of chronic failure. While the treachery of our technology may provide many occasions for disenchantment, the sum total of failures has the effect of increasing our dependence on technical expertise. After all, who is there better suited to repair the technology than the technicians? We may, indeed, begin to value them and defer to them more as repairmen than inventors—just as we are apt to appreciate an airplane pilot's skill more when we are riding out a patch of severe turbulence than when we are smoothly under way. Thus, when the technology freezes up, our society resorts to the only cure that seems available: the hair of the dog. Where one technique has failed, another is called to its rescue; where one engineer has goofed, another—or several more—are summoned to pick up the pieces. What other choice have we? If modern society originally embraced industrialism with hope and pride, we seem to have little alternative at this advanced stage but to cling on with desperation. So, by imperceptible degrees, we license technical intelligence, in its pursuit of all the factors it must control, to move well beyond the sphere of "hardware" engineering—until it

The Artificial Environment: Disneyland Is Better

begins to orchestrate the entire surrounding social context. The result is a proliferation of what Jacques Ellul has called "human techniques": behavioral and management sciences, simulation and gaming processes, information control, personnel administration, market and motivational research, etc.,—the highest stage of technological integration.[2]

Once social policy becomes so determined to make us dance to the rhythm of technology, it is inevitable that the entire intellectual and moral context of our lives should be transformed. As our collective concern for the stability of urban-industrialism mounts, science—or shall we say the scientized temperament?—begins its militant march through the whole of culture. For who are the wizards of the artificial environment? Who, but the scientists and their entourage of technicians? Modern technology is, after all, the scientist's conception of nature harnessed and put to work for us. It is the practical social embodiment of the scientific worldview; and through the clutter of our technology, we gain little more than an occasional and distorted view of any other world. People may still nostalgically honor prescientific faiths, but no one—no priest or prophet—any longer speaks with authority to us about the nature of things except the scientists.

The scientists are no doubt what John Ziman has called them: "humble practical men who have learnt from experience the limitations of their arts, and know that they know very little." Caution and diffidence are the classic characteristics of the scientific personality. But Ziman goes on:

> Why should we even want to decide whether a particular discipline is scientific or not? The answer is, simply, that, *when it is available,* scientific knowledge is more reliable, on the whole, than non-scientific.... In a discipline where there is a scientific consensus the amount of *certain* knowledge may be limited, but it will be honestly labeled: "Trust your neck to this,"...[3]

The humility we are confronted with here is not unlike that of Undershaft's son in Shaw's *Major Barbara*. The modest young man had nothing more to say for himself than that he knew the difference between right and wrong.

If we lose sight of the arrogance that lies hidden in Ziman's "humble practical men" of science, who simply happen to know the difference between what is reliably so and what isn't and who possess the only means of telling the difference, it is doubtless because their claim makes such perfect sense to us. All the metaphysical and psychological premises of that claim have become the subliminal boundaries of the contemporary mindscape; we absorb them as if by osmosis from the artificial environment that envelops us and which has become the *only* environment we know. The scientists simply work within this universally endorsed reality. They search for what that reality defines as knowable and—with strict professional integrity—feel obliged to entertain no alternative realities. Scientific knowing becomes, within the artificial environment, the orthodox mode of knowing; all else defers to it. Soon enough the style of mind that began with the natural scientists is taken up by imitators throughout the culture. Until we find ourselves surrounded by every manner of scientific-technical expert, all of them purporting to know as the scientist knows: dispassionately, articulately, on the basis of empirical evidence or experiment, without idiosyncratic distortion, and if possible by the intervention of mathematics, statistics, or a suitably esoteric methodology.

In urban-industrial society, nothing enjoys much chance of being dignified as knowledge which does not proceed from such an epistemological stance. Therefore, we must have objective and specialized scrutiny of all things, of outer space and the inner psyche, of quasars and of sex, of history and of literature, of public opinion and personal neurosis, of how to learn and how to sleep and how to dream and how to raise babies and how to relax and how to digest and how to excrete. Take up the catalog of courses issued by any of our great multiversities—the mills of our "knowledge industry"—and there you will find the whole repertory of expertise laid out. And does it not cover every inch of the cultural ground with scientized speculation? What is there left that the non-expert can yet be said to know? Name it—and surely it will soon be christened as a new professional field of study with its own jargon and methodology, its own journals and academic departments.

The Systems Approach

Given the extent of this empire of expertise, it might seem at first glance incomprehensible that urban-industrial society has any serious problems left at all. With such a growing army of trained specialists at work accumulating data and generating theory about every aspect of our lives, we should surely have long since entered the New Jerusalem. Obviously we have not. But why? The answer most commonly forthcoming in recent years is that our expertise has simply not been properly co-ordinated; it has been practiced in too narrow, disorganized, and myopic a way. So we begin to hear of a new panacea: the "systems approach"—and in years to come, we shall undoubtedly be hearing a great deal more about it.

The Artificial Environment: Disneyland Is Better

Systems analysis, derived from World War II and Cold War military research, is the attempt to solve social problems by ganging up more and more experts of more and more kinds until every last "parameter" of the situation has been blanketed with technical competence and nothing has been left to amateurish improvisation. The method is frequently mentioned nowadays as the most valuable civilian spin-off of our aerospace programs. We are told that the same broad-gauged planning, management, and decision making that have succeeded in putting a man on the moon can now be used to redesign cities and reform education. In brief, systems analysis is the perfection of Ellul's "human technique." To quote the military-industrialist Simon Ramo, one of the great boosters of the approach, it is the effort to create a "multiheaded engineer," a "techno-political-econo-socio" expert

> who must include in "his" head the total intelligence, background, experience, wisdom, and creative ability in all aspects of the problem of applying science, and particularly, who must integrate his total intelligence... who must *mobilize* it all—to get real-life solutions to real-life problems.
>
> In this sense, then, a good systems-engineering team—and we have begun in the United States to develop such groups—combines individuals who have specialized variously in mathematics, physics, chemistry, other branches of physical science, psychology, sociology, business finance, government, and so on.[4]

Notably, the good systems team does not include poets, painters, holy men, or social revolutionaries, who, presumably, have nothing to contribute to "real-life solutions." And how should they, when, as Ramo tells us, the relevant experts are those who understand people "as members of a system of people, machines, matériel, and information flow, with specific well-described and often measurable performance requirements"? Above all, the experts are those who can place "quantiative measures on everything—very often, cost and time measures."

One might well ask at this point: but is this "science"?—this doctrinaire mathematicization of the person and society, this statistical manipulation of human beings as if they were so many atomic particles? In later chapters, I will try to trace the genealogy which does, in fact, make Simon Ramo and the terrible systematizers legitimate offspring of Galileo and Newton. But as far as contemporary politics is concerned, the question is wholly academic. Such behavioral engineering is taken without question to be science by those who practice it, by those who subsidize its practice, and by most of those upon whom it is practiced. And the promised reward for such unquestioning deference to scientific expertise is glowing indeed; no less, in Ramo's words, than a "golden age."

> Once most people are wedded to creative logic and objectivity to get solutions to society's problems, the world is going to be a lot better. Then maybe we can say an important thing, namely, that science and technology are then being used to the fullest on behalf of mankind.[5]

The message is clear. The ills that plague urban-industrial society are not techno-genetic in essence; they are not the result of a radically distorted relationship between human beings and their environment. Rather, they result from an as yet incomplete or poorly co-ordinated application of scientific expertise. The province of expertise must, therefore, be broadened and more carefully administered. Urban-industrial society, having hopelessly lost touch with life lived on a simpler, more "primitive" level, and convinced beyond question of the omnipotence of technical intelligence, can do no other than trust to ever greater numbers of experts to salvage the promise of industrialism. Accordingly, the principal business of education becomes the training of experts, for whom progressively more room is made available in government and the economy. From our desperate conviction that the endangered artificial environment is the *only* livable environment, there stems the mature social form of industrial society: the technocracy.

The Citadel of Expertise

The continuities of tradition, ideology, and inherited institutions—severely ruptured though they have been by the urban-industrial revolution—may never allow a perfect technocracy to come into existence in any society. Certainly the pure technocratic system that Francis Bacon heralded in his *New Atlantis* and which Saint-Simon, Auguste Comte, and Thorstein Veblen were later to champion, is best regarded as no more than an ideal type of what Saint-Simon called "the administrative state"—the society beyond politics to which social engineers like Simon Ramo might aspire. If one were to declare oneself foursquare for such an institutionalized regime of the experts, the reaction from most citizens and politicians would doubtless be hostile. "Technocracy Incorporated," the American political movement of the 1920s and 1930s which openly advocated such an H. G. Wells government of the technicians, attracted only a meager following—and understandably so. Our national societies are committed one and all

The Artificial Environment: Disneyland Is Better

to the preservation of ideals, symbols, and rhetoric left over from the preindustrial past or from nineteenth-century ideologies which spoke the language of class struggle and political economy, not technics. These cultural remnants are still very much with us, and there is no denying that in various ways they retard the speedy and open maturation of the technocracy. Politics must still wear a political costume; it must have somewhat to do with elected representation, ethical debate over policy, and competition for office. At least for the sake of appearances, the hands on the levers of power must be those of politicians—and the politicians must continue to pretend they serve the general will of the citizen body as expressed by conventional electoral means.

Yet it is the essence of social criticism to enumerate the ways in which the democratic intention of such political principles has been utterly subverted with one degree of subtlety or another by those possessing money or guns, property or bureaucratic privilege. The language and iconography of democracy dominate all the politics of our time, but political power is no less elitist for all that. So too the technocracy continues to respect the formal surface of democratic politics; it is another, and this time extraordinarily potent means of subverting democracy from within its own ideals and institutions. It is a citadel of expertise dominating the high ground of urban-industrial society, exercising control over a social system that is utterly beholden to technician and scientist for its survival and prosperity. It is, within modern society, what the control of the sacramental powers was to the medieval church—the monopoly of all that people value and revere: material plenty, physical power, a reliable and expanding body of knowledge. To be an expert or (as is more often the arrangement) to *own* the experts in the period of high industrialism is to possess the keys to the kingdom, and with them a power that neither guns nor money alone can provide, namely mastery over the artificial environment, which is the only reality most people any longer know.

Contemporary politics tends toward technocracy wherever the mystique of industrial expansion or the zeal for development come to dominate the hearts and minds of people—and that means just about everywhere on earth. Whether we are dealing with societies as superindustrialized as America or as underdeveloped as many new African and Asian states, competent, realistic, responsible government means a crisp air of bustle and forward-looking dynamism, with a steady eye on the growth rate and the balance of payments. It means much confabulation with the statistical enigmas of the economic index and a sober concern for the big technological inevitables to which one's society *must* without question adapt. It means, therefore, being smartly in step with the best technical and economic expertise money can buy. "Like a national flag and a national airline," E. J. Mishan observes, "a national plan for economic growth is deemed an essential item in the paraphernalia of every new nation state."[6]

Even the surviving sheiks and shahs of the world must cloak their public relations in pretensions of technological glory; they must be able to pose confidently for pictures against a background of hydroelectric projects, freeways, air terminals. Everywhere it is the appearance (if not always the reality) of change, newness, efficiency that legitimates political control. Nothing else will qualify in the eyes of the world as a proper claim to power and privilege. Even where the state is barbarically repressive, the governing regime will seek to gild its image with the evidence of economic progress. Think, for example, of how our state department justifies its support for dictatorships in Brazil and Spain, Taiwan and South Korea. These, we are assured, are progressive, development-oriented governments. We need only be told that the growth rate in Brazil in recent years has been above 9 per cent, and enough said. Justice, freedom, dignity are almost unanimously understood to be the automatic results of successful development... in time. That is not true; but an intimidating world consensus believes it to be true. Even those who are clear-sighted enough to recognize that growth is of no guaranteed benefit to the wretched, for the most part limit their politics to the hard task of winning the excluded a secure place in the artificial environment. They do not question the desirability of that environment, and less so its necessity.

This is the sense in which contemporary politics is technocratic politics—even when what the experts themselves might define as optimal rationality is thwarted by those who enlist their services. Which is more often the case than not. In practice, the experts work within limits not of their own making. They are the chief employees and principal legitimators of power, not its possessors. The well-designed technocracy must have on tap experts who are eager and able to do *everything* technically feasible; but precisely what will at last be done is seldom of the experts' choosing—though their influence is far from negligible. Their chance to do their thing much depends upon their talent for being loyal employees. It is as Jean Meynaud has remarked in his study *Technocracy*. Successful experts know how to combine "technical boldness" with "social conservatism." Their guiding light is an ethically neutralized conception of "efficiency."

The Artificial Environment: Disneyland Is Better

Efficiency means putting facts before preconceived ideas.... Respect for the facts, which are always in advance of ideas, is one of the qualities often attributed to technicians, and one which they are always quick to claim.... Even when realities are irksome or ugly, the technician does not rebel against them. Technicians take the world as it is, without yielding to nostalgia or useless recriminations.... In a capitalist regime... technicians undoubtedly work towards consolidating the whole system of powers exercised by owners and managers of the means of production, although it is true that many of them make the public good their criterion.[7]

This does not mean that, like conscienceless automatons, the experts surrender all capacity to criticize or dissent. There may well be hot debate among them about options and priorities and, quite sincerely, about "the public good." It is simply that their debate does not reach very deep, nor is it heard far off. Their disagreements do not extend to advocating major reallocations of social power, nor do they challenge the cultural context of politics. Above all, good experts confine their dissent to the seminar table and the corridors of power where their peers can understand the jargon and where they can avoid the painful simplifications of popular controversy. It is the one overarching value all experts share: the preservation of intellectual respectability against those who would vulgarize the mysteries of the guild.

Notes

1. Alvin Toffler, *Future Shock* (New York: Random House, 1970), pp. 266-67. The book is a fast cruise over the glossy surface of the artificial environment, finishing with a facile endorsement of urban-industrialism.
2. Jacques Ellul, *The Technological Society* (New York: Knopf, 1964), pp. 338-39. Ellul is far too pessimistic in this work for my tastes. But as a sketch of the urban-industrial trap seen from its grimmest angle, the book is unsurpassed.
3. John Ziman, *Public Knowledge: The Social Dimension of Science* (Cambridge University Press, 1968), pp. 28, 74. An excellent little book on the professional practice of science.
4. Simon Ramo, *Cure for Chaos: Fresh Solutions to Social Problems through the System Approach* (New York: David McKay, 1969), p. 15.
5. Ibid., p. 116. For a critical study of systems analysis, see Robert Boguslaw, *The New Utopians: A Study of Systems Design and Social Change* (Englewood Cliffs, N.J.: Prentice-Hall, 1965).
6. E. J. Mishan, *The Costs of Economic Growth* (London: Pelican Books, 1969), p. 27. First published in 1967, this shrewd critique of "growthmania" is one of the first voices of the new zero-growth economics.
7. Jean Meynaud, *Technocracy,* trans. Paul Barnes (New York: Free Press, 1969), pp. 209, 248.

B. The Polity

The authors of the readings in this section anticipate key questions about how our political system does, and should, deal with technology. They take a hard look at the character of technologists and politicians and consider how these parties ought to interact to relate technology to political interests and social goals.

In Chapter 4, Don K. Price describes the intellectual approaches of professions along a spectrum from truth to power. Scientists search for knowledge in an abstract and precise realm; professionals, such as engineers, apply scientific knowledge for particular purposes; administrators draw on diverse sources of information in less formal ways; and politicians act on the basis of value judgments and power interests. These "four estates" are drawn together in that there is hardly any problem for which science cannot contribute to the solution. But no major public problem should be left to experts. Explicit checks and balances in the interaction of these groups are needed to accommodate technology and policy.

In Chapter 5, Roger Revelle contrasts the nature of the scientist or technologist (whom he sees as closely joined today) with that of the politician. He finds a complementarity conducive to mutual advising. He goes on to describe two examples of the way technology and politics necessarily intertwine in the energy crisis and the green revolution.

This section could well go on to consider the political processes and institutional arrangements for dealing with technology. Several of the readings listed in the bibliography extend in this direction. The following chapters investigate specific facets of this general issue.

4. The Spectrum from Truth to Power

DON K. PRICE

Eddington, as a popularizer of the new science, used to startle the layman by pointing out that the table in front of him was really two tables. One was the table of everyday experience, firm and solid. The other was the real table, the table that was known to the hard sciences, and it was made up of insubstantial atomic particles and a lot of empty space. The mere housewife, after listening to such a lecture, was a little inclined to be afraid that the dinner she set on the table might fall through it. At this level of experience, of course, not many people are unduly confused by the problem how to relate the abstractions of science to their concrete policy problems; Eddington's lay listeners went on eating substantial dinners, and took no more practical notice of his abstractions than his scientific students took of his religious views.

The ordinary citizen, in his everyday personal experience, does not need to worry about the abstractions of modern science. He can ignore Eddington's other table. The real table, for him, is the one he sees and feels; his test of the relevance of the new physics is the same as the one that Dr. Samuel Johnson applied to Bishop Berkeley's metaphysics when he kicked the stone in the street and proved to his own satisfaction that it was real. But the ordinary citizen today must be more discriminating in his thinking about abstractions and their relation to reality. For some practical purposes, the concepts of the new science are as irrelevant as were Eddington's theories to the housewife's dinner table; for others, they are the key to the future. So the ordinary citizen, with respect to the larger issues that face the nation and the world, cannot afford to dismiss the abstractions of modern science as ideas that are irrelevant to practical concerns; when properly

Excerpted from *The Scientific Estate* by Don K. Price. Copyright © 1965 by the President and Fellows of Harvard College. Reprinted by permission of Belknap Press of Harvard University Press.

applied they are obviously the source of tremendous power that can determine the success or failure of human purposes. How they are applied to those purposes may determine the future of humanity, or whether it is to have any future. This is a question that the ordinary citizen is interested in, no matter how ignorant he may be of science. So his problem is how he, or his representatives, can control the purposes to which science is applied, especially when he cannot understand science, and is not in the habit of electing representatives who know any more about it than he does.

To cope with this problem, it is tempting to turn to one of two alternatives, both of which try to preserve the ideal that an electorate, or its elected representatives, should make policy decisions by public discussion of the key issues, including their scientific aspects.

The first of these alternatives proposes to rely on giving the average citizen a deeper understanding of science. This is the classic formula of Jeffersonian democracy, and there is no doubt that if the electorate can have a better comprehension of science, the processes of politics can deal more rationally with the issues of a technological age. But this is a less plausible prescription for a complete cure than it once seemed. It is less plausible because, as we have seen, science has not only become so abstruse and so specialized that the great majority of men and women (including scientists) cannot be expected to understand very much of it, but it has also become so abstract that it is hard for the expert, let alone the layman, to tell just how it may be relevant to any given problem. While the physical sciences have for several decades been making themselves less comprehensible to the layman, the social sciences have been casting doubt on the concept of the rational voter, and on the idea that the rational debate of issues in a legislature is the main balance wheel of a democratic system. As a result, many are persuaded not merely that we cannot build a nuclear-age society on a Jeffersonian model, but also that government must be entrusted to those who are expert.

As a result, some are inclined to turn to the second idealistic alternative: to advocate the advancement of scientists to positions of high authority, and to urge the scientific community to organize itself more adequately to accept its social responsibilities. This too is desirable, but like the proposal to educate the average citizen in science it is subject to considerable limitations. For a basic approach of modern science has been to purge itself of a concern for purposes and values, in order to deal more reliably with the study of material phenomena and their causes and effects. Men and institutions dedicated to this approach may find it hard to convert themselves to an interest in the purposes of politics, and when they do, they are as likely to disagree about them as any other citizens.

These two ideas are not exclusive alternatives; both are desirable, and (within limits) feasible. But they are not required by the political theory on which the American (or British) constitutional system was based, and we fortunately do not in practice rely mainly on them. For even if we had an electorate and a Congress made up entirely of scientists, we should still have to face the difficult part of the problem: how, after we know the scientific aspects of an issue, we may proceed to agree on what we wish to do about it and then to apply science to effect that purpose. And scientists as such prefer to have no more to do with the idea of purpose in the practical affairs of government than in their own scientific methods. That part of the problem they leave to the professions and to administration.

The Professionals

Enginering, medicine, and law, in different ways, have the function of taking the abstractions of science (or other systematic knowledge) and applying them to the concrete and practical affairs of men. That is not only their function; it is their purpose. Science can insist on ignoring questions of purpose in order to be objective and precise; the professions cannot. So they sometimes take their purposes from generally accepted traditional values, which as professions they have to convert into working codes, as the physicians do when they start with the assumption that people generally wish to be healthy, and use to that end anything they can learn from biology. Or the professions take their purposes from the demands of their customers or employers, as an engineer does when he designs a bridge or an automobile, or from the decisions of duly constituted public authority. And that, of course, brings us to the kind of problem with which we are here concerned: the way in which a profession like engineering or medicine can serve as a bridge between the sciences and politics.

This function of the professions is not new: it has been of vital interest for a long time. When an engineer, following the safety regulations of the Coast Guard or the Federal Aviation Agency, translates the laws of physics into the specifications of a steamboat boiler or the design of a jet airliner, he is mixing science with a great many other considerations all relating to the purposes to be served. And it is always purposes in the plural—a series of compromises of various considerations, such as speed, safety, economy, and so on. Similarly, when a public health doctor or sanitarian translates the laws of biology

Don K. Price

into a system for the prevention or control of disease and epidemics, he is doing the same thing. With a little less precision, a budget officer or central banker does the same thing with respect to the science of economics.

Scientists who made the jump all at once from exploring the nucleus of the atom to thinking about the international control of atomic weapons—the long leap from science to politics—were tempted to transmute knowledge directly into political decisions without thinking much about the normal intervening stages, the stages of engineering and administration. To them, it seemed intolerable that decisions affecting the lives and welfare of themselves and their families should be made without full public understanding and debate of the issues, and especially of their scientific aspects. It is a safe guess that not many of them had the slightest interest in the technical procedures that governments were taking to safeguard the purity of the water that they drank every day, or to enforce building standards to make sure that their homes would not collapse or burn over their heads, or to make sure that the commercial airliners would carry them safely on their way to Los Alamos or Washington.

Each of these functions of government is carried on by engineering or medical professionals, applying scientific knowledge in the light of administrative and political considerations, and each is shot through with compromises (or, as weapons engineers like to say, trade-offs) among various purposes. We accept these compromises as a matter of course. It would be possible to eliminate nearly all cases of any particular epidemic disease, or nearly all transportation casualties, or for that matter nearly all professional crime, if we were willing to pay the price, in money or freedom or both. How far we go in any given case depends in part on scientific and technical considerations and in part on the opinion of the average citizen. But it depends, too, on the degree to which scientific and professional people are permitted to act on their own, and the degree to which they are subject to administrative and political control. And as the new science and new technology are now producing forces so powerful that they are matters of world-wide, rather than merely municipal, concern, the relation of the professions that are based on science—especially the engineering and medical professions—to the scientific community on the one hand, and to the world of administration and politics on the other, becomes a problem of some political and constitutional importance.

The role of these professions as intermediaries between abstract knowledge and political action is very like that of the profession of law (including the courts), which had to be created as an intermediary

The Spectrum from Truth to Power

between church and government. Many of the great constitutional issues from the Middle Ages to the nineteenth century, in England and the United States, grew out of the relation of religion and politics. Religion was accepted as the source of truth, but laymen had the right to ask by what institutional arrangements, and what procedures, was that truth to be applied to human affairs? The solution was to take away from the churches the power to enforce their ideas, or even to define the laws that were to regulate personal and public affairs; this led to the growth of the legal profession, the independent judiciary, and legislatures that were conceded the power to enact new law as well as define traditional law. The Constitution of the United States was not merely an abstract invention, creating a mechanical system of checks and balances among masses of undifferentiated voters. It was also an attempt to find a workable balance in society among institutions and types of people—among the churches, which were to be established or disestablished according to the tastes of the several states, and the judiciary, which would of course be dominated by the legal profession, and the various types of military and civil officials and politicians.

Now that science is accepted as a source of truth at least equal in practical significance to theology, the institutions that proclaim its truths and apply them to public affairs have to work out with administrative and political authority their constitutional relationships. And the constitutional history of the twentieth century will have to be as preoccupied with the university laboratory and security clearances and the status of systems-engineering corporations, as that of an earlier era was with episcopacy and the test oaths and the independence of the judiciary. The analogy is by no means perfect, if only for the reason that there is no way out of our difficulties by means of disestablishing science; without government money you can worship but you cannot run cyclotrons and radio telescopes. More important, science has wrought changes in our society—in its urbanization, its standards of living, and its increase of population—that can be dealt with only with the help of science and the professions.

In another way, however, the analogy is helpful. The effort to disestablish the churches, and thus to reduce the direct control of ecclesiastical authority over secular affairs, was not the work of scientists, and perhaps not mainly of men with skeptical temperaments, but of men with strong religious convictions. The dissenters, wishing to restrict the political power of churches in order to keep them free of political control, had as much to do with their disestablishment as did the skeptics. Similarly, scientists themselves are generally not eager to

Don K. Price

have scientific institutions identified too closely with political power and political responsibility.

In the previous chapter,* I considered the limitations—in practice and perhaps in theory—that science sets on its own ability to provide a complete knowledge of reality and to answer policy questions. The professionals who make it their business to relate scientific knowledge to political purpose are keenly aware of an analogous set of limitations on the competence of their professional skills.

The Limits of Professional Competence

In recent years, the most systematic professional experience that bears on this problem has been that in the field of operations research. In this very practical field, which has undertaken to apply the techniques of the exact sciences to the problems of almost all kinds of large organizations, public and private, the professional practitioners have been discovering limits on their techniques that are analogous to the limits that scientists have been discovering on their ability to acquire deterministic knowledge and to predict phenomena. Operations analysts, in manufacturing corporations as well as in government service, repeatedly observe that their scientific techniques work with greater power and precision on the problems encountered at lower levels of the administrative hierarchy. They observe that the higher the rank of the official needing answers to policy questions, the less exact and less reliable are the answers that scientific techniques can provide.[1] And these common observations correspond in some ways to the limitations—already discussed—on the ability of science to give a complete and predictable picture of the real world.

The first of those limitations comes from the very quality that has been responsible for science's great power: its high degree of abstraction, which has ruled out of consideration the biases and prejudices of human purposes, and rescued it from a confusion of ends and means. Because the sciences have moved ahead fastest in the most abstract fields, we as citizens (and our administrators and political leaders) have at our disposal precise techniques for answering questions before we know the questions we wish to ask. We like to make use of the skills about which we are sure, without always being willing to wait to consider with equal depth of study the more complex aspects of issues which are not susceptible to quantitative techniques. We prefer to

---

*Editor's note: Price is referring to the previous chapter in his book *The Scientific Estate*.

The Spectrum from Truth to Power

leave those aspects to be decided by snap judgment or the prejudices of politicians.

As one operations analyst has put it, we have too many "studies which try to determine the exact best way to perform an operation which shouldn't be performed at all."[2] We ask our scientists and engineers to tell us the best way to design a high-altitude high-speed bomber to attack Moscow, when we may need instead a low-speed low-altitude observation plane to help get the guerrillas out of the jungles of Laos, or vice versa. It is clear that the aeronautical engineers can develop either kind of plane for us; it is also clear that we cannot decide what kind of strategy to adopt until the engineers tell us what kinds of planes they can develop, and just what those planes can do, and about what they will cost; but it is, unfortunately, equally clear that aeronautical engineering cannot solve the strategic and political question what kind of planes we ought to have.

This is a painfully obvious point, perhaps, but at times it seems that we are determined to ignore it. Our columnists and Congressmen are not much interested in talking about whether we should put a man on the moon; it is much more interesting to argue about the various techniques of getting him there. You can readily invent chemicals to kill a pest like a fire ant, but it is harder to visualize the damage you may do to other animal and plant life if you use that chemical, and very much harder to decide how to staff the Department of Agriculture so as to get the right mixture of chemical and biological and political judgment involved in the decision whether to use it or not.[3] Chemistry is a more exact science than biology, and biology than administration, but the order of exactness may turn out to be the opposite of the order of importance for political decisions. For the importance and usefulness of a science may turn not on giving exact answers to the question that has been asked (or taken for granted) but on suggesting new or alternative questions. In public affairs as in physics, the problem never holds still; it is always being redefined.

This is in general the lesson learned by the social scientists, statisticians, and mathematicians who have worked on the theory of decisions. Science can contribute to the making of a decision only if someone has first specified a consistent set of objectives or values which he seeks to reach with maximum efficiency. And it is much easier to be consistent in thinking about small things that are closely related to each other, or in making a small additional improvement in a device or a procedure that is already well understood, than it is in thinking about a major problem of policy.[4]

Another limit on scientific precision that I discussed was the limit of size and complexity. In any major issue of policy or strategy, you cannot isolate and control all the things you are dealing with. And so in problems of military tactics and strategy, even though the weapons systems have been developed by engineers on physical principles, it has often been observed that some of the most effective operations research has been carried out not by the physical scientists, but by biologists or economists. The head of an Air Force operations research unit at the time of the Korean War had been trained as an ichthyologist. This surprised some of his colleagues who had not read what P. M. S. Blackett had written in 1943. Blackett had pointed out that military forces were not willing to operate under the controlled conditions that please the physicist, who likes to have a "great deal of numerical data about relatively simple phenomena," so that the problems of analyzing war operations are "rather nearer, in general, to many problems, say of biology or of economics, than to most problems of physics."[5] Here again, the range from the more exact to the less exact sciences runs in inverse proportion to their suitability for solving problems that seem important to the average citizen or the administrator.

In administration as in physics, velocity can alter mass. The faster a situation is moving, the more it is affected by the problem of the quantity of data available. When a decision has to be made in a short time, you cannot wait to gather all the data that are relevant and determine your course of action according to the results of objective research. So the responsible official and his scientific adviser have to treat some parts of the problem as matters of pure chance, and others as value judgments, even though both might be determined by exact study if they had world enough and time.[6]

This is illustrated by the negotiation of the agreement between the United States and Russia for the prohibition of nuclear tests in the atmosphere. Throughout the talks, some political leaders seemed to be looking for ways to push their difficult policy problems off on the scientists, while some scientists were eager to make clear the limits on the ways in which science could contribute to the problem. As Dr. Wolfgang Panofsky, one of the technical consultants, pointed out later, the balance of risks in the proposed agreement turned only in part on the technical and scientific data—although these data provided the essential foundation and starting point for the discussion. They depended also on such other factors as the relative military posture of the two nations, and the effects on that relationship of various degrees of compliance or evasion of an agreement; on the capacity of intelligence agencies, by traditional espionage as well as

The Spectrum from Truth to Power

technical means, to detect violations; on the effect of an agreement on world opinion and on the prospects for further arms limitations; and on the effect that control measures might have on international secrecy.[7]

It was the inherent uncertainty of such factors that caused the scientific witnesses before several committees of the Senate either to disagree with one another in their policy recommendations or to refuse to express any opinion at all on any questions but limited technical ones. This inability to give a scientific answer to a complex question was commented on with either cynical pleasure or despair by Senators of varying shades of opinion.[8]

Still another limit that physical scientists see on their ability to achieve precision and make predictions comes from the inadequacy of the instruments of observation. Bridgman despaired of getting a complete picture of the configuration of the atoms in any material system. How much greater is the problem of knowing the total picture in detail on any major political or administrative issue! Even before you get to the problem of human error and human motivation, you run up against the sheer cost of information in any operating system. The amount of time and effort spent in producing the data for the guidance of the executive and the legislature—the accounts, the personnel forms, the organization surveys, and the statistics—all this puts obvious limits on the use of a scientific method to govern large-scale economic or political systems.

But if the limits on predictability and control in the physical sciences have their analogies in public affairs, the limits of biology are even more relevant. The utterly random occurrence of significant events in biology, the uniqueness of every issue, the incalculable complexity of every system, and the way in which new issues and problems emerge when old ones are combined—all these points have their obvious parallels in the world of public affairs. Such factors put limits on the precision and the predictive capacity of the sciences but, it is necessary to emphasize, they do not by any means make them powerless. There are many ways in which the scientific method can operate in situations of uncertainty. The statistical approach, and the modern techniques of cybernetics and control theory, can provide essential guides to action in fields in which the individual case cannot be observed or predicted. Even so, the essential decisions regarding the purposes which are to be served by such techniques, and the cases to which it will be useful to apply them, and how much to rely on their findings, generally remain matters of judgment which the techniques themselves cannot supply.[9] When statistical models fail to provide an accu-

Don K. Price

rate forecast of the future, it may be because of scientific or statistical uncertainty, or it may be because the statistician has deceived himself about the essential nature of his problem.

In the professions that make use of the sciences, there is plenty of controversy regarding the extent to which science can guide professional judgment. The profession of medicine has obviously made tremendous advances as a result of scientific research; yet even among professors of medicine warnings are raised against excessive faith in science. As one senior statesman of the profession put it, "The dogmatic assumption of determinism in human behavior, fostered in large part by the sophomoric expectation of certainty in knowledge," has a pernicious effect on the practice of medicine, by leading the young doctor to "try compulsively by the unwise and neurotic multiplication of tests and superfluous instrumentation to achieve the illusion of certainty ... —a modernistic and expensive superstition."[10]

A similar distaste for excessive reliance on scientific methods is often shown by military officers, especially those who dislike the way in which the political executives of the Department of Defense use the findings of civilian operations analysts to refute recommendations from professional military men. As one military planner wrote, "Today's planning is inadequate because of its almost complete dependence on scientific methodology, which cannot reckon with those acts of will that have always determined the conduct of wars."[11] But in general, such critics have not been nearly as systematic and clear in explaining the limitations on scientific methods, and the necessity of subordinating them to a system of administrative judgment and political responsibility, as the leading operations analysts themselves have been.

Those analysts have a lively appreciation of the tremendous contributions science has made and is sure to make in the future to military affairs, but they also have a thoroughly professional understanding of the limits on the competence of scientific techniques to solve administrative and political problems. These limits do not derive from any traditional or mystical values, about which scientists are skeptical. Nor do they derive from any romantic conception of the importance of practical experience in actual combat, which they are inclined to resent—as one Pentagon scientist showed when, having been taunted by a noted general with his lack of battle experience, asked "How many thermonuclear wars have you fought?" They derive instead from the limitations, equally clear in scientific theory and in professional practice, on the possibility of applying abstract and systematic knowledge to the confused world of partial perception, in which the important problems are usually those that arise in unique and unpredictable situations.

The Four Estates

So it is abundantly clear that some types of decisions have to be made, not by professionals using scientific techniques, but by other kinds of individual thought and collective deliberation. The members of the professions based on the sciences thus have a role in public affairs that is different from the role of the scientists on the one hand and from the roles of administrators and politicians on the other. Though the distinctions among these roles are not fully defined in legal terms, they are clear enough so that the several types of thought and action can be related to one another in an ordered system of authority and responsibility—in short, in a constitutional system.

The professions (for example, engineering and medicine) make tremendous use of the findings of the sciences, but they add something more: a purpose. Science has advanced by getting rid of the idea of purpose, except the abstract purpose of advancing truth and knowledge. But the profession puts it back again; basic science could not cure a patient or build a bridge or an airplane, but the medical and engineering professions are organized to do so. Each is organized around a combination of a social purpose and a body of knowledge, much of it drawn from science. Each is organized as an almost corporate entity, with some control over its standards of admission. Its responsibility to an individual client, or to a corporate or governmental employer, is to serve within a defined and limited field; within that field, the professional has an obligation to standards of ethics and competence that his profession, and not his employer, dictates. The engineer has a sense of responsibility to build structures or systems that will work, and this responsibility is not merely to his employer. The doctor feels an obligation to the patient that goes beyond any contractual relation with an individual client, or any administrative instructions from the clinic or hospital that may pay him.

The general administrator is, in these senses, not a professional: his responsibility is not restricted to some special aspect of an organization's affairs that is related to a special body of knowledge or a special type of training, and it is more difficult for him to define a sense of obligation to a professional purpose that to some degree transcends the purposes of his employer. He is obliged to deal with all aspects of the concrete problems that his organization faces, and for that reason his education—no matter how thorough and useful it may be—cannot be reduced to a specific discipline or a restricted field. On the contrary, he must be prepared to understand and to use a wide variety of professional expertise and scholarly disciplines, as he helps his political superiors (or the directors of a business corporation) attain their general purposes. Administrators may organize themselves into associations of a quasi-professional type, but they cannot, like the true professions, undertake to determine the standards for entrance into their vocation, since it is too much subordinated to the unpredictable purposes of politics or business. They are obliged to maintain standards of objectivity and competence; without their enforcement of these standards the politicians would be powerless to control the machinery of government. But administrators cannot base their standards on precise scientific knowledge.

Still further away from the precision and abstraction of the sciences, and from the self-discipline and body of established principles of the true profession, are the politicians. The men who exercise legislative or executive power may make use of the skills of administrators and engineers and scientists, but in the end they make their most important decisions on the basis of value judgments or hunch or compromise or power interests. There can be no common discipline or body of established principles to guide them, for their business is to deal with problems in which either the inadequacy of scientific and professional data, or the conflict of expert opinion, makes it necessary or possible to come to decisions that are based on judgment and must be sustained by persuasion or authority. In government, the politician is apt to make every decision both to accomplish its ostensible purpose and to maintain or increase his power—just as in private business, the principal executive or owner is apt to make every decision both to produce some product or service and to make a profit.

These distinctions exist within any large institution whether it is public or private. In a business organization the owners, or the board of directors and the principal executives, have the function of making the general and ultimate decisions that correspond to political decisions in government. The larger the corporation, the more the function will be political in the sense of being related to the politics of the local or national government, as well as in the sense of being concerned with such incalculable elements as public and employee relations.

The four broad functions in government and public affairs—the scientific, the professional, the administrative, and the political—are by no means sharply distinguished from one another even in theory, but fall along a gradation or spectrum within our political system. At

The Spectrum from Truth to Power

one end of the spectrum, pure science is concerned with knowledge and truth; at the other end, pure politics is concerned with power and action. But neither ever exists in its pure form. Every person, in his actual work, is concerned to some extent with all four functions. The laboratory scientist is probably interested in some professional association with his colleagues, in the administration of his laboratory, and in the support for his work that comes only from money or influence. At the other extreme, the Congressman or the President must be interested in the ways in which new knowledge affects his status and his purposes, and in which professional skills and administrative competence can support them.

Yet men and institutions tend to associate themselves with one function or another, and many of the more interesting problems of politics arise from the ways in which these four types cooperate or conflict with one another. Their relationships have not made obsolete the classic concern of political science with the relations among branches or levels of government, and between them and competing political parties and economic or ideological interests. But they have added a significant and interesting complication to the study of contemporary politics, and one that will be of growing importance as long as science continues to increase its influence on public affairs.

In one sense, this is the revival of an old topic, and not the invention of a new one. The existence of political groups of institutions and individuals that are distinguished not by formal public office, nor by economic or class interest, but by the differences in the nature of their training and their skills, is no new phenomenon. To remind myself and the reader that this is not an altogether new phenomenon, but only an old one in new form, I am calling these groups by the old term "estates"—hoping that no one will imagine that I intend to suggest anything more than a very loose analogy indeed between these modern estates and their medieval counterparts.

The differences in function among the several estates are not merely based on tradition; they correspond in a measure to the nature of knowledge, and are found in every modern country. If a government is to act as an organized whole, it must make decisions on problems for which neither the sciences, nor the professions that are based on them, are prepared to give the complete answers. But although science can never provide a complete solution for any major political problem, no major problem can be dealt with today without its help, and its help can be effective only if it is given generous support, and a large measure of freedom, by government.

The scientific estate, of course, is merely a subdivision, although

Don K. Price

perhaps the most influential one, of the broader scholarly estate. Historically, the philosophers and the theologians, being concerned with another variety of pure knowledge, have played an analogous role in society. And the lawyers and the clergy, in taking over their basic ideas and applying them, with modifications, to human purposes, play important roles in the professional estate along with the engineers and the physicians. The difference between the older and the newer wings of each estate is that the Western world has had several centuries more experience in working out a satisfactory constitutional position for the lawyers and the clergy, while the scientists and engineers and physicians have risen so rapidly in influence as a result of the advancement of science that the politicians have not had time to develop a new set of relationships with them.

The Twofold Principle of Freedom and Responsibility

But the United States is clearly working out some crude principles as a basis for these relationships, or adapting some old ones, and these principles may help to establish a new set of checks and balances within the constitutional system. That system will be the subject of the next chapter, but it will be useful here to anticipate a general point. The most important principle seems to be a twofold one: (1) the closer the estate is to the end of the spectrum that is concerned solely with truth, the more it is entitled to freedom and self-government; and (2) the closer it gets to the exercise of power, the less it is permitted to organize itself as a corporate entity, and the more it is required to submit to the test of political responsibility, in the sense of submitting to the ultimate decision of the electorate.

It is true that in the United States the college or university is not organized in legal form as a self-governing community of scholars; it is either under the legal direction of a private board of lay trustees, or of a board of state officials. But in actuality, the American university, with respect to its academic affairs, is a community to which society concedes the right of self-government; the full members of the academic fellowship—the tenure professors—consider that they are entitled to control the admission of new members, and to protect them in their freedom. Similarly, the government concedes completely the right of scientists and scholars to organize any self-governing societies for scientific purposes that they please.

When it comes to the professions, whose work mixes a commitment of service to the public with its scientific interests, the government is

The Spectrum from Truth to Power

usually content to leave the control of standards to the professional society as a matter of practice, but in theory (and sometimes in practice) it insists on maintaining the right to exercise control.

The administrators, who can rely less on systematic and testable knowledge and are closer to ultimate power, are not permitted to exercise, through their quasi-professional societies, any influence on the terms of their admission to their vocations. They are appointed, on terms regulated by law, by procedures controlled by politicians.

And as for the politicians who exercise ultimate power, but can never be certain that their major decisions are justified by scientific knowledge or provable truth, they are permitted to hold office only by periodic re-election by the people.

This twofold principle about the spectrum between power and truth is not generally accepted in all other parts of the world, even though the practice is reasonably similar in other English-speaking countries and in some countries of Western Europe. The existence of the four estates seems to be common to all countries with advanced science and technology: the respective functions of the estates seem to derive from the way in which any large-scale institution has to be organized in order to translate abstract scientific knowledge into purposeful action. But the mutual independence of the several estates seems to be a quite different matter, that is, one that is not determined by the nature of science and technology (or if you like, by the mode of production) but by the way men think about political power in relation to truth.

For the twofold principle that seems to be accepted in the United States is by no means axiomatic. The idea that an institution concerned only with truth should be permitted to govern itself, and the idea that those who hold political power should be accountable to the people, are plausible only on the basis of certain historic political assumptions. If you think of a nation primarily as an organization that must have a coherent purpose, and think of its inner institutional relationships as a problem in communications and control, you will come out with a quite different approach. The logic of that approach is quite simple: If we have a reliable method for discovering the truth, namely, science, should we not use it to solve all our most important problems? If we propose to do so, should not those who make our most important decisions be selected for their understanding of the most relevant science? How can such a process of selection be carried on except by the judgment of their peers (as any university department chooses its professors) rather than by a popularity contest among ignorant and apathetic voters? And if they are controlling society on scientific principles, should they not direct scientific institutions to carry forward their research along lines that will be of the greatest service to society?

The logic of these questions forces us to admit that, although the way in which scientific knowledge is related to political purpose seems to require the existence of something like our four estates, it by no means requires the relationship among them that is conventional in the Western constitutional tradition. "The real brains at the bottom," as our science fiction hero put it, is an accurate observation if not a justified complaint. Is it really right that supreme authority should be vested in the Congress and the top political offices in the executive branch, which are filled with men who by any conventional test of abstract and theoretical intelligence are not the match of the scientists and professionals who work several layers down in the hierarchy? Even if the government must make decisions that cannot be determined completely by the rigorous processes of the natural sciences, many new techniques exist for dealing in a scientific way with situations of uncertainty and incomplete knowledge. Since science is producing the dynamic changes in our society, why should not our government be headed by men who will bring to its administrative and political functions the greatest possible ability to use science to the maximum?

The answer to these questions turns not on what you think about the nature of knowledge, but on what you believe about the nature of man and politics.

The Western, and particularly the American, point of view is determined by two traditional fears. The first is the fear that the scientist or the professional will never be guided completely by his desire for scientific truth and his professional ethics; the second is the fear that he will. In logic, these fears are contradictory; politically, they lead to the same end.

As for the first fear, neither the politician nor the administrator is prepared to accept any institutional arrangements that depend on the detachment, the unselfishness, and the purely scientific motives of the scientist. His distrust is warranted by what the scientists themselves have told him. For if limits to objective knowledge are set by the nature of things and the imperfections of the instruments that science can use for observing them, the limits are even more severe when men are serving as those instruments, and making deductions from the observations. For it is not merely that there are finite limits to what man can observe and understand, but that he has a capacity that is

The Spectrum from Truth to Power

nearly infinite for reading the evidence in the light of his own interests and passions.

To say this is not at all to deny that the sciences, within their own fields, provide a remarkably effective training in the virtues of objectivity and honesty. Even more important, they have the world's most effective policing system. Since the main prize in the game is the esteem of one's scientific colleagues for the results of one's research, and that esteem is based on the publication of the results of experiments which can be tested, the scientist has a most powerful incentive for objectivity and honesty. But the system works only when it deals with scientific data and logical patterns—abstractions from reality which can be dealt with uniformly in laboratories and studies all over the world. And to keep the system pure, scientists tend to frown on any colleague who mixes with such data irrelevant considerations like purposes or values that cannot be tested according to uniform standards by everyone else.

When scientists and engineers undertake to apply science to practical problems, however, they find it impossible to stick to quantitative data and rules of logic. The problems include too many other aspects, and in dealing with them they naturally behave like human beings. And so weary cynics like lawyers and political scientists take a certain malicious pleasure in showing that the natural scientists are led by confidence in their special scientific techniques to try to make judgments on political values or policy decisions, without realizing all the other components of a nonscientific nature that go into those decisions. The lesson is a valid and important one, and needs to be driven home to the general public much more than to the scientists, most of whom know it already even if they do not like for nonscientists to remind them of it.

If you feel obliged to document the lesson, the illustrations are plentifully available. Read the hearings published by the Atomic Energy Commission in the case of J. Robert Oppenheimer, and see how two groups of eminent scientists and engineers, differing very little indeed at any given time on specific scientific matters, mistrusted each others' political intentions to such a point that we had the closest thing to a heresy trial that modern American politics has provided. Or read the history of the disputes over nuclear fall-out, or the civil defense shelter program, or the use of insecticides, in each of which both sides used much the same statistics and drew widely different conclusions. Or the story of the Geneva disarmament negotiations with the Soviets, in which the scientists themselves must have found it difficult to tell when they were trying to sort out the technical evidence and when they were making judgments about foreign-policy priorities or estimating the intentions and good faith of the Russians.[12]

You can explain these difficulties in professional and scientific terms, if you like, by saying that the physicists and chemists all ought to realize that issues of this kind need to be worked on by social as well as natural scientists. This is obviously true, although the social scientists are at least as guilty as physicists of mistaking the abstractions of their respective disciplines for the sum total of all wisdom. Or you can add an extra bit of rational explanation by noting that the physical scientist is used to working with data that may be difficult but not malevolent—that is, the data have no will of their own to exercise against the experimenter—whereas in politics the data not only have wills of their own, but sometimes oppose the experimenter not because they dislike what he is doing but because they dislike him. Thus Einstein remarked, as an encouragement to those scientists who seek to find order in the complex universe, that "God is subtle, but He is not malicious."[13] But in an international negotiation on, for example, arms control, no matter how high a scientific content the subject matter may have, the contribution that science can make is a limited one because the Russians may be malicious. In the eyes of the Democrats, so are the Republicans, and vice versa. Maybe in the eyes of the Russians, so are the Americans.

And this type of conflict creates in politics, domestic or international, a problem quite different from those problems on which the scientific observer can take a detached and superior point of view, and ignore the elements of conflict. It sets up a system of relationships on any major political issue in which it is hard to judge the capabilities and intentions of your adversary, and to determine the exact degree to which he means exactly what he says, or is discounting your own reaction in advance, or is deliberately using conflict with you for other irrelevant purposes or out of irrational ill will—and equally hard to judge your own motives.

If we understand that the scientist and the engineer cannot be guided in their political actions entirely by what science teaches them, and that even when they think they are doing so they may be deceiving themselves and others, we are led to a realization that the fruit of the tree of knowledge is not always peace. Maybe this is the reason that, among professional political scientists, we read a great deal more today than we did a generation ago about the dogma—or to moderns, the myth—of original sin. Perhaps the fear that scientists and professionals will not be guided entirely by their scientific knowledge and

The Spectrum from Truth to Power

professional ethics in public affairs, and hence cannot be trusted with political power, is a political attitude derived directly from an old Puritanical prejudice. The Reformers began by observing the immorality of the higher clergy and therefore distrusting hierarchy in an ecclesiastical establishment, and ended by dividing and weakening the role of the church in political affairs. An analogous attack has begun in recent years on the ethics of scientists who use government funds, and on their conflicts of interest.

But the old dogma had a more subtle side to it: it recognized that it was not only the base or material side of man's nature that caused him the greatest trouble, but his spiritual pride. Men are more likely to fight with each other for noble than for base motives. The most powerful personal motive is not the gratification of the senses, or the acquisition of mere wealth or power, but the conviction that one's own skills and knowledge have a special contribution to make to the salvation of humanity, or that they have provided an insight which the rest of the world must be induced to accept.[14] The most tyrannical political systems are those built not on corruption, but on self-righteous fanaticism.

That is why, in regulating the constitutional relationships among the estates of our society, we should be less concerned that the scientists and professionals will yield to material interests or sensual temptations, than that they will be utterly and unselfishly devoted to their respective disciplines or professions. The two types of temptation of course can be related; the more importance society attaches to chemistry, the more income it will let individual chemists earn; the more it is persuaded that doctors can cure its ills, the more likely it is to concede to the American Medical Association the right to prescribe the terms of public insurance for medical care. At any rate, so the chemists and doctors are tempted to believe. But if we tend to think of the problem as one in which the main incentive is material gain, we not only fail to do justice to the motives of the contending parties, but we fail to understand the more difficult part of the problem.

Because of the rapid advance of the sciences, and because of the contribution that they are making continuously to engineering and medicine, there can clearly be no set limit on the contribution they may be asked or permitted to make to our social problems. They do not merely produce things that the public asks for; what they discover determines the range of new possibilities that are open to us. Invention is the mother of necessity, as a British observer recently remarked.[15] Since scientists and mathematicians and engineers are extending their ability to solve problems in which there is a high degree of uncertainty, the administrators and politicians would be stupid not to encourage them to go as far as they can. But the administrators and politicians would be even more stupid if they failed to note the aspects of those problems on which the scientists and professionals are making judgments not required by the nature of their subject matter—judgments that have the effect, and perhaps the purpose, of extending the power of their particular estates.

The Case of "Command and Control"

This problem may be illuminated, perhaps, if we take as an illustration the field in which the most fateful social consequences are mixed with the most advanced scientific techniques. This is the current problem that has become the subject of countless sermons, novels, and speeches: the problem that the engineers call "command and control." It has to do, of course, with the way in which the nation prepares to control its intercontinental missiles and aircraft, armed with nuclear and thermonuclear weapons. This is the most threatening problem with which science and technology have confronted society. The nature of the relation of political to military power has always been one of the most difficult constitutional problems, and here it has been complicated by the vast increase in the power and velocity of the weapons at hand. The problems of command and control have been thoroughly studied from the point of view of their effect on international politics and diplomacy; somewhat less attention has been paid to the way in which they affect the relationship, within our own constitutional system, of the several estates. If we can identify the main outlines of the problem it may help us to deal with other problems in which scientific advances may alter the balance of power within our system of government.

Command and control is the problem of dealing with intricate, lethal, and nearly instantaneous weapons systems through the exercise of fallible human faculties, and even more fallible systems of organization.[16] When the British air defense forces, in the Second World War, found that the old system of airplane spotters telephoning reports to a command center was sure to break down under mass attack, they began to develop something that evolved into a highly automated system, in which radar and the most elaborate electronic computers were used for spotting and tracking the attacking planes (or missiles), cataloging the weapons at the disposal of the defense, and directing them against the attackers. All this now goes on over distances that stretch across seas and continents, and it involves most

The Spectrum from Truth to Power

massive calculations, at speeds made possible only by electronics, of the velocity and distance and altitude of each of a myriad of weapons. The general type of system has now, of course, been developed not only for defense against air attack, but for control of the long-range bombers and intercontinental ballistic missiles—the ICBM's, in military jargon—that would be used in retaliation.

So we have a lot of ICBM's scattered around our deserts and mountains, each one subject to command from an electronic system that identifies an enemy attack by means of long-range radar observation, and that can technically be made to respond either absolutely automatically or by waiting for orders from authorities at any level you choose. It is possible to have each ICBM aimed at a single predetermined target, or (at greater cost) made ready to be ordered to alternative targets.

Now observe the effect on our constitutional system of this particular bit of technology. A generation ago, the outbreak of war abroad gave the United States, with its ocean shelters, a year or so to start making its weapons and training its troops. A decade ago, the time scale had shortened, but it still would take a good many hours for a bomber plane to travel from Russia to America or vice versa, and this left some leeway for decisions; after the alarm, planes could be moved from vulnerable bases, and retaliatory planes could be started off, and recalled at will after some hours of flight. But today, an ICBM could travel in either direction between the U.S.S.R. and the U.S.A. in half an hour, and a ballistic missile (though it may be blown up in flight) cannot be recalled. If you add to the problems caused by this compression of time the possibility that most or all of the top political executives of the government might be killed by a single unexpected attack, you can readily visualize the difficulty of making plans to decide who will be entitled (and by what procedure) to give the ICBM's the command when to attack, and what to attack. The simple Constitutional principle of the supremacy of civil over military authority is meaningless unless all this can be worked out in technological and administrative detail.

Since no system can be infallible, people have worried about the danger that this kind of intricate and sensitive system will get out of hand. Can we start a nuclear war that nobody intended, because someone mistakes a radar signal, or some piece of equipment fails, or someone in authority is drunk or crazy? Such a disaster is not inconceivable, but we should worry much more about a less spectacular type of threat. For this kind of worry before the fact is the equivalent of the typical clamor for an investigation that goes on after an event:

everyone seeks to pin on some unhappy culprit the responsibility for a disaster that was caused not by the criminal intent or delinquency of an individual, but by the negligence or inadequacy or lack of responsibility built into a system, sometimes as the result of the misguided zeal of the most honorable and patriotic men.[17] The system itself is what we should worry most about. The professional engineers and military officers have as urgent a motive as any professional pacifist for preventing the obvious potential failures of men or equipment.[18] But they do not have the motivation to inquire into the limits imposed by their own professional biases.

The United States, shortly after the bungling victory in the Spanish-American War proved its military incompetence, undertook to create a professionalized system of top military command. In order to reconcile this with the traditional distrust of a standing army, Congress did not set up an independent command headquarters, but based the system on staff advice to a civilian Secretary. True, the Chief of Staff has always been in some senses a commanding general; nevertheless the distinction has been an important and significant one,[19] because it made it clear that in constitutional theory military operations were under civilian command, a theory that could be translated into reality by any strong Secretary. But the whole theory of advice depended on the premise that it was in fact possible for the civilian authority to refuse military advice.

It has long been true that the President and the Secretary of Defense (or War) were almost powerless if they waited until a crisis and then were presented with advice in the form of a proposed operational order; if they wished to have influence, they had to have a hand in ordering the assumptions on which were based the war plans and the research and development plans. The quicker the reaction time involved in the potential military operations, the slower the process of translating policy decisions into effective policy. As technology requires us to be prepared to take in a very few minutes decisions affecting the survival of the nation and perhaps the world, it also requires the Secretary of Defense and the President to make decisions five or ten years ahead of the possible event if they wish to influence it. The problem that the nation should be worrying about is not the dramatic possibility of a war caused by the insanity or malevolence of an individual, but the much more difficult problem of building an organized system to make the engineers and the military officers as responsible in fact, as they are in law, to political authority.

If we look at the broad outlines of the command and control problem in this context, they suggest to us three lessons about the relation

The Spectrum from Truth to Power

of science and the professions to administrative and political authority.

First, the scientists and professionals, in order to do their own jobs, must be involved in the formulation of policy, and must be granted wide discretion in their own work.

Second, politicians and administrators must control the key aspects of technological plans if they are to protect their own ability to make responsible decisions.

Third, the ability of a free society to make effective use of science and technology depends on some workable (though probably informal) system of checks and balances among the four estates.

1. *The professional role in the formulation of policy.* The surest way for the top administrator (in this case, a general or a general staff) to lose control over his essential purposes is to try to tell his professional subordinates exactly what to do, and make them stick to it. His problem is not to control things that he fully understands, but to control the development of new possibilities that he cannot understand. As Brockway McMillan, Assistant Secretary of the Air Force, put it in a speech in 1962, the function of engineering is to start with a problem, which is of course a statement of purpose, or an expression of political values that the top administrator and his political superiors must define. But then the engineer must translate that purpose into specific objectives, requirements, and criteria of performance. This is not a matter of pure science; that term, Secretary McMillan pointed out, is too precise and rigid. It is a matter of engineering doctrine, which must be developed on the basis of scientific concepts, but it must also govern the allocation of resources—in short, a kind of budget—to make the system work. You cannot state the requirements for a moon rocket, he pointed out, without using the concepts of physics, such as the laws of conservation of energy and momentum, but you cannot do so either without deciding what you want to do, and what various parts of the job are worth in terms of money and scientific effort.[20] In this process, the best way for the top administrator to ensure failure is to prescribe a specific plan of action in advance.

In an age of slow technological development this fact was not so clearly apparent; the engineer moved to novel systems slowly enough so that his boss could keep up with him. But the more abstract sciences, by greatly speeding up the velocity of development, have changed all that. And this is why the more scientific engineers have been so stubborn about insisting that they not be asked merely to make improvements that the administrator can foresee and state in a set of prescribed requirements, but that they be permitted to play a role in the interaction between ends and means, between strategic purpose and scientific possibility. This kind of stubbornness showed up in the process of developing the first radar for immediate use in combat. Sir Robert Watson-Watt quoted with pleasure the exasperated military comment that the radar engineers "won't tighten a nut on a bolt until they have had the whole strategic plan . . . explained to them." And this attitude is what led the Royal Air Force to become the first military service in which scientists became an integral part of the military planning process.[21]

The process of planning, insofar as it involves the use by politics of the sciences, is not a one-way street. In the case of the command and control system, the engineer cannot know what kind of help he can use from the scientist until he knows what the military officer needs. But the military officer cannot know what he needs unless the engineer, with the benefit of new knowledge from the scientist, can tell him what is becoming possible. No military officer in 1939 could have written an official requirement for the atomic bomb, or a little earlier for radar. In the politics and administration of large-scale engineering enterprises, as in the basic sciences, it is necessary to guard against a kind of teleology—the desire to make what we can learn conform to the way we think things ought to be.

2. *The role of the politician and administrator in technological planning.* As Secretary McMillan warned in the speech already quoted, "a command and control system exists to support the commander, not to supplant him." What reason is there for concern on that score?

Science makes it possible, of course, to develop gadgets that are so effective in doing specific things that they outrun the possibilities of human control. A fighter airplane or a communications system can operate too rapidly for the human nervous system to follow. If your problem is to do exactly the same thing over and over again—to handle larger quantities of standardized operations, or even to vary the job according to conditions and rules you can foresee, you might as well put a computer on the job and make it thoroughly automatic. But the limit on the usefulness of this approach is that the machine does not care what it does; it has no sense of values or purpose.[22] (By values or purpose in this context we do not *need* to mean ideas coming from supernatural sources; it is enough to refer to those of the responsible political authorities—including those aspects of a decision on which adequate knowledge is lacking, either because they relate to things that are unpredictable by the present methods of science, or about which it would take too long or be too costly to get the information.)

The Spectrum from Truth to Power

So the man for whom the machine has been designed may find that he has invested resources in an elaborate system that cannot be changed to do the new job required by new conditions or new discoveries. Or he may find that it is a system that commits him to do things he never wanted to do, and that the engineer who built it failed to warn him that he was putting all his eggs into one electronic basket. Accordingly the better computer experts are now warning the customers that if the purposes of an organization are to be best served, in the light of future uncertainties, it is often better to resist the temptation to install the biggest and fanciest computer and the most complete system of automation based on the most elaborate concepts. The purpose of the organization has to be protected against the desire for technical elegance, and a simpler and slower system may make it possible to attain the desired ends of the organization more effectively in a changing environment. It helps, in determining that purpose, to look at the problem from the perspective of various scientific disciplines and various professions, but no one of them can define that purpose, and each one may be tempted to try to do so for its own professional interests.

The administrator's need for flexibility is least in some types of business operations where the product is standardized and nearly sure to remain so. But it becomes much greater in military matters for the reason that a commander is dealing with an adversary whose responses he cannot know in advance. He may not want, after all, to fire at the targets he had first picked; the enemy may have moved them. Or there may be a different enemy. Or he may need to change the rate of his fire to hold a missile supply in reserve, or to make negotiations with allies or enemies possible. This is why the experts warn us that "strategies should not be inadvertently built into the command and control system."[23]

Any reader with an interest in politics, of course, should ask just what this warning means. It is stated in the passive voice, and therefore covers up the issue. *Who* might be tempted to build an undesirable strategy into a system? Is anyone likely to do so entirely by inadvertence? The real danger is probably that the top political authorities of the government might find themselves in charge of a system which gave them too little flexibility, and too little range of choice in difficult diplomatic or strategic situations, because it had been based on decisions by engineers and military officers which the politicians did not fully understand. If this should happen, it might be very hard indeed to tell whether the engineers or the generals had caused the trouble, and for what motives. The engineers might be tempted to do so, either for the purpose of developing the most advanced technology possible in the interest of national security, or so as to help their companies sell something fancier and more costly to the government. Or the generals might be tempted to do so, either in order to get the quickest and most reliable response to an enemy threat, or in order to put their political superiors in a position where they would have to leave the conduct of a war more exclusively to military commanders and would be unable to compromise the chances of all-out victory by diplomatic and political bargaining.

If we choose to make use of the techniques of automation to accomplish things wer are sure we wish to do, it is important to recognize two things. First, although an elaborate system of command and control may let you get a particular thing done much more rapidly and more cheaply and more reliably, it will very probably force you to do it in one particular way, and cut off other choices. Second, unless you are very careful to understand the whole business before you start, your lack of understanding may conceal a number of ways in which other people, with quite different biases and preferences, have committed you to do things you never meant to do. Or, in more technical language, "A machine has the advantage of making its synthesis without personal bias ... But it also tends to eliminate any allowance for the personal biases of the inputs which it uses."[24]

3. *Checks and balances among the four estates.* We have to look at this problem of automation—or any problem involving the large-scale use of new technology by society—not merely as one of providing information to make possible the unified action of a lot of people with a common purpose. Instead it is the problem of providing a system of authority that can reconcile widely different purposes—and with the differences resulting not merely from conflicting material interests or general ideology, but from the differences in intellectual discipline or professional background that distinguish the several estates in modern society. Consequently, as the government has developed the system of command and control of air defense and intercontinental missiles, it has in practice shown an awareness that the relationship among the several estates is one that must be handled on the basis of constitutional principle, even though it may be a principle unknown to the courts and the Constitutional lawyers.

One might have thought that air defense and missiles offered the greatest potential scope to the unbridled development of technology and automation. The military nature of the programs meant that legal restraints on discretionary power were weak. The purpose was comparatively clear, and had almost unquestioning public support.

The Spectrum from Truth to Power

The scientific techniques were advanced and complex, and well beyond the full comprehension of the layman.

Yet it is obvious that there has been a lively system of checks and balances at work. The conflicts of interest have been identified and fought over at length; the most important issues have been debated by competing scientific and professional groups, and brought urgently to the attention of administrative officers (both military and civilian) and political authority (both executive and legislative). If the competing interests of the four estates have been able to bring a measure of responsibility and restraint into the command and control of guided missiles, it seems probable that there is nothing in the nature of advanced technology and automation that will inevitably bring about a highly centralized system of political authority.

We have been establishing new kinds of checks and balances within the governmental system that depend not on formal legal provisions, but on a respect for scientific truth and for professional expertise. This respect is stronger, no doubt, among the scientists and the experts—including the engineers, the generals, and even the administrators—than among the general public. But even the general public has come to distrust political authority and to support those who defend the scientists and experts against political interference. As a result, it is possible to defend pure knowledge, even though it is legally powerless, at each step along the gradation from science through professional work and administration into politics.

At each step, one estate distrusts the next for compromising objective truth in the interest of political purpose. Thus, in a problem like command and control, the scientists are likely to distrust the engineers for not being alert to the great discoveries that could be made if basic research were not restricted for pragmatic or economic (or other irrational) reasons, and to resent the limitations that are put on the laboratories for reasons of short-run expediency. The engineers distrust the generals and admirals and budget officers for restricting their full development of technological possibilities on account of operational or traditional—usually reactionary—reasons. The generals distrust their political superiors for wishing to interfere in strategic matters on nonmilitary, and presumably immoral, grounds; they suspect politicians of putting a concern for dollars ahead of the lives of soldiers, or of being soft on Communism, or of being unreliable when life-and-death issues are at stake, or even of trying to make decisions that military men know more about.

There can be no completely hierarchical relationship among the four estates. Even within government itself, the collective pride and ethical beliefs of a profession and the firm faith of the scientist in intellectual freedom are by no means overridden easily by political authority. An organization chart is conventionally drawn in terms of a pyramid of power, with the chief political officer on top, his administrators just below, and his professional or scientific subordinates lower still. But if that chart were drawn in the order of scientific precision, or demonstrable truth, it would have to be turned upside down.

This point of view, which sees the spectrum from science to politics as one involving a steady decline in demonstrable truth, is naturally held most often by the scientists. But if they were alone in that view, it would not have much weight in practical affairs. It has weight because it is shared quite generally, at least at various times and for different purposes. Even politicians themselves will appeal to it in cases in which the apparent majority of scientific testimony supports their side of a policy debate, or in the perhaps less frequent cases when it genuinely persuades them to accept a new policy.

But this general view of the moral superiority of science and the professions is counterbalanced by another way of looking at the spectrum from science to politics. For it is possible to see it as ranging not from pure truth to naked power, but from inhuman abstraction to moral responsibility. And this is the way the spectrum looks as the chain of distrust runs in the opposite direction to the one we noted above. The political executive distrusts the generals and admirals for wishing to have such tight control over their weapons systems that statesmen and diplomats are given no room to maneuver, and for using the argument of military necessity to demand so large a share of the nation's resources that its total strength is actually weakened. The generals and admirals are impatient with weapons engineers for developing endless technical refinements, rather than being willing to supply the combat units with simple and reliable weapons when they need them. The weapons engineers are even more impatient with the basic scientists, who are too much interested in fundamental knowledge to think very much about the feasibility of their ideas, or their practical and economic development.

There is hardly any social problem on which science cannot make some contribution. Science may help not only in the development of particular techniques for reaching the goals that others have determined; it may help refine our value judgments, and determine the nature of the goals themselves. For this reason, the infusion of science, as a way of thinking, should be welcomed in the professions, in administration, and in politics. On the other hand, the motives of the scientist (and of anyone else who may seek to apply science to any

The Spectrum from Truth to Power

given problem) may be tainted by various forms of self-interest, and he may be defining the problem, and making assumptions about the relevance of his science to it, in ways that are unjustified from various points of view—from the point of view of other sciences, or of perfectly valid value judgments of the layman. For this reason, there is no major public problem that does not have its political aspects, and none that should be left entirely to the judgment of the experts.

But if this is true, one might say, we have a set of conflicts of interest built into every issue, with the four estates involved in perpetual civil war. Quite obviously that does not happen; in some way, in every society, a realization of common purposes and values transcends these differences of approach, and unites the estates at least partially in a common effort. The extent to which there must be union, and the circumstances under which division and disagreement are tolerated or encouraged, are crucial questions in the development of any constitutional system. No matter what common purposes unite us, we must still work out a number of practical questions about the relationship of the four estates. We must ask, for example, just how government should decide to have a particular issue (or a part of a particular issue) decided by one estate or another; how each estate selects its leaders and is governed; how much interchange there is among them, and on what terms; and how the checks and balances in this system are related to those of our formal Constitution.

Even if all Americans agree on the twofold principle suggested above—that the more a function or institution is concerned with truth, the more it has a right to freedom, and the more it is concerned with power, the more it should be subjected to the test of political responsibility—those practical questions must be faced. I will take up such questions in the next chapter, but first I will try to be clear about the general approach that in the end will probably determine the answers to them. This general approach is summed up in the twofold principle, and it is quite different from that of the Marxist dialecticians.

The Western Tradition

In our domestic politics, we have gradually been getting accustomed to the necessity of working toward common ends with people whose theoretical philosophic views or religious beliefs we disapprove, but whose human qualities we respect, and whose practical purposes we share. In international politics, the power of modern weapons now may be forcing us to recognize similar elements of common purpose with those we consider our adversaries—if only a common interest in survival as separate nations. On this subject, it may be even more dangerous for us to be cynical than to be naive.

Some therefore are tempted to imagine that the scientific-technological revolution (a term that Communist Party theoreticians are beginning to use more frequently) has been making the social systems of the advanced industrial nations more compatible with one another. Advanced science and technology, so the argument runs, require higher education and critical minds, and thus lead toward a more liberal and open kind of political system. On the other hand, industrial technology has been centralizing economic power in the United States, and unintentionally bringing it more and more under political control. As the two systems come more nearly to resemble each other, why should they not get along better? Quintin Hogg, British Minister for Science, illustrates this train of thought. In 1964 he told the Second Parliamentary and Scientific Conference of the Organisation for Economic Cooperation and Development and the Council of Europe he believed that current technical developments would drive the economic systems of all developed nations closer together. Speaking of Soviet Communism, he remarked, "I fancy that it may not turn out to be so different from the American way of life as the inhabitants of the Kremlin or the White House may expect."[25]

The same view has often been expressed, though usually in private, in the United States, in conservative as well as liberal circles. Many scientists are tempted to share that view by the old utopian dream of a future in which technical and industrial progress would let political authority wither away, or at least greatly reduce its influence. This still seems to be the hope, in spite of recent evidence, of many conservative businessmen.[26]

There is no doubt, too, that the common requirements of technology have brought about great similarities in the organization of factories or laboratories or military forces, regardless of the nature of the political system. This leads not only scientists, but even administrators and politicians, to develop a certain amount of international sympathy on the basis of common problems, and may soften the harsh outlines of ideological conflict. Thus Khrushchev, in a major speech to the Communist Party Central Committee in 1962, reviewed his economic and technological problems, and admitted that, "having destroyed class and social barriers in October, 1917, we later, in the course of economic construction, in many cases erected departmental and local-interest barriers that, intentionally or unintentionally, restrict the possibilities of our development."[27] Americans may well take a

The Spectrum from Truth to Power

sympathetic pleasure in such a bit of realism, with its general theoretical implications, and in other comments on technical policy in Khrushchev's speech. For example, he admitted that the Soviet Union, for all its central party authority, finds it hard to centralize science for civilian purposes as it has managed to do for military purposes. He warned that capitalism has the advantage of competition to spur the adoption of new technology. And he summed up some of the common human problems of technological innovation in all societies as he described his visit to an old blacksmith who told him directly to get that girl efficiency expert with her stop watch away from his forge and cut out those time-and-motion studies.[28]

The official Communist Party doctrine, of course, has painted a quite different picture of the way in which science and technology contribute to the planning of a new society. A party theoretician put the argument thus in an article in *Kommunist,* the theoretical journal of the Bolshevik Party: "The present scientific-technological revolution is unfolding at a time when there exist two opposite social systems in the world. Its most important elements are as a rule not a secret of this or that industrially highly developed country. The struggle for the fullest, fastest, and most economic utilization of the possibilities offered by this revolution is one of the most important aspects of the economic competition of the two systems ... The general realization of the fruits of the scientific-technological revolution requires a level of socialization, and of a planned coordination of science, technology, and production, that are unthinkable under the capitalist system."[29]

As the Secretary of the Central Committee of the Communist Party of the Soviet Union put it, "The Party has restored a Leninist and profoundly scientific approach to the *solution of economic problems.* Just as Lenin put forward the task of establishing the material and technical foundation for socialism through industrialization of the country, so the Party has worked out, at this new stage, a plan for the creation of the material and technical basis of communism. One of the distinctive features of the past decade has been the fact that the Party initiated and led the scientific and technical revolution and thus placed the Soviet Union in the forefront of world scientific and technical progress, making socialism its standard bearer."[30]

This doctrinaire argument has been immensely persuasive to many underdeveloped nations, which wish to shape their political institutions so as to take full advantage of technological progress. Considerable doubt, to say the least, has been cast on the argument, not only by Khrushchev's earthy and practical comments, but by many intellectual leaders in the Communist countries who have discovered, with Milovan Djilas, that conflicts of interest can arise among the estates in society without having any basis in private property. This discovery, and the parallel discovery that in the United States a new mixture of governmental and private enterprise has been invented (partly as a result of scientific and technological programs) to modify or replace the classical form of capitalism, have reduced the apparent difference between the two systems with respect to the national planning of technological progress. In that respect, the two systems are now close enough for most people to be willing to have them judged on the basis of peaceful competition.

But there is a fundamental aspect of Communist theory regarding the relation of science to politics that makes it more difficult for the rest of the world to confront the Communist world on terms of friendly rivalry. One party theoretician put the issue in the form of a question: "Is it really possible to call scientific the concepts of contemporary bourgeois sociology . . . if all of them question the existence of objective laws of social development, and the possibility of predicting the paths of this development?"[31] Although this quotation pertains to a rather academic issue, it seems to me of much greater importance. It sums up the demand of Communist theory not only that the social sciences and history be included with the natural sciences in a single philosophical system, but that this system predict the future development of society.[32] Is this a scientific approach to the problems of society?

Half a century ago there may have been some reason for the ordinary citizen to think so, and to fear that the progress of science would bring about a drift toward some form of dogmatic materialism as the basis for a dictatorship. But this does not seem to be in accord with the way science itself has been working, in either its theoretical or practical developments. The "scientific prediction" of the future development of society is not really a prediction; it is an assertion of purpose. The several sciences can combine to make such a prediction only if they are asked to conform to a predetermined theory, and if the request is supported by the authority of a small and disciplined and indoctrinated party. To combine in such a corporate body control over both ends of the spectrum from political authority to scientific knowledge is not a decision that is encouraged by science. It comes instead from a very old and uncritical faith: the belief that an organization of human beings can know the ultimate purpose of the world, bring a select group of its members to a state of superior knowledge and power, and provide the doctrine to guide an authoritarian government to rule the rest.

The Spectrum from Truth to Power

The Marxists have not been the first to fasten such a notion on Eastern politics. Their belief has its ancient analogy in the belief of the Orthodox church that it could make perfect the faithful; a vision of perfection can justify the use of any ruthless means. The Roman church never shared this belief in the possibility of human perfection, and the Reformed churches, of course, were at the opposite extreme with their dogma of total depravity. And no matter how self-righteously and rigorously Calvinists may have tried to rule their cities and countries, it was pretty hard to set up a permanent authoritarian rule, or even to maintain a unified church, when their basic doctrine put rulers and ruled at a common level of imperfection.[33]

If the inner logic of the sciences themselves does not bring all of them—including the social sciences—together in a uniform methodological system, there is no reason to think that anything in the political habits or traditions of the West will force them to do so. The several sciences may indeed acknowledge allegiance to a common set of very general principles; they all believe in truth and objectivity, in the publication and verification of a scholar's findings, and so on. But this is like the agreement by the dissenting Christian sects on a number of basic doctrines. Agreement on organization or action is a different matter. The countries in which theologians have long insisted on organizing themselves in separate sects, and on not letting the government set up an orderly and unified establishment for them, seem to be the ones in which scientists do the same.

We may well have faith that there is, in some ideal sense, one system of truth comprehending the approaches of all the sciences—physical, biological, and social—but they cannot be unified to produce a governing ideology or social plan without closing off a tremendous variety of possible future choices, and without being used as an instrument by which a new governing class may maintain a monopoly on power. Like an excessively automated system, this may produce impressive results for a time in a predetermined direction. But there is no logical reason to suppose that it is more efficient as a general political system, whether you prefer to think of the main purpose of such a system as keeping open the opportunities of technological and economic progress, or of protecting the human freedoms that are fundamentally based on even more important values.

The most fundamental disagreement between the nations of the Western political tradition and those of the Communist world does not turn on their attitudes toward private property. The greatest mistake in Western political strategy consists in committing itself to the defense of property as the main basis for the preservation of freedom. Private property is indeed a useful and important means to that end, but it is not an absolute end in itself, and the effect of scientific advance on a technological civilization has made property less and less important as a source of power, and as a way of limiting political power. Far more fundamental is the way men think about the desirability of organizing truth in the service of power, and using power to determine truth. Whether truth is conceived in the old terms of religion, or the new terms of science, the greatest source of tyranny is the conviction that there is a single way of determining truth, and that it should be interpreted by a single disciplined organization.

Having rejected this conviction, Americans need to base their constitutional system on a division of power that will take fully into account the newer institutional forms of power as well as the old—the new estates of the scientific era, as well as branches of government and business corporations. And as we do so, a clear understanding that science alone cannot solve political problems will be the surest safeguard for the protection of the sciences and professions in a proper degree of independence from political authority.

Notes

1. Robert Dorfman, in his survey of "Operations Research" for the *American Economic Review* in September 1960, noted types of problems that were not susceptible to solution by operations research, and went on to say, "Fortunately, this galaxy of unsolvable problems is less obtrusive in narrow operational contexts than in broad strategic ones. It is easier to ascertain the objectives of a department than of a firm, of a section than of a department, of a particular phase of the work than of a section" (p. 611). Much the same distinction is apparent in Herbert A. Simon's discussion of programmed and nonprogrammed decisions, and their relation to hierarchical organization, in his *The New Science of Management Decision* (New York: Harper, 1960), pp. 5-7, 49-50.

2. Albert Wohlstetter, "Defense Decisions: Design vs. Analysis," abstract of paper presented at Second International Conference on Operational Research, Aix-en-Provence, September 1960.

3. The story of the fire ant is only one of many interesting cases that led to the concern exemplified in Rachel Carson's *The Silent Spring* (Boston: Houghton Mifflin, 1962), and in the subsequent report of the President's Science Advisory Committee on the use of pesticides. For the fire ant story, see W. L. Brown, Jr., "Mass Insect Control Programs: Four Case Histories," *Psyche: A Journal of Entomology,* June-September 1961, p. 75, and Edward O. Wilson, "The Fire Ant," *Scientific American,* March 1958, p. 36. Brown said in his 1961 article that some of the insect control programs amounted to "scalping in order to cure dandruff." The articles by Wilson and Brown show how the scientific journals may be ahead of the popular authors, and even further ahead of political action.

The Spectrum from Truth to Power

4. There is an interesting analogy here between the politician's use of science and the scientist's use of the discoveries of other scientists, for example by using mechanical or electronic aids in searching for citations. H. Burr Steinbach observes that "the importance, to a scientist, of *knowing the literature* is (i) inversely proportional to the size of the idea (many people think of many small things) but (ii) directly proportional to the proximity to technology (finding out someone else did something saves time—sometimes)." "The Quest for Certainty: Science Citation Index," *Science*, July 10, 1964, p. 142.

5. P. M. S. Blackett, "Operational Research," *The Advancement of Science*, April 1948. This paper had been written and circulated privately during the war.

6. Similar testimony as to the extent to which operations research is progressively less able to deal precisely and completely with problems that have to be faced by higher levels of authority may be found in Bernard Brodie, *Strategy in the Missile Age* (Princeton, N.J.: Princeton University Press, 1959), p. 388, and Charles J. Hitch and Roland N. McKean, *The Economics of Defense in the Nuclear Age* (Cambridge, Mass.: Harvard University Press, 1960), pp. 125-128.

7. Unpublished lecture by Dr. Panofsky at Tufts University, March 7, 1964.

8. See for example Senator Henry M. Jackson's comment in *Arms Control and Disarmament,* Hearings before Senate Committee on Armed Services, Preparedness Investigating Subcommittee, 87th Cong., 2d Sess., 1962, p. 26. Also Senator J. W. Fulbright's comment in *Nuclear Test Ban Treaty: Hearings*, Senate Committee on Foreign Relations, 88th Cong., 1st Sess., 1963, pp. 639, 642. Also Representative Jack Westland's comment in *Technical Aspects of Detection and Inspection Controls of a Nuclear Weapons Test Ban,* Hearings before the Special Subcommittee on Radiation and the Subcommittee on Research and Development of the Joint Committee on Atomic Energy, Part 1, 86th Cong., 2d Sess., 1960, p. 75.

9. Richard Bellman, for example, warns the mathematician of the difficulties of determining which problems will or will not be "susceptible to the collection of devices, tricks, and legerdemain called mathematics." He goes on to note that the "classical techniques are ineffectual as far as decision making in the face of complexity and uncertainty is concerned," and emphasized the value of the "engineering attitude," which is "partial control with partial understanding." *Challenges of Modern Control Theory,* RAND Corporation, Santa Monica, Calif., RM 3956-PR, January 1964.

10. John C. Whitehorn, M.D., Professor Emeritus of Psychiatry, Johns Hopkins, "Education for Uncertainty," a lecture at the Massachusetts General Hospital, 1961, printed by the University of Rochester School of Medicine and Dentistry.

11. Col. Francis X. Kane, USAF, "Security Is Too Important to Be Left to Computers," *Fortune*, April 1964, p. 146.

12. U.S. Atomic Energy Commission, *In the Matter of J. Robert Oppenheimer*, transcript of hearing before Personnel Security Board, 1954. See esp. pp. 384-394, 709-727, 742-770. See also Robert Gilpin, *American Scientists and Nuclear Weapons Policy* (Princeton, N.J.: Princeton University Press, 1962).

13. "Raffiniert ist der Herr Gott, aber boshaft ist er nicht." The remark, taken from a conversation, adorns the professors' lounge in the Princeton

Don K. Price

University mathematics department. *Albert Einstein: Philosopher-Scientist,* ed. Paul A. Schilpp (Evanston, Ill.: Library of Living Philosophers, 1949), p. 691.

14. If the reader's taste runs to science more than theology, he may compare the following by Bernard Berelson and Gary A. Steiner, in their *Human Behavior* (New York: Harcourt, Brace & World, 1964): "But there is another way in which man comes to terms with reality when it is inconsistent with his needs and preferences; and it is here that the behavioral-science model departs most noticeably from the others. In his quest for satisfaction, man is not just a seeker of truth, but of deceptions, of himself as well as others" (pp. 663–664).

15. Nigel Calder, "Parliament and Science," *The New Scientist,* May 28, 1964, p. 534.

16. Thornton Read, in his *Command and Control* (Center of International Studies, Princeton University, 1961) gives a very lucid summary of the problem. The difficulties that result from too much reliance on the computer are summed up in Anthony G. Oettinger, "A Bull's Eye View of Management and Engineering Information Systems," *Proceedings of the 19th ACM National Conference,* Association for Computing Machinery, New York, 1964, pp. B.1-1 to B.1-14.

17. Roberta Wohlstetter, in her *Pearl Harbor: Warning and Decision* (Stanford, Calif.: Stanford University Press, 1962), points out that the blunder at Pearl Harbor was not the result of lack of warning, but of poor strategic analysis, and that this in turn came less because we did not understand the Japanese than because we did not understand the deficiencies in our own system of responsibility.

18. Sidney Hook criticizes the political and fictional critics of the command and control system in *The Fail-Safe Fallacy* (New York: Stein & Day, 1963).

19. For an impression of the significance of the distinction, see the way in which the Army Chief of Staff system was contrasted with the Navy's Commander in Chief of the Fleet system by Henry L. Stimson and McGeorge Bundy, *On Active Service in Peace and War* (New York: Harper, 1948), pp. 32-33, 450-451, 506.

20. Brockway McMillan, address before the First Congress on the Information System Sciences, Nov. 19, 1962.

21. Sir Robert Watson-Watt, *The Pulse of Radar* (New York: Dial Press, 1959), pp. 277-278.

22. "In this setting there exists a considerable danger that complex decision systems involving human parameters will be broken down into routine segments which are more or less independent of human reaction, and that the combination will then be called a credible simulation of the total system... Also, there is an attractive but dangerous precedent for restricting value parameters in the interest of simplicity and neatness; the result... may be satisfactory in a 'game' situation but is disastrous in application." David L. Johnson and Arthur L. Kobler, "The Man-Computer Relationship," *Science,* Nov. 23, 1962, p. 876.

23. Read, p. 21; Johnson and Kobler, p. 877.

24. Read, p. 15.

25. Calder, p. 534.

26. Zbigniew Brzezinski and Samuel P. Huntington vigorously disagree with the theory that the systems of the U.S.A. and U.S.S.R. are converging;

The Spectrum from Truth to Power

one of their reasons is that in the U.S.A., unlike the U.S.S.R., the policy-making process is dispersed among groups whose power depends not on legal power or on property but on special knowledge. See their *Political Power: USA/USSR* (New York: Viking Press, 1964), p. 413. See also Sheldon S. Wolin, *Politics and Vision* (Boston: Little, Brown, 1960). In his last chapter Wolin discusses the ways in which modern business has believed that it could assume essentially political functions.

27. "Khrushchev's Report to the Party Plenary Session," Nov. 19, 1962, in *The Current Digest of the Soviet Press*, vol. 15, no. 47, p. 5.

28. *Ibid.*, pp. 5, 13.

29. S. Kheinman, "Sozdanie material'no-tekhnicheskoi bazy kommunizma i nauchno-tekhnicheskaya revolyutsiya," *Kommunist*, No. 12 (August 1962), 47–48. The title in English: "The Establishment of the Material-Technical Basis of Communism and the Scientific-Technological Revolution."

30. Boris N. Ponomaryev, "Leninism Is Our Banner and All-Conquering Weapon," *Current Soviet Documents*, 1, no. 8 (May 13, 1963) (Crosscurrents Press, New York). The italics in the quotation were in the original.

31. L. Ilichev, "Nauchnaya osnova rukovodstva razvitiem obshchestva; nekotorye problemy razvitiya obshchestvennykh nauk," *Kommunist*, No. 16 (November 1962), 34. The title in English: "The Scientific Basis of the Management of the Development of Society; Certain Problems of the Development of the Social Sciences."

32. The theoretical problems of the relation of science to materialist dialectic, which denies positivism and reductionism almost as vigorously as idealism, are summarized by Maxim W. Mikulak, in "Philosophy and Science," *Survey: a Journal of East European Studies,* July 1964, pp. 147–156. See also Richard Pipes, "Foreword to the Issue 'The Russian Intelligentsia,'" *Daedalus*, Summer 1960, and other essays in that number, especially those by David Joravsky and Gustav Wetter.

33. To illustrate the difference between Orthodox and Roman Catholic theology with respect to perfectibility, as seen from the Roman point of view: Gerhart B. Ladner argues that the idea of reform (as distinct from revolution or renaissance) was inhibited by the Eastern Orthodox belief, as typified by St. Gregory of Nyssa, who held that "the reform of man had been purification of the soul which would thus mirror God more and more clearly; a resulting vision of God is not excluded even on earth." By contrast, "Augustine did not think that the terrestrial condition of even the holiest man warrants the expectation of a vision of the fullness of God this side of heaven." See Ladner's *The Idea of Reform* (Cambridge, Mass.: Harvard University Press, 1959), pp. 190–191. On the issue of perfectibility and its relation to ecclesiastical government, see also Reinhold Niebuhr, *The Nature and Destiny of Man* (New York: Charles Scribner's Sons, 1947), Part II, p. 133.

5. The Scientist and the Politician

ROGER REVELLE

In 1889, John Wesley Powell, explorer, geologist, and ethnologist, was the retiring president of the American Association for the Advancement of Science. Major Powell was the founder of the U.S. Geological Survey and the Bureau of American Ethnology; he was an enthusiastic advocate of the creation of a Federal Department of Science. In his monograph on the lands west of the 100th meridian he was the first to show that a shortage of water would limit the development of the West. One might have expected him to use this presidential platform to present his views on some broad issues of science and society, but the traditions were different in those days. His presidential address consisted of a dissertation on "Evolution of music from dance to symphony." It was only in the 1920's that retiring AAAS presidents began to talk about broader issues, usually in terms of science as a great human enterprise and the duty of our Association to defend it and spread its benefits among mankind.

Recognition of the social responsibility of scientists and technologists for the uses of their discoveries came after World War II when the terrifying threats posed by nuclear and thermonuclear weapons became clear to all. But science and technology were still thought of as essentially neutral. Their work could be used for good or ill, depending on the decisions made by other sectors of society. Scientists could try to influence these decisions, but, in their capacity as scientists mainly on technical grounds. Scientists and politicians should maintain an arm's length relationship with each other.

Copyright © 1975 by the American Association for the Advancement of Science. Reprinted from *Science*, Vol. 187, March 21, 1975, pp. 1100–1105, by permission.

The Scientist and the Politician

The Necessity for Cooperation

Today this comfortable arm's length relationship will no longer do. The threats to civilization are too real and too immanent for anything other than the closest kind of cooperation among politicians and scientists in the search for ways out of our present dilemmas, in increasing public understanding of the issues involved, and in mobilizing the confidence and will of the people.

Many events have converged to create our present precarious situation. Among its outward signs are environmental decay; the depletion of resources; the rise of ever more pervasive, ever more unresponsive, bureaucracies, both corporate and governmental; the economic contradictions of capitalism, as they manifest themselves in double-digit inflation, combined with economic stagnation or decline; the tragic gap between the rich and the poor countries with the resulting threats of rapid population growth, starvation, and social collapse; above all the malignant insanity of the arms race, and the strategy of mutual terror. But the real crisis of the West may exist within ourselves—in a failure of nerve, a loss of self-confidence and a sense of purpose, a widening disillusionment with technology, and with economic growth based on technology—in short, a loss of faith in the inevitability or even the possibility of human progress, the great idea that has powered our civilization for 300 years.

When I speak about science it must be understood that I am talking about both science and technology, for in our times they cannot be separated. This was not always true. The most important physical inventions in human history—fire, fermentation, farming, and the working of metals—all occurred before the natural sciences were born. Likewise, the most important social inventions—birth control and cities—were made without benefit of the social sciences. But ever since some people began to realize they could learn about nature through the combination of theory and experiment which we now call science, the possibilities of applying this knowledge have been in the forefront of Western thought. Francis Bacon said that we seek knowledge of nature to extend our dominion over things. Three hundred and fifty years earlier, Roger Bacon had written, "Machines may be made by which the largest ships, with only one man steering them, will be moved faster than if they were filled with rowers. Wagons may be built which will move with incredible speed and without the aid of beasts. Flying machines can be constructed in which a man may beat the air with wings like a bird." And Descartes said, "We can have useful knowledge by which, cognizant of the forces and actions of fire,

Roger Revelle

water, air, the stars, the heavens, and all the other bodies that surround us, knowing them as distinctly as we know the various crafts of the artisans, we may be able to apply them in the same fashion to every use to which they're suited, and thus make ourselves masters and possessors of Nature."

Today, science and technology might be described as having a kind of incestuous relationship because technology is both the mother and the daughter of science. Most modern technological developments, including jet airplanes, computers, contraceptive pills, and hybrid corn, not to mention such mixed blessings as the control of nuclear energy, could not have occurred without the knowledge and understanding gained by science. And scientific discovery depends more and more on our ability to supplement our eyes and ears with complex instruments based on new technologies.

Some of the Difficulties

In discussing the need for cooperation between scientists and politicians, we must start by being aware of the difficulties. First, scientists and politicians have little mutual empathy. The personalities, methods, motivations, roles, and orientation of scientists and politicians are each foreign to the other. The politician is publicly egotistical, gregarious, garrulous, and has a strong gambling instinct. The scientist, at least in his own image, is publicly modest, introverted, relatively inarticulate, and seeks certainty rather than risk. In the past, the best science has been conducted within a narrow discipline, whereas the politician's methods are multidisciplinary in a way those of scientists can never be; they include the ancient arts of rhetoric and myth-making, appealing to the emotional and the irrational in other men as well as to their calculating self-interest. The scientist is motivated by the need to explain, predict, and control phenomena. The politician is motivated by a desire for power. Or to make the contrast more exact, scientists and technologists seek power over nature; the politician seeks power over men. The scientist's role in society is to gain knowledge and understanding; the politician's is to decide and to act. Indeed, in our democracy he alone has the obligation, as the people's elected representative, to make decisions as to what society shall do and to take responsibility for those decisions. In his search for truth, the scientist is oriented toward the future; the politician's orientation is usually here and now. He desires quick visible pay-offs for which he often seems willing to mortgage the future. For

The Scientist and the Politician

the politician in a democratic society, infinity is the election after the next one.

Many barriers must be overcome before an effective working relationship can be established between politicians and scientists.

1. Most politicians are lawyers, and lawyers feel most at ease in adversary proceedings in which conflicting viewpoints and evidence are presented as a basis for decision. Scientists generally shun such proceedings, preferring to work in committees or otherwise cooperatively, to examine facts and hypotheses in order to find agreement or to clarify disagreements.

2. Scientists like to categorize themselves in professional disciplines or subdisciplines. Politicians and lawyers traditionally resist classification and tend to think of themselves as generalists rather than specialists.

3. Politicians and lawyers are interested in the particular rather than the general. They do not seek universal truths or broad generalizations, but are content with determining as best they can the facts that bear on a particular situation. Scientists are much more interested in regularities that underlie or explain an entire class of phenomena.

4. Politicians and lawyers need the best possible answers as soon as possible. They are not interested in the scientist's search for certainty and his patient testing of hypotheses. They want to know and use the present state of the art, and in this respect they are more closely akin to engineers than scientists.

5. Technological advances all too often outpace political institutions. In his speech at the centennial celebration of the National Academy of Sciences, President Kennedy said, "Whenever you scientists invent a new technology, we politicians have to make a new political invention to deal with it."

6. With regard to the social sciences, politicians tend to believe, with some justification, that they know as much or more about human behavior from a practical point of view as the psychologist or the sociologist. They are unsympathetic with the tentative nature of the social sciences and the wide divergences of opinion among social scientists concerning the existence and character of regularities or laws of human and societal behavior. Moreover, like other shamans, they do not like to have their secrets exposed.

7. The main task of the politician is to mediate among competing social pressures—to arrive at compromises that are most acceptable or least unacceptable to most people. The scientist tends to take an un-

Roger Revelle

compromising position which reflects the truth as it is known at a particular point in time.

On Advice-Giving

How can the politician and the scientist work more closely together? One way is to give each other advice. Scientists have long accepted the idea that they should advise politicians, but they are liable to react with alarm and incredulity when it is suggested that politicians should advise *them*. The fact is, of course, that nowadays politicians advise scientists in the most forcible and direct possible way—by granting or withholding support for research and development. I would argue only that this process of mutual advice-giving should contain a better feedback mechanism. The scientist should advise the politician concerning the advice in the form of financial support that the politician gives the scientist. And similarly, the politician should advise the scientist concerning the kinds of scientific and technical advice and the conditions under which it is given that will be most useful to him.

Limitations of scientific advice. When scientists advise politicians both groups need to recognize the limitations of the process.

Most scientific questions of interest to politicians contain a nonscientific, and often what Alvin Weinberg has called a "transcientific element." This is in part because the politician needs answers under time and action constraints that are incompatible with scientific certainty, and in part, as Victor Weisskopf has pointed out, because a complete description of a phenomenon in scientific terms may not contain the elements of the phenomenon that human beings consider most relevant. Political decisions in such situations must be based on judgments that are outside the realm of science. Tran-scientific questions that go beyond scientific certainty necessarily involve large areas of judgment in which individual scientists disagree. This disagreement is hard for the politician to deal with because he has difficulty in recognizing what, in the scientist's advice, represents scientific certainty and what are elements of judgment, estimation, and uncertainty. In other situations, the politician may accept the scientist's word that there is not enough information for scientific certainty without recognizing that nevertheless there is enough for political action.

The scientist can help the politician evaluate the costs of making different kinds of mistakes, but insofar as these costs depend on value judgments, the scientist can give no more help than any other citizen.

The Scientist and the Politician

It is often stated that scientists and technologists can give useful advice to politicians about the character and time scale of future technological developments, but this is true only to the extent that the basic scientific discoveries or technical inventions have already been made. Otherwise, the scientist can only limit himself to predictions that physical and biological laws will not be violated. By pointing out to the politician the intractable nature of many scientific and technical problems, he can help the politician suppress his natural tendency to solve these problems by political action. He can help him to avoid legislating a cure for cancer.

The same difficulty exists, a fortiori, in forecasting the technical developments that society will want or should want in the future. So-called normative technological forecasts depend on today's values and ideologies, which in a world of change may change at least as rapidly as technology itself. Normative technological forecasts are likely to be useful only if they are self-fulfilling prophecies.

The scientist can advise the politician on how to cure human ills, but not on how to produce human happiness. He must resist the tendency of the politician and the public to look upon him as a member of a remote omniscient priesthood even though public faith in science and scientists may be essential for the health of both science and society.

Two kinds of scientific advice. Politicians need and can utilize scientific and technical advice given in several different ways. These can be conveniently classified as inside and outside, or, if you will, in-house and out-house. The scientific adviser who is inside the system must be completely inside it. He must accept the politician's rules of accountability and responsibility, which always mean discretion and loyalty and often anonymity. In other words, he should be a professional rather than an amateur. If he differs with his boss, he should resign, but not go public.

Scientific and technical advice from outside the political system is also useful especially if it serves to educate the public as well as politicians. It may be given by supposedly impartial committees formed by academies and scientific associations or even by scientific and technical pressure groups. Scientists can often give the most effective advice by "taking their case to the people."

Does the President need a science adviser? It is clear from what I have been saying that the presence within the Executive Office of the President of a group of full-time knowledgeable specialists, containing a mix of social and natural scientists from different disciplines, is essential to maintain the long-range effectiveness of science and technology in the United States, to develop broad strategies for science policy as

Roger Revelle

well as ad hoc solutions for immediate issues, and to resolve conflicting claims on scientific and technical resources.

Whether advice from these technical experts should be presented directly to the President or should be filtered through a group of generalists in the White House will depend on the President's own needs and style. Some presidents will want completed staff work in which scientific and technical factors are weighed against economic, political, and social ones in the alternatives presented to them for decision. Others will prefer to judge for themselves among the clash of opinions and to formulate their own options.

Political Directions for Science and Technology

The giving and receiving of advice is only one aspect of the need for new relationships between technology and politics. Dean Harvey Brooks of Harvard University recently gave a speech in which he asked the rhetorical question, "Are scientists obsolete?" He immediately answered in the negative by describing 11 major problem areas for future research and development. Most of these problems have arisen in part from the legislative or administrative actions of politicians or will depend for their solution on cooperation among politicians, scientists, and technologists; for example, energy supply and conservation; pollution control; technology assessment, that is, the side effects and secondary effects of new technologies; efficient, humane, and cost-effective health services; world food supplies and nutrition; the communications revolution as it will be affected by the breakdown of the traditional monopoly of communications by common carriers; problems of increasing the productivity of the public sector, particularly at state and municipal levels, and the productivity of the service industries in general; and the attempt to regain a comparative advantage for the United States in international trade, particularly in capital goods technology. To these I would add the stopping and reversal of the arms race, the highly inequitable distribution of income in the world's poor countries, which probably lies at the root of population problems, and the social and economic transformations being brought about by multinational corporations, which may no longer be responsibe to the traditional economic laws of supply and demand.

Beside their political nature, these problems have three common characteristics. (i) They involve the study of complex systems—the ecology of the natural world and of human societies, and systems of information, communication, transportation, and control—that can-

The Scientist and the Politician

not be studied by the conventional reductionist methods of the natural sciences but demand instead a synthetic approach. (ii) They require intimate collaboration between social and natural scientists—consider, for example, the relationships between population distribution and the impact of environmental changes, and communal responses to risks created by environmental hazards. (iii) The responsibility for solutions must be broadly shared among the people who will be affected. To exercise this responsibility the public needs much more understanding of the issues, including the technical and scientific aspects. One of the major tasks of all concerned must be to increase public understanding of the potentialities and limitations of science and technology and the socioeconomic changes they both create and require.

Two Case Studies

To understand the difficulties of joining scientific knowledge with political action, let us take two case studies: our national failure to anticipate the energy crisis and the slow progress of the "green revolution" in India.

The energy crisis. Although scientists concerned with natural resources knew and publicly stated for many years that the potential oil reserves in the United States are limited, and at least an order of magnitude smaller than our coal reserves, the energy crisis in the special form it has taken during the last 2 years was not and could not have been predicted from geological considerations, because, in its immediate aspects, it is economic and political. Nor would it have occurred to traditional economists that the Arab countries and the other oil producers could organize an effective cartel. The actual costs of oil production in the Middle East, both tangible and intangible, are about 20¢ a barrel, a fiftieth of the present price, and the production potential is so much greater than world demand that the principal worry of the international oil companies before 1973 was the possibility of a break in prices due to oversupply.

Development of our own energy resources and a reduction of our profligate use of energy seemed to conflict with other social and economic goals, although from a long-range point of view this conflict was more apparent than real. In the short range, however, the new awareness of the environmental damage caused by energy production, transportation, and use delayed construction of the Alaska pipeline, virtually prevented the burning of high-sulfur coal and oil, inhibited exploration and production of off-shore oil, and mandated energy-consuming antipollution measures. The national fixation on economic growth and the political pressures of vested interests determined government pricing and import policies which exacerbated the crisis of supply and prevented the adoption of energy conservation programs. High interest rates and a shortage of capital, together with the opposition of environmentalists, slowed the construction of nuclear reactors which could have taken some of the load off fossil fuels. The automobile manufacturers stubbornly clung to making monstrous gas guzzlers because of their high profit margins rather than smaller cars that would economize on fuel.

There were sharp disagreements among scientists and technologists on the size of U.S. reserves of petroleum, natural gas, and uranium and on the feasibility and timing of new energy sources. Since 1920 some petroleum geologists have been predicting that our oil reserves were about to run out. These cries of wolf, in the face of increased oil supplies that resulted from new technologies for petroluem exploration and recovery, lulled the public and the politicians into complacency. Optimistic estimates of uranium resources by the Atomic Energy Commission led to a weak program of research and development for breeder reactors. Scientists and engineers differed widely on the likely time scales or even the feasibility of development of thermonuclear fusion and other new energy sources.

Political action was inhibited by public disinterest. The average American, like his political representatives, has a high discount rate concerning the future. Public opinion about long-range problems rarely crystallizes into a sense of urgency. The newspapers, popular magazines, and television reflect this lack of interest by giving little coverage, particularly when the problems concern such hard-to-understand subjects as energy and fossil fuels which are full of numbers. Scientists and engineers have contributed to the lack of public interest by cultivating, or at least not abjuring, the public's faith that science and technology will always come up with a miracle in time to avert a problem. John Sawhill, formerly the Federal Energy Administrator, said in a recent interview, "We can't move too fast on science and technology. The President can't introduce a program until the people are ready to support it, and the people won't be ready until they are in a crisis situation. Once we are in a crisis we can shape a crash program to deal with it. I believe in the efficacy of crash programs. It's only when you marshall all your talents and resources on a crash basis that you get good hard results." In the light of the record in the energy crisis, this viewpoint can be most charitably described as trying to produce a baby in 1 month by putting nine men on the job.

The Scientist and the Politician

Finally, there were structural difficulties in the federal government and the energy industry that grossly retarded effective federal action. More than a dozen federal agencies were charged with regulation of the energy industry or with energy research and development. The economic structure of the industry itself, as a partially regulated free enterprise, makes government intervention difficult and a cause for resentment. The industry raises formidable obstacles to government action and its structure makes it hard to predict the effects of federal intervention.

We can't go home again. The energy crisis will not go away. Indeed, it is likely to persist for the rest of our lives and perhaps that of our children. It could be the immediate cause of the collapse of Western civilization as we know it. A considerable degree of energy conservation is both possible and desirable, but it is mindless to suppose that we can reverse our dependence on nonhuman energy. We have gone too far in raising life expectancy, and hence the numbers of people, and in lifting the burden of physical labor from the backs of farmers and city dwellers. Without mechanical energy our cities would be uninhabitable and many people would starve.

An old saying has it that "slavery will persist until the loom weaves itself." All ancient civilizations, no matter how enlightened or creative, rested on some form of slavery, because human and animal muscle power was the principal energy available for mechanical work. It is not because we are enlightened that we have abolished slavery but because we have discovered a cheaper source of energy. A man can produce in a day about a kilowatt-hour of mechanical work; to keep him working on the meagerest of diets costs 15¢. Even at present oil prices, a kilowatt-hour of electrical power, or the equivalent in gasoline, costs only about 2¢. By its discovery of less expensive energy than human muscles, Western civilization, unlike all others, has been able to make men free.

Put in other terms, once a society has climbed onto the treadmill of technology, it can never get off again. No solution of the energy problem is possible without far-reaching technological advances in both the production and the conservation of energy. But the lesson up to now is that such technological advances will not occur in time, or may not take place at all, without enlightened, far-seeing political action, courageous political leadership, and clear public understanding of the issues and possibilities. Our future welfare and perhaps our survival will depend on the closest kind of cooperation between politicians, technologists, and natural and social scientists. This, in turn, will depend both on an enlightened public support and on the politician's sensitivity in recognizing what the people would want if they had a chance to want it, that is, the choices they would make if those choices were actually available. Social scientists have a special role to play in defining and appraising the possible range of public choices.

How to stop a revolution. Why has the "green revolution" progressed so slowly in India? When the new high-yielding varieties of wheat were first introduced in the middle 1960's they caught hold with great rapidity. From a few demonstration farms the new varieties spread to millions of acres within 5 years and the results were spectacular. India's wheat harvest doubled from 1967 to 1971. In 1971, India produced a large surplus of food grains, more than enough to feed the millions of refugees who poured over the borders from what is now Bangladesh. But from 1972 onward, food production has hardly increased at all, nor has the area planted to high-yielding varieties of wheat and rice.

A drought in 1972 and poor weather conditions over large regions during the 1974 monsoon, combined with the worldwide rise in petroleum and fertilizer prices, are partly responsible, but a major share of responsibility must be assigned to governmental actions and inactions. Farm prices of cereals have been kept low in order to placate the urban masses while fertilizer prices have steadily risen. It takes two or three times as much wheat or rice to buy a pound of fertilizer in India as in Japan and considerably more in India than in Pakistan.

To buy fertilizers the farmers must have credit on reasonable terms and this has not been available. Most small farmers are still in the grip of the traditional village moneylenders. Many of these farmers are sharecropping tenants with little or no security of tenure; they do not have much to gain from planting new high-yielding seeds which require expensive inputs of fertilizer, irrigation water, and plant protection. Land reform has been virtually nonexistent even though many larger landowners have failed to intensify their farming practices. The state governments, which are mainly responsible for agricultural development, have been dominated by the richer farmers and they have neglected the interest of the small farmers and the incentives they need to increase their production.

Insufficient resources have been allocated to the agricultural sector, with the result that development of irrigation has been very slow even though in most regions irrigation is required for the new varieties. Because of the great uncertainties in rainfall, farmers are reluctant to invest in fertilizers in unirrigated areas, and consequently the growth in the use of fertilizers has lagged far behind the expectations of the Planning Commission. Extension services, which could provide instruction to the farmers on proper techniques of fertilizer application, and soil testing services, which could indicate the required mix of fertilizer and needed soil amendments, are completely inadequate, and consequently crop responses to fertilizers are much lower than they should be. At the same time there have been short falls in fertilizer supply because of the slow rate of development of domestic fertilizer production.

Because the central and state governments have neglected the development of seed multiplication farms, the seeds of the new varieties have been in short supply. Many of the seeds, supposedly of high yielding varieties, purchased by the farmers are adulterated with seeds of the older varieties or even with weed seeds. The agricultural research establishment has been sufficiently remote from the realities of farming that new rice varieties adapted to the special situations in many regions of the vast country have not been developed.

The future of the green revolution. An enormous increase in agricultural productivity based on new agricultural technology is still possible in India, but it can occur only if social and economic reforms are carried through by vigorous government action and if the government can find the will to divert greater resources from other sectors to agriculture. Here we find another clear-cut example of the issues I have tried to stress: biological and physical science and technology can be usefully applied on a large scale only in the context of changes in social and economic institutions, illuminated by the insights of the social sciences, and carried out by politicians who are able to mobilize an effective public support.

A National Policy for Science and Technology

What can we do to improve the effectiveness of cooperation between politicians and scientists and technologists? One step would be the formulation and adoption by Congress of a National Policy for Science and Technology as a guide to legislative and executive action, just as 25 years ago the Congress adopted a national economic policy. Such a policy would have three clearly expressed goals: (i) to maintain the health and effectiveness of scientific research and technological development in the United States; (ii) to assure the maximum usefulness of scientific and technological advances in serving the people's interests; and (iii) to provide for evaluation and assessment of unforeseen or undesirable effects of new technology and advances in applied science.

Dealing with long-range problems. The policy would recognize that government support is essential to deal with long-range problems at an adequate level of effort. It would emphasize the development of ways to recognize and define problems that will arise in the future and to improve the government's capacity to deal with them. Only the federal government can maintain a sufficiently low "social discount rate" to give a significant "present value" to problems that may take decades to solve.

Maintaining the health of the scientific enterprise. The National Policy for Science and Technology should ensure a continuing flow of highly educated, dedicated, and promising young people into the nation's scientific and technical effort. It should aim at the widest possible base for identifying and educating these young people without regard to sex, ethnic origin, geographical regions, or socioeconomic status. At the same time, the policy should state the national intent to maintain a full range of research and development facilities which will ensure that the nation's scientific and technological community can perform at high effectiveness.

Because experience shows that science and technology form a seamless web, a national policy established by Congress should state the nation's intent to support both pure and applied science, both free research and mission-oriented research and development, both "big science" requiring expensive installations, and "little science" requiring modest equipment, and science in all its disciplines. Science and technology depend on a variety of institutions—universities, government laboratories, industry, and nonprofit institutions—and the National Policy should encompass all of them.

Scientific freedom and responsibility. A National Policy for Science and Technology should provide for a broader public understanding of the nature of the scientific enterprise and the possibilities and limitations of technological development. To further such understanding, scientists and engineers should be guaranteed the freedom to express their ideas about the probable consequences for society of their discoveries, and their sense of responsibility for the potential social effects of their research should be encouraged.

Uses, priorities, and impacts. A National Policy for Science and Technology should aim to ensure that new or prospective technological developments are taken fully into account in the budgeting and programs of federal agencies. At the same time, it should encompass mechanisms to ensure that priorities for scientific and technological research and development set by different federal agencies make sense in terms of national goals, and are realistic in terms of feasibility, scientific and technical manpower requirements, timing of expected

The Scientist and the Politician

results, and available funds. It should recognize governmental responsibility for evaluating the probable impact of new products on human health and welfare and the natural environment.

Science, technology, and foreign policy. Finally, one aim of a National Policy for Science and Technology should be that the best scientific and technical information is fully utilized in making and implementing the nation's foreign policy, and that our unique scientific and technical capabilities are both an instrument and an object of foreign policy. Attention should be paid to means of increasing imports and exports of technology and to the "balance of trade" in technical exchange with other countries. International cooperation and cost-sharing in scientific research and technical development should be encouraged and technology transfer as a major element of assistance programs for less-developed countries should be facilitated. International constraints on oceanic, atmospheric, and space research should be avoided.

What room does this talk of the applications of science and technology leave for the strength and integrity of science itself? These do not lie in the impressive products or the powerful instruments of science but in the minds of the scientists and the system of discourse they have developed to seek a more perfect understanding. The strength and integrity of science can be protected by its uses in education, its international character, and its uniquely human quality.

Scientific research in its broadest sense is the solving of problems to which no one knows the answers. This is the essence of higher education and it is why teaching and research should be inseparable. Scientific education must include the learning of facts and unifying principles, but it is far more important that students learn how to discover, recognize, and use the truth.

It is a truism to say that neither the work nor the results of science can be confined within national boundaries. Many people in many nations can contribute to it. All human beings everywhere may benefit or be harmed by its application. Science unifies men.

The search for an ever-growing but never complete understanding is that uniquely human activity which distinguishes human beings from all other living things. Indeed, I would claim that man is the first step in the evolution of a new form of matter—new because it can understand the world and itself. We look at the stars with awe and wonder. The stars do not look at us at all because they have no eyes to see with and no minds to be struck with wonder.

Understanding can be sought for and obtained in different ways. But one of the most powerful, because it builds on the past and

Roger Revelle

combines the efforts of many individuals, is the method of science—that method of free inquiry, theoretical construction and empirical testing which is the special heritage of Western civilization. Science is more than the handmaiden of technology or the abstract possession of a few. It does more than allay men's fears and inform their hopes. It helps to give them their true dignity as men. Aristotle said it in simple words more than two millennia ago:

> The search for Truth is in one way hard and another easy. For it is evident that no one can master it fully, nor miss it wholly. But each adds a little to our knowledge of Nature, and from all the facts assembled there arises a certain grandeur.

C. The Economy

In this section our emphasis shifts to the influence of technological innovation on economic affairs. The previous chapters have dealt with the problems and disruptions induced by technological change. By contrast, economists typically view technology as a vital ingredient, together with labor and capital, for producing our society's needed goods and services.

In Chapter 6, Donald Schon ponders the proper role of the federal government in prompting technological innovation. In fact, the federal government has been a leading participant in inducing new industrial technology in many, but not all, areas. Yet in many respects, federal actions raise the cost of private innovations. Technological innovation underlies our economy; federal policies, ranging from regulation to procurement and general fiscal policy, affect it greatly—but we have little understanding of the relationships.

Michael Boretsky (chapter 7) perceives a slippage in American technology's international position which threatens our well-being. He worries that social opposition toward technology threatens our economy. A technology-intensive group of industries has been the source of international trade surpluses, but these surpluses are slipping. Data cast doubt about the adequacy of our research and development efforts. Transferring advanced technology to other nations undermines our competitive position. According to Boretsky, we are losing our lead in technology at our own peril.

6. The National Climate for Technological Innovation

DONALD SCHON

In our discussion of problems of innovation and approaches to change in corporation, it was possible to identify someone responsible for change. Corporate management bears the principal, though not the only, responsibility.

But when we come to industries the picture clouds. Though we speak glibly of "leaders of industry," it is difficult, if not impossible, to identify the leader of any industry. We can perhaps identify spokesmen for steel and automobiles, but these men serve for the most part to reflect the view of men in the industry, not to lead in the sense in which a corporation manager may lead. Although there are industry associations—many hundreds of them—they are loose confederations which serve as lobbies, as centers for propaganda campaigns ("Buy Leather," "Buy Wool," etc.) and occasionally for the support of research judged to be of sufficiently general and nonproprietary interest to be worthy of common support. They are not, as a rule, effective vehicles for change.[1]

Even if it were possible to identify industries and industry leaders as agents of change, it is by no means obvious that industries are the most desirable units for initiating change. We have just seen that traditional industries, like textiles, machine tools and building, are in process of metamorphosis. It is in the very nature of technical change that industries merge, separate and take new forms. We ought not to consider change in the traditional textile industry, for example, without also taking into account the paper and chemical industries. We have also seen how individual corporations, partly as R & D-based companies and partly through simple pressure for diversification, may function in a variety of industries.

Excerpted from the book *Technology and Change* by Donald A. Schon. Copyright © 1967 by Donald A. Schon. Reprinted by permission of DELACORTE PRESS/SEYMOUR LAWRENCE.

The National Climate for Technological Innovation

It would appear that the only sensible unit for change is industry (rather than "an industry") or, at the very least, larger aggregations of what we ordinarily call industries.

But it is even more difficult to identify an agent of change for industry or for the chemical-paper-textile complex than for textiles. There is no single institution or set of institutions to turn to.

There is only the whole complex of institutions: companies, industry associations at varying levels, universities, research institutes and governments—municipal, state, and federal. At each level, and in each relationship, change in the direction of innovation is possible.

—Companies within an industry may come together to promote innovation for the good of the industry.

—One industry may, as we have been seeing, invade another, with innovation as a consequence.

—The industries, universities and research institutes of a region may come together to promote the economic development of the region.

There is also the relationship between industry and the Federal Government, a relationship in some ways more important than those listed above because it is important for all segments of industry, not for just a few, and because change in the government-industry relationship may spur change elsewhere. It is appropriate to consider this relationship, moreover, because government as a set of institutions exists and is amenable to certain kinds of change; this is a relationship it may be possible to do something about.

In this chapter we will consider briefly the Federal Government-industry relationship as part of a national climate for technical innovation, focusing on those issues which seem to come most directly from the patterns and problems of innovation so far discussed.

Government-Industry Relations: The Starting Point

Observers of the relation between Government and industry in the United States and in other industrialized nations of the world are bound to be struck by the contrast. In countries like Britain, Germany, Japan and Soviet Russia, government speaks and acts as though it had a position of responsibility and leadership in the introduction of new technology in industry. To a greater or lesser degree, the institutions of those countries reflect this view. Each of these nations has a vehicle—like Britain's Department of Scientific and Industrial Research (or, more recently, its Ministry of Technology), Holland's TNO, Germany's AIF—for joint Government-industry collaboration in industrial research and development. In each of these nations a substantial part of technical innovation in the civilian economy is Government-financed and encouraged. Government's leadership in research and development in these nations usually goes along with a leading role in export-import matters, the provision of risk capital, industrial standards and codes and other issues essential to technical innovation. The situation of the building industry in France is a useful example; here Government provides financial support for a substantial fraction of the technical development; codes and standards develop through a Government institution; the Government provides a vehicle (the so-called Agrément system) for testing and certifying innovations, and Government itself undertakes most of the new building. Government, in short, is a leading participant in the game of technical innovation.

In the United States, on the other hand, an observer gets the impression that Government plays the role of umpire and groundskeeper while private industry plays the game.

This impression is partly a matter of official view and partly a matter of observed fact. It is difficult to disentangle the two. There is a strong conventional wisdom about the appropriate and actual roles of Government and private industry which takes the form of advocacy of the Free Enterprise System. No one can have lived very long in American industry without encountering the view and without being called upon to sustain it. Men who are considered leaders of industry regularly espouse it—gain the right to speak for it, in fact, as a consequence of their leadership. Government leaders violate it at their peril. Reforms or changes are proposed in its name, never in opposition to it. But like other strong tenets of the conventional wisdom—Democracy, for example—the Free Enterprise System is a rather elastic concept and difficult to define in precise terms.

In its strongest sense, Free Enterprise, applied to the problems of technical innovation, requires that *all* technical innovation in our society be the province of the private sector and that any initiative by Government be considered an illegitimate intrusion. Among businessmen, however, even the staunchest supporters of Free Enterprise usually require certain inputs from Government. What, exactly, is viewed as an area for appropriate Government contribution varies from group to group. There is little widespread agreement on the point. The following are examples of Government functions accepted as legitimate by some business groups:

—Collecting and publishing statistics about the industrial process (Gross National Product, numbers of manufacturing firms, imports, etc.)

The National Climate for Technological Innovation

—Exercising control of tariff and quota regulations to protect American business from certain forms of foreign competition

—Employing monetary and fiscal policy to encourage economic growth and discourage depressions

—Subsidizing certain industries (aircraft and maritime, for example)

—Undertaking major technical innovations, requiring great capital investment, in the public interest: for example, weapons systems and systems for national defense; the technology of space exploration; peaceful uses of atomic energy; desalinization of water and, most recently, the supersonic transport

The list is by no means complete. At the strong end of the Free Enterprise spectrum are those who would accept little or nothing of the above as a legitimate function (perhaps reluctantly accepting some functions as necessary "because of the crises Government has created"). At the weak end of the spectrum are those who would accept all of the above and a good deal more, excepting only Government's direct intervention in product and process development in areas where private industry is capable of supporting the work itself.

If we turn from the beliefs held by businessmen to observation of what has actually happened in the United States over the last fifty to one hundred years, we may observe the following:

—Throughout our history, the Federal Government has taken initiative in seeing to major technical innovation believed to be in the public interest. The line of examples includes bounties offered for certain new developments, such as aircraft; granting of land rights for the early expansion westward of the railroads; direct Federal support of the development of new transportation and communications systems (ranging from the telegraph to new satellite communications systems); Federal support of the development and first models of new technology which later came to play a major role in the civilian economy (for example, the development of the aircraft and computer industries and much of the electronics industry; atomic energy; numerically controlled machine tools; operations research).

The Federal Government has supported programs of increase in agricultural productivity, which began with the Morrell Act in mid-nineteenth century, and culminated in the network of agricultural research and experiment stations, land grant colleges and county agents which has helped to make us the most agriculturally productive nation in the world and has created the problem of agricultural surpluses.

The fact is that the Federal Government has acted regularly as a

Donald Schon

leading participant in the game of technical innovation when, under certain conditions, the public interest seemed to require it.[2] It is also true that the mainstream of new products and processes in industry—the growth of the American chemical industry, for example, and the proliferation of new consumer products and methods for making them—all occurred with little, if any, direct Federal involvement. There has been, as we have seen, at least over the last roughly fifty years, a process of technical change in American industry which has much more to do with interactions among industries than with interaction between any industry and the Federal Government. The process of competition between individual firms, and, more significantly, the process of invasion of one industry by another, has been widely responsible for technical innovation in the civilian economy.

The Federal Climate

Starting from this array of attitudes toward Government's role in technical innovation, and this brief historical summary, we may now ask what Government's role *ought* to be.

When we consider what has been said about problems of innovation in the corporation, it is apparent that there is one major aspect of innovation to which Government can make no direct contribution. The entrepreneurial problems of the firm, the dynamics of innovation in the corporation, are at the root of the problem. They are very much internal to the corporation. No Government policy or program can or should affect them directly. There is a very great need, at the corporate level, for education, training and experiment in the problems of innovation, but these are already the objects of considerable private effort; and what private interests do not do, in this field, it is unlikely that Government can do well.

On the other hand, there is a kind of indirect Federal contribution which is very much to the point. For better or worse, the Manhattan Project and the whole wartime scientific effort provided a model for the conduct of postwar research and development in industry. TVA and military construction in World War II provided models of construction as a mass production process. Government's way of undertaking technical innovation for its own needs has given, and can continue to give, a powerful model of processes of innovation. In spite of the very mixed American attitudes toward Government—attitudes which treat Government as evil, just as in the time of the American Revolution—these Federal models are perhaps more influential than any others and provoke a kind of mimesis which is more effective than deliberate Federal efforts to affect corporate behavior.

The National Climate for Technological Innovation

The provision of such models is related to a kind of national setting of tone which is also peculiarly a function of the Federal Government, and especially of the President. The President can convey a sense of satisfaction with the *status quo,* which discourages innovation. He can also convey a sense of urgency about change, an excitement about the process of innovation, and impart to private industry a sense of responsibility for constructive change which supports forces already working in the direction of innovation.

When we turn from the individual corporation to the level of technical change in industry, direct, constructive Federal action seems more feasible.

In principle, this action could take the form of direct Government involvement in the industrial game. It might proceed then, under the following model:

> Let Government, as the most powerful national institution and the one most closely associated with the public interest, take over the job of technical innovation for the nation—determining in which areas support and encouragement are required, undertaking and supporting the development of new products and processes, regardless of the competitive inclination of industry.

This is, in effect, a Great Man theory of Government which gives to Government, in relation to industry, a form of power and responsibility similar to those of the corporate Great Man in relation to his subordinates. There are certain areas of public sector need, described by John Galbraith in his *Affluent Society,* which private industry for a variety of reasons has been unable to fill.[3] Here there may be room for a new kind of direct Government involvement. A full discussion of this issue lies beyond the scope of this book, however. As far as the Great Man theory of Government might apply to innovation in areas of present private concern, it is sufficient to remark how that theory attributes to Government, on no particular grounds, greater wisdom in the choice of need and greater ability in the innovative task than are to be found in the private sector.

But we have seen in previous chapters something of the processes of industrial change involved in large-scale innovation. It is appropriate to ask how Government, by deliberate policy and program, could create conditions favorable to those processes of change, freeing private enterprise to become more effective in innovation. Government's role here would be analogous to the role of the corporate leader in what we have described earlier as the second model of corporate change.

Donald Schon

We have seen that the principal form of innovation in mature industry is innovation by invasion. Such invasions dislocate companies, workers and regions of the country at the same time as they tend to create new jobs, new companies and new industrial regions. The economic problems of Appalachia, for example, have their roots in the displacement of coal by petroleum as a source of energy and in the introduction of automatic coal-mining machinery. The economic problems of the midwestern (not the Great Lakes) states come from the displacement of the small family farm by the large industrialized farm, capable of producing well beyond current demand. The economic problems of New England are in part the problems of the traditional paper, textile, leather and shoe industries affected by the new synthetics, by foreign competition and by the economic pressure drop which has driven much of this industry south. Dislocation of workers in many industries and many regions results from the introduction of more productive equipment, including computerized equipment. Technology is by no means the only, or perhaps even the major source, of such dislocations. Market shifts, such as the shift in Federal market now being experienced in places like California and Long Island, can be severe in their effects. Shifts in other economic factors, such as labor costs and industrial concentration, can be equally powerful.

In all these instances, however, our way of coping with the effects of dislocation is critical not only to the well-being of those affected but to the processes of change, themselves. This is true in at least two ways:

—Industry is highly sensitive to public opinion, at present. A company will tend to introduce more productive methods and equipment unless it can foresee a feasible way of dealing with those affected by the change. Companies tend to defer introducing new equipment, which reduces labor content, until they can foresee a period of growth which will enable them to employ displaced workers in other jobs, permitting them to reduce their labor force by attenuation, or unless those displaced are likely to be able to find jobs elsewhere.

—If our way of coping with technological dislocations is to support those hurt by the change, the effect may well be to impede technological change.

Although the Federal Government has no formal plan for coping with technological dislocations, its actions tend to follow a consistent strategy. It waits for a crisis and then moves in to protect and support those who are adversely affected by change. Its protection may take a variety of forms.[4] It may give support through tariffs and quotas,

The National Climate for Technological Innovation

loans and subsidies, special procurement policies, tax relief or special supporting research. In one or another of these ways, Government has responded to the problems of industries like railways, shipping, agriculture, textiles, and to the problems of small business in general. It responds to the needs of affected regions through depressed-area programs which concentrate on public works and loans to local corporations, both designed to provide local jobs for the unemployed. Until recently, its principal response to the needs of displaced workers was in the form of unemployment insurance and welfare payments.

While the pros and cons of this kind of support need to be debated in each instance, it is highly likely that such a policy—a policy of stability—tends to stand in the way of technical innovation. By supporting invaded industry and invaded companies—where, as we have seen, invasion is a dominant pattern of technical change—Government tends to raise the cost of technical innovation. By offering a cushion to those who are threatened, it creates a climate reinforcing the disposition of traditional industry to resist innovation. Most seriously, once a policy of protective support has been initiated, it is difficult or impossible to stop. The outside threat does not disappear. The invaded industry or region does not develop independent resources, and it becomes accustomed to the helping hand.

A quite different strategy in the face of threat from new technology would be that of encouraging mobility and adjustment. Such a strategy would include:

—A labor mobility policy, with an effective job information system, support to workers in moving to new positions, improved training and retraining programs and a data base suitable for judging the effectiveness of the program

—Encouragement to firms threatened by new technology to diversify into more profitable areas of the economy

—Assistance to depressed areas in the development of new industry based on new technology

Such a strategy would require measures of its own effectiveness (measures which are now generally absent); a new level of adequacy in information about the dislocations produced by technical change; and an effective co-ordination of agency activities in such diverse fields as taxation, loans and subsidies, regional assistance, research and development and manpower adjustment. The key to such a policy would be the effort not to support the injured but to place them in a position to support themselves. It would offer support not to specific industries but to the process of technological change throughout the economy.

Donald Schon

A New Technology of Government

At this point, the problem of developing a policy of industrial and worker mobility in response to the dislocations produced by technical change merges with the problem of developing a national policy climate favorable to technical change.

We can look vainly for those specific Federal policies related to technical innovation while missing the point that there is scarcely any Federal policy which does *not* affect innovation. Federal policy provides the framework, partially obvious and partially hidden, within which the industrial game is played:

—Antitrust policy may encourage competition and the continued existence of small firms with capacity for innovation, or it may have other effects adverse to innovation.

—Regulatory policy (the policy of agencies such as the FDA, ICC, FCC, FPC and CBA) may inhibit technical innovation, encourage it or channel it in certain directions.

—There are both obvious and hidden effects on technical innovations as a result of patent policy, policy on the granting of subsidies and loans, Federal support of research and development, and import-export policy as reflected in tariffs and quotas.

—Federal and local standards and codes may serve to free the development of new technology, or because they are based on materials specifications derived from old technology, they may freeze technological advance. Uniform, performance-based codes encourage new solutions to technical problems; fragmented codes may raise the cost of introducing new technology to an impossible level.

—Federal procurement policy may make opportunities for pilot attempts at new technology, or it may stamp in the use of procurement specifications based on old technology.

—Federal policy related to the over-all state of the economy—particularly monetary and fiscal policy—has an enormous and well-established effect on innovation, serving to encourage or discourage expansion and investment in new equipment. Decisions to invest in new technology hinge on expectations concerning economic growth which affect levels of optimism about paying off such investment and about ability to provide opportunities for employees dislocated by new technology.

In short, the Federal Government provides *government* for the process of technical innovation in industry in the most literal sense. Government and industry represent a closed-loop system in which the industrial process (including the process of technical change in indus-

The National Climate for Technological Innovation

try) gives rise to modification of policy, which in turn affects the system. It is clear that this relationship holds. In the areas of fiscal and monetary policy it has become increasingly clear *how* it holds. It is nowhere near so clear in other areas of policy. Policies interact and their interaction is generally obscure.

Patent policy may stimulate or inhibit technological innovation by offering more or less protection for innovation. Federal patent policy may serve either to attract industrial enterprise to an area of technology or to keep it away from an area by removing the attraction of exclusivity. Antitrust policy may stimulate innovation by permitting the continuing formation of new small firms based on new technology; it may, on the other hand, prevent concentration of industry essential to technological innovation in industries where high capital investment in innovation is required. Tariff and quota policy may protect a domestic industry which requires protection for its early development, or it may lull a traditional industry to sleep. Federal procurement policy may offer opportunity for experiment in new technology, free of regulatory restrictions, or it may stamp in restrictive specifications based on old technology.

In most instances the Government policy in question has as its primary goal (or, at any rate, had as its originally intended goal) quite a different set of objectives. It may have been designed to protect the public safety, to procure goods at least cost, to prevent excessive control of an industry by one company, and the like; but it nevertheless affects the process of technological innovation. In policy as in mechanical design, it is impossible to make a move which has only the effect intended for it. These side effects of Government policy may be at least as important as the originally intended effect. In some instances, they may turn out to be the principal benefits or evils of the policy.

For good or for evil the effects of Government policy on technological innovation are poorly understood. We understand only in primitive terms how fiscal policy affects incentive to investment in industry. We understand far less how antitrust policy affects willingness to invest in innovation. Still less do we understand the interlocking effects of tax and antitrust policy. A tax policy aimed at encouraging investment in new capital equipment (and, consequently, at encouraging investment in new technology) may run head on into an antitrust policy administered in such a way as to discourage such investment, or a patent policy which deprives corporations of the incentive to be first among competitors in development.

The single and interlocking, primary and secondary effects of Government policy on innovation are the basis of a technology of Government designed to create a climate favorable to innovation. Such a technology would have to do with the deliberate exercise of instruments of Government policy in order to give maximum freedom to industrial processes of technical innovation, or at any rate, the ability to anticipate the inhibiting effects of such policy when it is dictated by values other than technical innovation. Such a technology would require looking at the relation between Government and industry as though it were a complex servo-mechanism, much as we now look at the interaction of Government and industry through the medium of fiscal policy.

Such an enterprise requires new or extended activities, which include:

Sensors: means of collecting data about relevant social and economic variables, such as the structure of the unemployed population, rate of re-employment, corporate profitability and productivity, on a scale greater than the one now available. Most important, data must be made available about the effect of Federal policy and program. Government frequently has little, if any, information about the side effects of its policies on technical innovation—although it is clear from the variety of strong and usually conflicting reactions that it has *some* effect—and often lacks the very feedback which would make possible the evaluation of its policies from the point of view of their main intended effects. What are the effects of tax policy on corporate capital investment? What are the effects of manpower adjustment policy on the match between workers and jobs? The difficulty of obtaining such data on a regular and trustworthy basis are enormous because of both the complexity of the system and the reluctance of industry to give up the required information. The collection of such information would be a major enterprise, comparable to the current efforts of the Census of Manufacturers and the Bureau of Labor Statistics.

Analysis: means of interpreting information in terms of a theory of the relationship between Federal action and the economy. In these matters, we are currently theory-poor. But there are encouraging beginnings in the development of models of the economy and models of the interaction of Federal policy and the economy which would enable us to anticipate more effectively the effect on the economy of a course of Federal action.

Decision: mechanisms for rationalizing policy decisions across agency lines. Nearly all major fields of Federal policy have an effect on technical innovation in industry, and many have interacting effects

The National Climate for Technological Innovation

which may either complement or conflict with one another. There would have to be vehicles for perceiving such relationships and for permitting reasoned co-operative policy-making where appropriate. The prospect of a technology and a style of operation which would make Government a more effective governor of the economic process, including technical innovation, is by no means pleasing to everyone. Behind it lurks the specter of a managed economy with increased Governmental control.

But it is not clear that by gaining greater insight into the effects and side effects of its policies by improving the base of information in which it acts, and by co-ordinating the administration and planning of policies which may otherwise conflict, Government would thereby exercise greater control of industry. Although it may be argued that Government should abandon or reduce the control it now exercises in its regulatory, tax, antitrust or import-export policies, the fact is that in these fields it now exercises considerable control. By learning more about what it does here and resolving some of its internal conflicts Government may become more effective in creating an environment of greater freedom for innovation in private enterprise. The choice is not so much between greater and lesser control as between arbitrary and more nearly rational control.

Nevertheless, this fear of control raises an issue central to the whole problem of Government-industry relations concerning innovation: the issue of trust. We have lived for generations in a climate of prevailing mutual distrust between industry and Government. This distrust has its roots in our Revolutionary origins and has grown in the climate of increased Federal responsibility and regulatory control which began in the time of the New Deal. Because of it, Government and industry have tended to sleepwalk into new situations. Industry has accepted Federal manipulation of monetary and fiscal policy designed to assure economic growth, for example, and has moved noticeably toward an acceptance of Keynesean economic policy while continuing to enunciate the old doctrines of anti-Federalism and individualism.

Establishing a basis for Government-industry trust is central to all those policies which might create a more favorable climate for innovation. It is as important to the Government-industry relationship as it is to the relationship between boss and subordinates or between marketing and technology in the firm. But, because it is a trust between institutions rather than between individuals, it is far more precarious and difficult to achieve. Confrontations between individual leaders may be necessary to it but are not sufficient. The establishment of this

Donald Schon

trust, as an objective, and of the institutional arrangements which would permit it to be sustained, ranks highest on the list of measures aimed at creating a national climate favorable to innovation.

Notes

1. The growth of industrial associations has usually been in response to a perceived external threat, either from competing industries, foreign or domestic, or from Government policy. In the case of some American industrial associations—the American Gas Association and the Structural Clay Tile Products Institute are examples—such threats have led to association-sponsored research and development.

In Britain there is a longer and fuller history of industrial association-sponsored research, centering on the Department of Scientific and Industrial Research (DSIR), founded shortly after World War I and now part of the Ministry for Technology. The history of this effort suggests some of the potentials and limitations of industrial association research as it has been practiced to date. In general, such efforts tend to concentrate on areas such as product improvement, increase in productivity, cost reduction and better understanding of materials or processes central to the industry. Highly significant technological innovation, in the sense described above, tends not to come from this sort of activity, for reasons that have to do with (a) the problem of conducting research under the direction of multiple "bosses," (b) the desire of individual firms to capture or hold secret any development of potential proprietary importance and (c) the unresolved marketing and entrepreneurial problems associated with commercializing research results.

2. A. Hunter Dupree has developed this point at far greater length in his *Science in the Federal Government* (Cambridge, Mass.: Harvard University Press, 1957).

3. "Failure to keep public services in minimal relation to private production and use of goods is a cause of social disorder or impairs economic performance.... By failing to exploit the opportunity to expand public production we are missing opportunities for enjoyment which otherwise we might have had. Presumably a community can be as well rewarded by buying better schools or better parks as by buying bigger automobiles. By concentrating on the latter rather than the former it is failing to maximize its satisfactions. As with schools in the community, so with public services over the country at large" (Galbraith, 1958, p. 204).

4. This protectivist policy coupled with the longevity of Government bureaucracies helps to account for the fact that many Government agencies now appear as institutions designed to cope with problems thirty, fifty or even seventy-five years old. Where the policy is protectivist, the Government action may help to perpetuate, if only in vestigial form, the problem it is designed to solve. The continuation of many such institutions then raises major problems for an organizational response to the need for new programs or new missions. Something of this sort underlies the problem of Government response to change, which is in many ways analogous to the industrial problems we have been discussing.

7. Trends in U.S. Technology: A Political Economist's View

MICHAEL BORETSKY

Until the Civil War, the United States was what we would now call an underdeveloped country, and its technology, with few exceptions, was no more than a poor offset rendition of that of Europe. Abundant natural resources, however, coupled with Alexander Hamilton's policy of industrialization (*1*), which was vigorously pursued from the time of the Revolution until about the end of World War II, plus an otherwise positive attitude toward technological and industrial development on the part of government and society at large, soon made the U.S. a rapidly developing country. The most relevant historical evidence suggests that American technology reached parity with Europe by 1870 or thereabout; by the turn of the century, it was in most respects higher than Europe's; and by the end of World War II it had become a literal "wonder" to the rest of the world. Parallel with this technological development, and because of it, America became the greatest world power—economically, politically, and militarily.

Toward the end of the 1960s, and especially since 1971, however, a number of societal attitudes toward technology, on one hand, and quite a few unfavorable economic trends implying similarly unfavorable trends in technology, on the other, have induced considerable speculation about America's economic and political future. The present concern is not with the country's technology relevant to defense and the conquest of space, which occupied the last two decades, but with technology relevant to the quality of life in society at large, as well as, more specifically, productivity and commercial markets at home and abroad.

The debate in question has been underway for some three years or so, but there is still considerable confusion as to the real state of affairs

Reprinted by permission, *American Scientist,* journal of Sigma Xi, The Scientific Research Society of North America, vol. 63, January–February 1975, pp. 70–82.

Michael Boretsky

in U.S. technology, and an almost unbelievable amount of confusion as to what precisely the problems are, the reasons for these problems, and what can and should be done about them. There are people who argue that the country's rate of technological progress is heading downward, that our technological leadership in the world is rapidly disappearing, and that in technological prowess, they predict, the United States will become "just another industrialized country" in a matter of a few years.

Others would seem to hold the opposite view: There is absolutely nothing wrong with U.S. technology. To "prove" this theory, they invariably cite the fact that Japan, all European countries, and Canada have been buying and otherwise importing U.S. technology en masse, and that more recently the U.S.S.R. has started doing the same. If there were something wrong with U.S. technology, they contend, foreign countries would not buy it.

There are still others who argue that even if the U.S. were losing technological preeminence, there is nothing to worry about because (1) the trend is probably inevitable and, to some degree, desirable, (2) U.S. technology is so much more advanced than that of other countries that it will take quite some time for them to approach the U.S. level, and (3) the U.S. economy has strengths other than technology. In an attempt to resolve these conflicting viewpoints and to assess just where the U.S. stands and seems headed on the technological tote board of comparative economic advantage, I shall try to give a brief analysis of facts bearing on the current state of U.S. technology, the problems to be faced, and the apparent causes of these problems.

Current State of U.S. Technology

The only meaningful way to define the state of a country's technology at any single point in time is by reference to the country's own historical performance and the state of relevant affairs abroad at the time in question. In reference to its own historical performance, there is no question that the current level of U.S. technological development is higher than it has ever been. For example, in 1972, U.S. consumption of Btu's for productive purposes per civilian employed in the private economy, which is probably the single most comprehensive indicator of overall relative technological advancement—or at least relative technological intensity—was about 62% higher than in 1950 (1950 = 100, 1972 = 162); and private GNP (in constant prices) per person employed in the private economy was 74% higher.

The reason for the apparently close correlation between these two

Trends in U.S. Technology

indicators is that, historically, at least to date, the essence of most innovations in civilian technology has been the substitution, usually in all kinds of equipment and mechanical implements, of Btu's for human and animal energy. The ratio between the two figures, 0.93 to 1, might not be a precise measure of the dependence of the output per man on the consumption of Btu's per man, and hence technology, but this dependence is unquestionably very high.

Similar indicators show that U.S. technology is also still much more advanced, or at least much more intensive, than the technology of any other country (see Table 1; although the estimates refer to 1970, the relationships could not have changed enough in three years to invalidate this conclusion). Based on a variety of other data, which because of space limitations cannot be presented here, I also conclude that, as of now, there are hardly any economically important product or service lines in which the quality of U.S. technology does not equal or excel that anywhere abroad. This conclusion might be reassuring to most Americans and incite envy in most foreigners, but numerous economic and social trends show clearly that U.S. technology faces many serious problems. The most far-reaching among these problems are the decline in productivity growth, both of labor and capital, and the deterioration in the U.S. foreign trade position.

Decline in productivity growth. The importance of the decline of productivity growth of both labor and capital derives from the fact

Table 1. Comparison of consumption of primary energy sources and GNP per civilian employed for selected countries in 1970

Country	Ratio of U.S. to foreign country's consumption of primary energy sources, other than motor gasoline, per civilian employed	Ratio of U.S. to foreign country's GNP per civilian employed[a]
France	2.8	1.4
West Germany	2.2	1.8
Belgium	1.4	1.6
Netherlands	1.6	1.7
Italy	3.2	2.1
United Kingdom	2.1	2.1
Unweighted average for the above countries	2.0	1.7
Canada	1.2	1.2
Japan	3.6	3.1
U.S.S.R.	3.2[b]	3.4

SOURCES: Department of Commerce, OECD, and individual country data.
[a] GNP valued in dollars of comparable purchasing power.
[b] 1969, relative consumption of energy for productive purposes.

163

that they are practically the only sources of growth in the country's material well-being. Without growth in productivity, money income might increase but real income, or real material well-being, will not. The critical dependence of labor productivity growth on technology has been alluded to in the preceding comparisons of consumption of Btu's with output per man. The same is by and large true for capital productivity. At least conceptually, the only time capital productivity can grow is when the efficiency of new capital goods (plant and equipment) grows faster than the prices of these goods, and this can happen only when the economic value of resource-saving technological innovtions embodied in new capital goods exceeds the cost of the innovations. Both are generally recognized by most economists (2).

Viewing the information on trends in labor productivity in Table 2 and Figure 1 from the perspective of U.S. historical performance, we should first note that for about 20 years following World War II the U.S. maintained about the same rate of labor productivity growth as the average of the preceding 80-plus years (2.4 to 2.5% per year). This greatly contributed to our becoming the world's foremost economic power. In the 1965–71 period, however, our growth rate slipped to about one-half of this long-term average. From 1965 through 1973, which included the two most recent boom years, when labor productivity presumably should have shot up, our growth rate averaged 1.7% per year, slightly better than in 1965–71, but still 30% or so less than the preceding long-term average. Table 3 shows that the decline in the growth of labor productivity in 1965–73 was paralleled by a decline in capital productivity, and Table 4 shows how dramatically it has lagged compared to other nations.

Table 2. Rate of growth in GNP per civilian employed, average percent per year

Country	1870–1950	1950–1965	1965–1971
United States	2.4	2.5	1.3
France	1.7	4.6	4.9
West Germany	1.6	4.8	4.3
Belgium	1.6	3.0	3.7
Netherlands	1.1	3.7	4.7
Italy	1.5	5.5	5.7
United Kingdom	1.6	2.2	2.5
Unweighted average for the 6 European countries	1.5	4.0	4.3
U.S.S.R. (Russia)	1.7[a]	4.2	4.3
Japan	1.4[a]	6.8	9.6

SOURCES: See Ref. 16.
[a] Growth in per capita GNP.

Trends in U.S. Technology

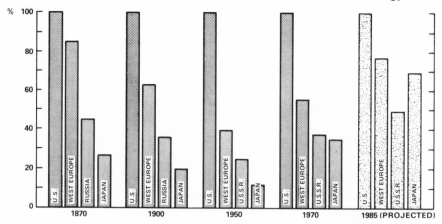

Figure 1. Comparative GNP per civilian employed, 1870–1985. The figures for Western Europe are an average of France, West Germany, Belgium, Netherlands, Italy, and United Kingdom. (Data from the sources listed in Ref. 16.)

Most economists argue that this decline was strictly cyclical, induced by policy measures that restrained the growth in output in 1969 and 1970. They also assert that the long-term rate will resume once the growth in total output resumes its "normal" path. As William D. Nordhaus (3) has demonstrated, however, this slowdown contains a substantial long-term element, and one that is not automatically reversible. The most persuasive prima facie observation favoring the Nordhaus argument is the fact that no labor-productivity-growth slowdown

Table 3. Estimates bearing on trends in capital productivity and relative capital requirements in U.S. private economy

	1947–65	1965–73
1. Average annual growth of private GNP in constant dollars	3.9%	3.9%
2. Average annual growth of employment in private economy	0.9%	1.9%
3. Average annual growth of private non-residential, business capital stock (gross value of plant and equipment) in constant dollars	3.3%	4.2%
4. Average annual growth of private GNP per person employed	3.0%	2.0%
5. Average annual growth in value of private nonresidential capital stock per person employed in the private economy	2.4%	2.3%
6. Ratio of average annual growth of private nonresidential capital stock (item 3) to growth of private GNP (item 1)	0.85	1.08
7. Ratio of average annual growth of private nonresidential capital stock per person employed (item 5) to growth of private GNP per person employed (item 4)	0.80	1.15

SOURCES: U.S. Department of Commerce and the Bureau of Labor Statistics.

Table 4. Data bearing on productivity growth of fixed nonresidential capital investment for selected countries

Country	A. Average annual growth in fixed nonresidential investment, % per year		B. Average annual growth in GNP, % per year		C. Average annual growth in GNP per civilian employed, % per year		D. Ratio of B to A		E. Ratio of C to A	
	1953–63	1963–69	1955–65	1965–71	1955–65	1965–71	1955–65 to 1953–63	1965–71 to 1963–69	1955–65 to 1953–63	1965–71 to 1963–69
United States	3.1	6.4	3.5	3.2	2.2	1.3	1.13	0.50	0.71	0.20
France	9.6	8.4	4.9	6.3	5.1	4.8	0.51	0.75	0.53	0.57
West Germany	11.1	5.4	6.2	4.3	4.5	4.3	0.56	0.80	0.41	0.80
Belgium-Luxembourg	5.9	5.2	3.5	4.4	2.7	3.7	0.59	0.85	0.46	0.71
Netherlands	7.1	7.9	4.4	5.0	3.1	4.3	0.62	0.63	0.44	0.54
Italy	8.6	1.2	5.4	5.3	5.6	5.5	0.63	4.42	0.65	4.58
Common Market countries, total	8.9	5.8	5.3	5.0	4.6	4.8	0.60	0.86	0.52	0.83
United Kingdom	6.3	5.2	2.9	2.3	2.1	2.8	0.46	0.44	0.33	0.54
Canada	3.0	6.0	4.3	4.8	1.9	1.8	1.43	0.80	0.66	0.30
Japan	14.8	14.2	9.6	11.2	8.0	9.6	0.65	0.79	0.54	0.68
U.S.S.R.	8.7	7.9	5.8	5.5	3.5	3.6	0.67	0.70	0.40	0.46
Unweighted average for the 10 foreign countries	8.3	6.8	5.2	5.5	4.1	4.5	0.63	0.81	0.49	0.66

SOURCES: See Ref. 16.
NOTE: In comparing growth in GNP and GNP per civilian employed with growth in fixed nonresidential investment posted in columns D and E, respectively, a 2-year lag for the effects of new investment to materialize is assumed.

in U.S. history for which we have statistics, including the Great Depression of 1929-39, was greater than 20% of the long-term average, compared with about 30 to 50% in the 1965-71 or 1973 period, most of which was not recessionary. Moreover, even a resumption of the long-term rate, however impressive it might have been in the past, would be far from sufficient for the country's comfort in the years to come.

As is evident in Figure 1, the U.S. gained the peak of its economic preeminence by about 1950, largely because its labor-productivity growth in the preceding 80-plus years was a mere .09% higher than the average abroad (2.4% in the U.S. versus about 1.5% abroad). Continuation of essentially the same rate in the 1950-65 period, however, produced a decline in the relative position of the U.S. vis-a-vis the other countries at about the same rate as the rate of our gain in 1870-1950, because the other countries more than tripled their rate of productivity growth. The decline in U.S. labor-productivity growth to about 1.3% per year in the 1965-71 period, coupled with a further speed-up of this growth abroad, accelerated the U.S. loss of eminence at a rate more than 1.5 times faster than it took to obtain the preeminent position in the years prior to 1950.

All these unfavorable trends have tremendous implications for the country's future growth in well-being, its relative economic and political position in the world, future inflation pressures, and the like. The decline in capital productivity (which is the same thing as progressive growth in relative capital requirements) implies, it should be noted, that the progressively growing shortages of "investible funds" and interest rates hitting double-digit levels, which we have been witnessing since the middle 1960s, are not short-term phenomena. Rather, they are a result of long-term forces at work in the economy—namely, private propensity to save and business profit rates remaining essentially at pre-1965 levels, while the capital requirement per unit of additional capacity grows. And these trends are more than likely to continue unless the general forces causing them are reversed. Moreover, the new capital-intensive programs that have been launched quite recently, such as the intensified environmental pollution control and "Project Independence," are likely to worsen these trends dramatically.

There is hardly any question that the principal generic force behind all these unfavorable trends, so far, is the decline in the country's rate of "civilian-life-oriented" technological advance relative to both our own past performance and the current performance of other industrialized countries. The only way to reverse these trends is to reverse the decline in the rate of U.S. technological advance.

Michael Boretsky

Deterioration in U.S. trade position. To most economists, and the government officials and politicians who rely on their advice, the deterioration in the U.S. trade position, which became visible in 1971 and 1972, represents a part of the productivity problem. Technology, in this view, is assumed to have contributed to this deterioration only to the extent to which it adversely affected the productivity of industries engaged in the production of internationally traded goods. Though seemingly in line with the established theory of international trade, this view is untenable for at least three reasons.

1. The true trade-executing mechanism is not comparative productivity and changes therein, but their inverse—changes in comparative costs and/or relative price levels. Although growth in productivity, as was shown in the preceding section, has been much faster abroad than in the U.S. throughout the post-World War II period, the rate of inflation, however defined, has been much lower in the United States than abroad (*4*).

2. Technology affects a country's foreign trade not only through productivity, and hence prices, but also more directly—via exports and imports of products embodying superior or unique technological know-how, the demand for which tends to be a function not so much of the product's formal prices as of the relative quality of the expertise in question (for example, such U.S. products as the Boeing 747 aircraft and enriched uranium). Conceptually, a superior technology embodied in a particular product is equivalent to a tangible surplus value which the buyer gets, either in the form of the product's greater economic efficiency compared with similar products embodying less advanced technology in the case of capital goods, or greater satisfaction on the part of the consumer in the case of consumer goods. This surplus value is also the reason people buy products embodying advanced technology—mostly at prices much higher than the price of similar products embodying inferior know-how. Indeed, I am told that, for example, the greater reliability of U.S.-made industrial equipment alone, compared with otherwise identical foreign-made equipment, frequently commands a price premium of 20% or more.

3. Apart from relative price levels and quality of technological know-how other than that working through productivity and prices, a country's foreign trade is also affected by its endowment of natural resources relative to its needs. The more abundant the resources—such as oil—relative to its needs, the more of these resources it is likely to export and the less it will import.

The commodity detail and time scale of Table 5 maps out the minimum role of technology in U.S. trade and the principal forces

Trends in U.S. Technology

Table 5. U.S. merchandise trade by major commodity group 1951–1972 (values in billions of dollars)

Commodity group	1951–55 (average)	1962	1965	1971	1972
Agricultural products					
Exports[a]	3.2	5.0	6.2	7.7	9.4
Imports	4.4	3.9	4.1	5.8	6.2
Balance	−1.2	1.1	2.1	1.9	3.2
Minerals, fuels, other raw materials					
Exports[a]	1.3	2.1	2.6	3.8	4.3
Imports	3.3	4.5	5.4	7.9	10.1
Balance	−2.0	−2.4	−2.8	−4.1	−5.8
Nontechnology-intensive manufactured products					
Exports[a]	3.7	3.5	4.4	6.3	7.1
Imports	1.9	5.1	7.4	14.6	17.8
Balance	1.8	−1.6	−3.0	−8.3	−10.7
Technology-intensive manufactured products					
Exports[a]	6.6	10.2	13.0	24.2	26.6
Imports	0.9	2.5	3.9	15.9	19.9
Balance	5.7	7.7	9.1	8.3	6.7
All commodities[b]					
Exports[a]	15.5	21.7	27.5	44.1	49.8
Imports	10.9	16.5	21.4	45.6	55.6
Balance	4.6	5.2	6.1	−1.5	−5.8

SOURCE: U.S. Department of Commerce.

[a] Exports include "noncommercial" shipments, such as military grant/aid shipments of agricultural commodities under Public Law 480, etc.

[b] Includes the four commodity groups listed above plus "goods and transactions not classified according to kind" and reexports.

behind the deterioration in the U.S. trade position. Of primary interest to this analysis are, of course, what I define as *technology-intensive manufactured products*. This product group includes chemicals; nonelectric machinery; electrical machinery and apparatus, including electronics; all types of transportation equipment, including aircraft and automobiles; and scientific and professional instruments and controls. The chief criterion in designating these products as technology-intensive is the relative intensity of the new-technology-generating inputs used in their production—research and development, scientific and engineering manpower used in functions other than R & D (design, production supervision, customer services, etc.), and the relative level of the workers' skill.

In the U.S., the industries manufacturing technology-intensive products spend about 11 to 12 times as great a proportion of their value-added on R & D as do industries manufacturing nontechnology-intensive products. The employment of scientists, engineers, and technicians in functions other than R & D relative to production workers in industries manufacturing these products is almost 5 times as great as in the other manufacturing industries; and the employment of craftsmen relative to "operatives and laborers" is some 70% or so greater than in the other industries.

Moreover, the industries manufacturing these products are not only the primary users of the new-technology-generating inputs, but also, unquestionably, the primary domestic originators of technological innovations both for their own use and for all other sectors of the economy through the equipment, instruments, and synthetic materials embodying innovations they supply. Internationally, the disparities in technological prowess among the industrialized countries are largely concentrated in these industries.

The technology-intensive group is clearly the only one that has consistently yielded surpluses that have covered deficits in trade in other commodity groups as well as deficits arising from other U.S. financial transactions with foreign countries. Table 5 shows that the gross surplus yielded by trade in this commodity group increased from 1951 to 1970; in 1971, however, the surplus started slipping and in 1972 had dropped to about $6.7 billion compared with $9.1 to 9.4 billion in 1965–70. From this it must be concluded that technological know-how has been, and still is, *the* force behind U.S. trade successes and behind the strength of the dollar's external value. This force, however, is getting progressively weaker, and this weakening has been one of the principal factors in the deterioration of the total U.S. trade position.

Moreover, contrary to general belief, the comparison of growth rates of U.S. imports and exports of these products implicit in Table 5 indicates clearly that the weakening of U.S. technological advantage in trade has not occurred suddenly; rather, as in productivity performance vis-a-vis other countries, it has been underway since the early 1950s. Indeed, the ratio of the growth of U.S. imports to the growth of U.S. exports of these products has been an almost-constant 2.5 to 1 throughout the entire 20-odd years covered in the table. With the higher rate of inflation abroad throughout the entire period, as noted earlier, the only explanation of the faster growth of U.S. imports of such products is the concomitantly faster growth of technological know-how abroad.

The two other forces that brought about the deterioration in the overall U.S. trade position are (1) U.S. losses of competitiveness in manufactured products other than the technology-intensive, where the evolution of general "industrial art" rather than R & D or advanced engineering plays a major role, and (2) the growing inadequacy of domestic supply of minerals, fuels, and other raw materials relative to the economy's needs. Overall, the deterioration could hardly have been more dramatic—almost a 180° turn-around of the overall balance in just seven years, from a $6.1 billion surplus in 1965 to a deficit of about $5.8 billion in 1972.

By August 1971 the deterioration process had forced an unprecedented devaluation of the dollar (because of lower rates of inflation in the United States than abroad), followed by another in February 1973, at which time the dollar was also set "afloat." From the time just prior to the initial runs on the dollar in April 1971 until the middle of May 1974, the dollar was effectively devalued against all foreign currencies by 9 to 14% (the figure depends on whether U.S. exports or imports are used in weighting individual changes) and by 14 to 24% against the currencies of the industrialized countries.

Analysis of the developments in U.S. trade from the middle of August 1971 to January 1973 leaves hardly any doubt that the primary effect of the first devaluation was a further deterioration in the trade balance, largely because of a worsening in the U.S. terms of trade (the ratio of U.S. import to export prices), by just about the amount of the actual worldwide devaluation. The second devaluation, and the subsequent floating of the dollar, might have had the potency to produce some improvements in the situation, but the data on what actually transpired in U.S. trade in 1973 (Table 6) and the reasons why, plus the skyrocketing prices of foreign oil, suggest that this improvement might only be temporary.

The only encouraging indications of a possible improvement in the U.S. trade situation implicit in Table 6 are a sizeable turnaround in the overall trade balance—from an overall gross deficit of $5.8 billion in 1972 to a gross surplus of about $2.2 billion in 1973—and a substantial increase, by $4.0 billion, in the trade surplus in technology-intensive manufactured products. However, the reasons for skepticism regarding the durability of this improvement are simply overwhelming:

1. The bulk of the overall improvement in the trade surplus came from trade in agricultural products, which also produced shortages in the supply of the products to domestic markets and a drastic accelera-

Michael Boretsky

Table 6. U.S. merchandise trade by major commodity group, 1973, compared with selected years and periods in the past

Commodity group	Billions of dollars			Average annual growth rate, in %		
	1971	1972	1973[c]	1962–71	1972	1973[c]
Agricultural products						
Exports[a]	7.7	9.4	17.7	4.0	22.1	87.9
Imports	5.8	6.2	8.4	3.7	12.1	37.3
Balance (gross)	1.9	3.2	9.3	—	—	—
Minerals, fuels, and other raw materials						
Exports[a]	3.8	4.3	6.0	5.5	13.2	41.1
Imports	7.9	10.1	14.1	5.2	27.8	40.0
Balance (gross)	−4.1	−5.8	−8.1	—	—	—
Nontechnology-intensive manufactured products						
Exports[a]	6.3	7.1	9.9	5.5	12.7	39.1
Imports	14.6	17.8	20.7	10.0	21.9	16.2
Balance (gross)	−8.3	−10.7	−10.8	—	—	—
Technology-intensive manufactured products						
Exports[a]	24.2	26.6	34.8	8.2	9.9	30.6
Imports	15.9	19.9	24.1	18.3	25.4	20.7
Balance (gross)	8.3	6.7	10.7	—	—	—
All commodities[b]						
Exports[a]	44.1	49.8	71.3	6.7	12.9	43.3
Imports	45.6	55.6	69.1	9.7	21.9	24.4
Balance (gross)	−1.5	−5.8	2.2	—	—	—

SOURCE: U.S. Department of Commerce.

[a] Exports include "noncommercial" shipments, such as military grant/aid, shipments of agricultural commodities under Public Law 480, etc.

[b] Includes the four commodity groups listed above plus "goods and transactions not classified according to kind" and reexports.

[c] Preliminary.

tion of inflationary pressures in the economy. This implies that we have now reached a magnitude of exports of agricultural products that is unsustainable unless the nation acquiesces to a continuation of all its ill effects on the economy.

2. The bulk of the increase in the surplus of technology-intensive products was due to the substantial increase in shipments of military equipment to Israel and the much greater than average growth in

Trends in U.S. Technology

GNP and investments, requiring U.S.-made capital equipment, in most foreign industrialized countries in 1973. This is not likely to continue (5).

3. There has been a substantially greater increase in the rate of inflation in all other industrial countries than in the United States (6) which, if it continues, might probably lead either to a worldwide depression or to drastic changes in the socioeconomic systems of these countries—or to both.

4. All the devaluations have completely failed to slow down the rate of growth in U.S. imports in any product group, and especially manufactured products, whether technology-intensive or nontechnology-intensive.

5. There was a small reflection in the 1973 trade statistics of the huge increases in foreign oil prices posted toward the end of the year. The adverse impact of these prices will increase dramatically in 1974 and thereafter.

Causes of Loss of U.S. Technological Advantage

How did all this happen? Undoubtedly, the causes have been numerous. In a generic sense I would attribute, one way or another, most of these problems to a change in American society's historic posture with respect to technology and long-term economic self-interest, on the one hand, and a lack of long-term economic policy making within the government, on the other.

The most important avenues through which these generic causes have exerted themselves have been (1) a lower growth of investment in new industrial plant and equipment in the U.S. than in other industrialized countries since the early 1950s, (2) an underinvestment in economically relevant R & D relative to other industrialized countries since the beginning of the 1960s, and (3) a worldwide and practically one-sided diffusion of existing U.S. advanced technology in a "naked" form since the end of World War II, especially since the end of the 1950s.

Investment in plant and equipment. It is generally assumed that investment in new industrial plant and equipment is the principal vehicle for the internal diffusion of new technology in the economy and, hence, the principal mechanism through which new technology raises factor productivity. This assumption, it should be noted, is the reason why many governments, including ours, use all kinds of investment tax credits and favorable capital depreciation allowances as their principal tool for fostering productivity growth. The assumption is unquestionably correct, but new investment in plant and equipment raises factor productivity only if the new capital goods continue to incorporate new resource-saving technology.

The comparisons given in Table 4 show that in the 1953-63 period, U.S. nonresidential fixed investment grew on the average only about 37% as rapidly as the average of the ten foreign countries listed in the table, and U.S. output per person employed grew at a rate about 54% of the foreign average. In the 1963-69 period, however, the U.S. growth in nonresidential investment was virtually the same as the average of these countries, but the growth in output per man slipped to only 29% of the foreign average. This implies that the smaller U.S. growth in nonresidential investment has been a factor in the loss of U.S. labor productivity advantage throughout the post-World War II period, but it was far more important in the 1950s than in the 1960s. In fact, in the 1960s it was, for all practical purposes, only of marginal significance.

This also implies, significantly, that our reliance on the investment tax credit and manipulation of capital depreciation allowances as the sole tool of productivity enhancement is far from an optimum effort and might actually be amiss. Indeed, the comparisons given in Table 4 show clearly that a serious attempt to improve the productivity performance of the U.S. economy would require not so much a greater rate of growth in new investments as a substantial lift-up in the rate of growth in technological improvements to be embodied in the newly produced capital goods. But to achieve this, the investment tax credit and manipulation of depreciation allowances as such can hardly be effective. Needless to say, lifting up the rate of growth of technological improvements in newly produced capital goods is also a sine qua non for enhancing the direct technological competitiveness of domestic industry in world markets, and for this the investment tax credit is almost irrelevant.

Lagging economically relevant R & D effort. Analysis of the comparative international R & D effort is still at a stage of infancy, and much of what has been done so far is misleading for at least three reasons:

1. Most of the comparisons of various nations' R & D expenditures made so far include all R & D for defense and space purposes. This produces a huge bias in favor of the U.S. Its inclusion is justifiable only to the extent that it produces civilian spinoff.

2. These comparisons are almost uniformly based on individual-country expenditures in local currencies converted into U.S. dollars by means of official exchange rates. The recent wave of exchange rate realignments has clearly demonstrated that the use of official ex-

change rates for purposes of objective analysis is meaningless. In the case in point, it also produced a huge bias in favor of the United States.

3. Whenever comparisons were made in terms of employment in R & D, the prevalent focus has been on so-called "qualified scientists and engineers." This, too, has tended to produce a heavy bias in favor of the U.S. because of international differences in educational systems. For example, a U.S. engineer with a B.S. degree is classified as a "qualified engineer," but a German graduate of an *Ingenieur Schule*, who has completed as many years of education, is classified as a "technician."

These three purely analytical misconceptions, I might note, played a large role in the so-called "technological gap" issue which was widely debated in the highest international policy circles in the second half of the 1960s. By avoiding these misconceptions, the data in Table 7 and Figure 2 realistically compare the economically relevant R & D effort in the United States during the 1960s with that of other industrialized countries. The data show that, though the U.S. was spending on and employing more professional manpower in economically relevant R & D than any other single country, the relative intensity of the U.S. R & D effort can hardly be classified as that of a leader. Indeed, in terms of expenditures, the U.S. relative intensity—per dollar's worth of available resources (GNP)—was only about 80% as great as that of the Common Market countries and Japan, and only about 60% as great as that of the United Kingdom. Measured by the number of professional manpower employed in this R & D, the relative intensity of the U.S. effort was almost 30% smaller than that of the Common Market countries, less than half that of the United Kingdom, and only about 35% of that of Japan.

Moreover, it is almost certain that by now the U.S. lag in economically relevant R & D compared to other industrialized countries has increased rather than narrowed. Table 8 shows that in 1969–71 the real growth in U.S. civilian (company-financed) industrial R & D, which represents the bulk of what I define as "economically relevant," slowed to no more than 1.4% per year (implicit in the employment of R & D scientists and engineers) but that growth of defense and space R & D declined by about 10% per year. The growth in total U.S. civilian-equivalent industrial R & D in that period might, therefore, be assumed to have been about zero.

Data for the R & D effort of all other countries in that period are still spotty, but we do know that, in the same period, Japan's head count of full-fledged researchers (net of technicians) working in in-

Table 7. Comparative civilian-equivalent R & D effort[a] for economic development[a] in the 1960s, U.S. and selected other industrialized countries

Country	Expenditures (average for 1963–67)			Employment of professional manpower[c] (1967)		
	$ millions (U.S. cost-equivalent)[b]	% of U.S.	Ratio to U.S. per $ worth of GNP	Thousands (full-time equivalent)	% of U.S.	Ratio to U.S. per $ worth of GNP
United States	7,992	100	1.00	342.5	100	1.00
France	1,750	22	1.35	78.5	23	1.37
West Germany	2,098	26	1.44	100.7	29	1.61
Belgium	195	2	0.69	10.8	3	1.00
Netherlands	482	6	1.82	26.9	8	2.42
Italy	420	5	0.50	27.3	8	0.79
Common Market	4,945	62	1.22	244.2	71	1.38
United Kingdom	2,132	27	1.69	116.3	34	2.17
Western Europe, total	7,972	100	1.14	402.5	118	1.34
Canada	406	5	0.64	24.2	7	0.88
Japan	1,667	21	1.21	197.9	58	2.86
Western Europe, Canada, and Japan, total	10,045	126	1.12	624.6	182	1.56

SOURCES: See Ref. 17.

[a] R & D for agriculture, civilian nuclear technology, civilian industrial technology infrastructure, and civilian-equivalent R & D performed for defense and space, assumed to amount to about 10% of the latter (about twice the percentage implied in Habridge House, Inc., *Government Patent Policy Study*, vol. 1, Final Report, p. 1–10).

[b] Expenditures in currencies of individual countries converted into U.S. dollars by means of official exchange rates in force in 1965 adjusted for the relative cost of R & D in dollars in the United States vis-a-vis the other countries.

[c] The concept of professional manpower includes qualified scientists, engineers, *and* technicians.

Trends in U.S. Technology

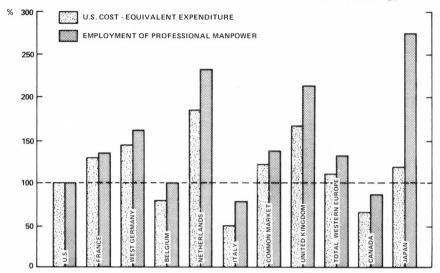

Figure 2. Comparative intensity in R & D effort for economic development in the 1960s. (Data from the sources listed in Ref. 17.)

dustrial R & D increased by about 35%, or a cumulative increase of about 16% per year (7). West German R & D expenditures by industry alone increased in the two years by 20.3%, or 9.7% per year (cumulative) in current prices, and, probably, some 2.2% per year (cumulative) in real terms. Total West German R & D expenditures in the 1969–71 period, which are largely civilian-market oriented, however, increased by about 14% per year (cumulative) in current prices and, probably, some 6% per year in real terms (8).

The R & D effort, however measured, constitutes only an input in a country's "inventiveness," which is the thing that makes the difference for the country's growth in technological know-how and, hence, growth in productivity and international competitiveness. Data on patents issued by the U.S. Patent Office to and applications received from U.S. and foreign residents (Figs. 3 and 4) strongly imply, however, that the trends are probably similar in actual relative inventiveness.

In interpreting the ratios in Figures 3 and 4, I should note, the more revealing statistics are the relative changes in the number of patents issued and/or applied for by foreign residents versus U.S. residents, rather than the absolute number, because U.S. residents tend to file patent applications for all types of inventions, whereas

Table 8. U.S. industrial R & D effort in selected years, 1953-71

Year	Total industrial R & D effort		R & D effort financed by federal government (largely DoD and NASA)		R & D effort financed by private company funds[a]	
	Expenditures $ billions (current)	Employment of R & D scientists and engineers, thousands (full-time equivalent)	Expenditures $ billions (current)	Employment of R & D scientists and engineers, thousands (full-time equivalent)	Expenditures $ billions (current)	Employment of R & D scientists and engineers, thousands (full-time equivalent)
1953	3.6	132.0[b]	1.4	45.0[c]	2.2	87.0[c]
1957	7.7	236.6	4.3	112.1[c]	3.4	124.5[c]
1963	12.6	333.8	7.3	161.7	5.3	172.1
1969	18.3	385.6	8.5	154.2	9.9	231.4
1971	18.3	362.4	7.6	124.6	10.5	237.8
Average annual growth rates, % per year						
1953-57	21.0	15.7	32.5	25.5	11.5	9.4
1957-63	8.6	5.9	9.2	6.3	7.7	5.6
1953-63	13.3	9.7	18.0	13.7	9.2	7.0
1963-69	6.4	2.4	2.6	-0.8	9.3	5.1
1969-71	0.0	-3.0	-5.5	-10.1	3.0	1.4

SOURCES: See Ref. 17.

[a] The company expenditures and employment do not include small company-financed R & D contracted to outside organizations, such as research institutions, universities, and colleges, etc. In 1970, for example, industrial firms contracted $243 million of R & D to outside organizations, or 2.4% of the R & D performed by the companies themselves.
[b] Average for 1952 and 1954.
[c] Estimated by the author on the basis of historical differences in expenditures per scientist and engineer in government-sponsored R & D vis-a-vis projects funded by the companies themselves.

Trends in U.S. Technology

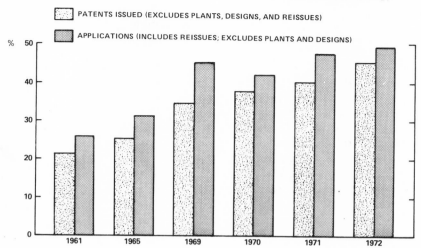

Figure 3. Patents issued by the U.S. Patent Office to and applications received from foreign residents as percentage of patents issued to and applications received from U.S. residents in selected years from 1961–1972. U.S. = 100%. (Data from U.S. Patent Office.)

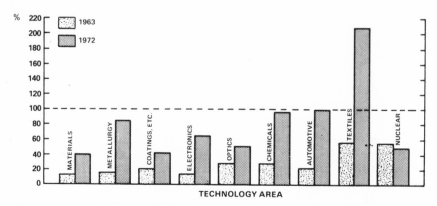

Figure 4. Patents issued by the U.S. Patent Office in nine selected technology areas to foreign residents as percentage of those issued to U.S. residents in 1963 and 1972. U.S. = 100%. *Materials* include paper, insulation materials, artificial wood and leather, etc. (exclude mineral and agricultural raw materials); *metallurgy* includes metalworking; *coatings* include bonding and molding; *electronics* includes electrical engineering; *optics* includes photography; *chemicals* include high polymers and other chemical compositions; *automotive* includes all automotive vehicles; *textiles* include textile machinery. (Data from U.S. Department of Commerce, Technology Assessment and Forecast, December 1973.)

foreign residents usually seek the protection of U.S. patents for only the most important inventions.

Export of "naked" technology. The issue of U.S. technology transfers—especially what they do or do not do for U.S. companies, on the one hand, and the U.S. at large, on the other—is terribly complex and subject to even greater controversy and confusion (9) than the issues discussed earlier. Since, in my judgment, this issue is of central importance for the U.S. situation under consideration here, I shall try to give as systematic an account of the conclusions of my analysis as space permits.

The international flow of advanced technology might take place in a great variety of ways, but the two and by far most important channels are transfers of advanced technology embodied in products produced in the U.S. and transfers in what, for the sake of brevity, I call a "naked" form. By naked transfers I mean sales of patent rights and licenses together with appropriate instructions, blueprints, and other technical assistance on the part of the seller which permit the buyer, independent foreign company, or foreign subsidiary of the "selling" company a quick and full exploitation of the know-how either for a fixed fee or for the more usual "running" fee (proportional to sales or cost of relevant products).

The relationship of the international transfer of advanced technology embodied in products produced in the U.S. to the international competitiveness of the companies, to the advantages of foreign countries, and to U.S. trade as such should have become obvious in the preceding analysis of the deterioration of the U.S. trade position. Briefly stated, if the companies have products embodying more advanced technology (compared with that being employed abroad), this technology makes the products in question either totally new or superior to similar products embodying conventional (inferior) technology produced in other countries. For this reason, foreign demand for such products with respect to price tends to be either inelastic or, at most, to have a fairly low elasticity (less than unity). This assures companies trying to sell such products in international markets a stronger competitiveness in these markets, which tends to make their overall sales and profits greater than otherwise would be the case. The advantages to foreign countries lie in their obtaining products which enable them to do things heretofore impossible or impractical and/or which yield an economic or utilitarian surplus value.

For the U.S. at large, finally, transfers of technology in such a form represent practically the only means of overcoming the disadvantages of high wages and high prices in its trade with other countries, as well as coping with market-induced employment opportunities and a growing tax base, and achieving reasonable equilibrium in balance of payments. The foreign trade advantages that any country derives from comparatively higher-quality technical know-how embodied in its products are generally of a monopolistic nature, since they can rarely be nullified by measures other than similar know-how. The point of emphasis is that, for a high-wage, high-price country like the United States, this advantage is a sine qua non of its international competitiveness.

The nature of the consequences of U.S. exports of advanced technology in a naked form, however, depend first on the nature of economic activity abroad in which the technology in question is used and, secondly, on whose interest is considered (10). Transfers of advanced technology in a naked form usable in areas of foreign economic activity which are supplementary or non-competitive with U.S. economic activities (such as oil and other extractive industries, utilities, and retail trade) tend to benefit the importing country, the companies exporting it, and the U.S. in general. The reason these imports benefit the importing country is that they tend, by the definition of "advanced technology," to enhance the productivity and, hence, the real income of the importing country and are frequently crucial to the economic activity in question. The reason the U.S. benefits is that such transfers command fees that otherwise would not be forthcoming to the companies and to this country, and if the technology in question is used in the production of products which the U.S. is importing, this technology tends to keep the cost of these products down and thus to secure better "terms of trade" for the United States than would otherwise be the case, unless, of course, the product in question is priced not according to cost but according to the buyer's "opportunity cost"—as is currently the case with Arab and other foreign oil—or simply what the market will bear. But even then the royalties obtainable from such transfers might cause them to be in the private as well as national interest.

However, transfers of advanced technology in naked form that are usable in areas of foreign economic activity *competitive* with the activities of the U.S., such as the manufacture of internationally traded goods, tend to be advantageous to the importing country and to the exporting companies, but detrimental to the general well-being of the U.S. The reason for the importing country's advantages are that these transfers speed up its rate of technological progress, frequently by great leaps and at a cost usually much lower than would be the case if the country were to develop the know-how by itself (11). Such trans-

Michael Boretsky

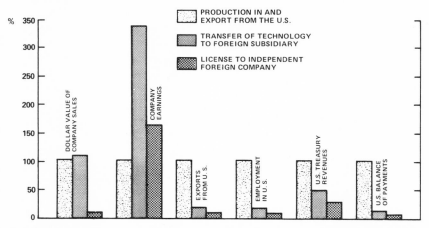

Figure 5. Comparison of the implications of export of advanced and market-tested manufacturing technology in "naked" form versus production in and export from the U.S. Note that for each item production in and export from the U.S. equals 100%. The middle bar of dollar value of company sales is only slightly larger than 100% because of unit price differential abroad.

fers represent an almost immediate substitute for imports, a potential base for export expansion, and an overall blessing to its balance of payments. And, of course, such transfers are the basis for growing market-induced employment opportunities as well as the general material and psychological well-being of the country. The advantages to U.S. companies lie simply in enhanced profitability and a potentially larger piece of the action abroad than they could get by exporting from the U.S. Finally, the U.S. gets the short end of such deals because the technology in question is usually new and thoroughly market-tested, the demand for products embodying this technology tends to be highly or at least fairly inelastic (by the definition of new and market-tested technology), and its export in a naked form, though commanding fees and in many instances conducive to some exports of products from the U.S. (capital goods and some components impractical to produce abroad), tends on the whole to be a substitute for exports of products from the U.S. as well as (at least in the long run) a promoter of competitive imports. The value of "displaced" exports plus promoted imports tends to be much greater than the returns of fees and promoted exports of products, even in cases where the would-be exports of the products embodying technology from the U.S. are appreciably smaller than the sales of these products when produced abroad. The discrepancies in the relative magnitudes

Trends in U.S. Technology

in question have implications not only for trade and balance payments but also for employment, treasury revenues, and many other variables (see Fig. 5).

The extent and growth of actual transfers of U.S. technology in naked form by major type between 1960 and 1972 might be judged by U.S. companies' receipts from and payments to foreign countries for technological royalties and license fees by industry (Table 9). The estimates include the companies' transactions with foreign companies as well as their foreign affiliates, but they do not include the value of the exchanges of technology between U.S. and foreign companies or the value of U.S. transfers in exchange for equity participation in foreign companies.

The impact of exports of naked manufacturing technology on U.S. trade of a specific industry, or rather a group of closely interrelated industries, is most readily identifiable in the case of the electronics and communications equipment industry. The technological basis of these products comprises about sixteen pivotal innovations that were developed after World War II. Of these sixteen innovations, twelve were developed exclusively in the U.S., two in the U.S. and, independently, in other countries, and only two exclusively in other countries

Table 9. U.S. companies' approximate receipts from and payments to foreign countries for technological royalties and license fees in selected years, 1960–72, in millions of dollars

Industry and item	1960	1965	1970	1972[a]	Average annual growth rate, % per year
Manufacturing industry					
Receipts	275	525	1,097	1,437	14.7
Payments	38	71	132	166	13.1
Balance	237	454	965	1,271	—
Ratio of receipts to payments	7.2	7.4	8.3	8.7	—
Industries other than manufacturing					
Receipts	89	141	273	362	12.4
Payments	12	20	41	42	10.0
Balance	77	121	232	320	—
Ratio of receipts to payments	7.4	7.1	6.7	8.6	—

SOURCE: U.S. Department of Commerce.
[a] Preliminary

(*12*). With such an overwhelming technological advantage we would expect the U.S. trade balance in the products based on this technology to be comfortably in the black. The worldwide dissemination of these U.S. innovations in a naked form, however, has been, by historical standards, almost instantaneous, and the U.S. trade balance in these products not only failed to grow, but gradually turned into the red, from a $206 million surplus in 1965 to almost a $1 billion deficit in 1972 (Table 10).

A series of alternative calculations, analogous to those presented in Figure 5, based on U.S. companies' aggregate receipts and payments for the transfers of manufacturing technology given in Table 9 and other relevant information imply that, had the U.S. done nothing more than it did in promoting new technology at home, but had fostered exports of its products with the same force as it fostered the export of naked technology, the kind of deterioration of U.S. trade and balance of payments that we have been witnessing since 1971 might have come eventually, perhaps in the twenty-first century, and not in 1971.

Thus far, the incidence of U.S. technology transfers has been spread fairly widely over most U.S. industries, but the heaviest concentration—some 80%—is in technology-intensive industries. The only major U.S. industry in the category of technology-intensive that has not done much of this exporting so far is the aerospace industry, which throughout the post-World War Ii period has been responsible for the largest U.S. trade surpluses of manufactured goods of any single industry. By now, however, according to numerous press reports and speeches of key industry people, even this industry seems to be determined to do the same thing (*13*).

In conclusion, I think my analysis makes it unmistakably clear that the choice of solution to the problems the U.S. faces in technology could radically change the country's socio-economic system and the course of its history. Continuation of current trends would lead, on the international front, to a decline in its economic and political position, pressure on the external value of the dollar resulting in periodic formal or informal devaluations (*14*), and the gradual worsening of its terms of trade, causing a lowering of the standard of living. On the domestic front, these trends would result in a continued lag in productivity growth, mounting inflationary pressures, lasting high interest rates, pressures for the redistribution of income, and lagging improvements or even a decline in the present level of the quality of life.

Obviously these conclusions have rather pressing policy implications, but to discuss them here would take this paper beyond its in-

Table 10. Trends in U.S. trade, production, and domestic consumption of electronics and communications equipment,[a] 1965–1972, in millions of dollars

Year	Trade			4. Industry shipments of primary products, inc. multiple counting	5. Industry shipments of primary products, net of multiple counting[b]	6. Apparent domestic consumption (Col. 5 and Col. 2 minus Col. 1)	7. Imports as % of domestic consumption (Col. 2 to Col. 6)	8. Imports as % of domestic consumption adjusted for price differences[c]
	1. Exports	2. Imports	3. Balance					
1965	670	464	+206	15,526	12,421	12,215	3.8	4.8
1966	815	749	+66	18,681	14,945	14,879	5.0	6.3
1967	891	793	+98	19,269	15,415	15,317	5.2	6.5
1968	1,085	1,154	−69	20,484	16,387	16,456	7.0	8.8
1969	1,395	1,559	−164	21,221	16,977	17,141	9.1	11.4
1970	1,547	1,788	−241	20,328	16,262	16,503	10.8	13.5
1971	1,491	2,108	−617	20,298	16,238	16,855	12.5	15.6
1972	1,876	2,829	−953	21,640	17,312	18,265	15.5	19.4

SOURCE: U.S. Department of Commerce.
[a] Industries included: SIC 3651 (radio and TV receiving sets, except communications tape); SIC 3661 (telephone and telegraph apparatus); SIC 3662 (radio and TV transmitting, signaling, and detection equipment); SIC 367 (electronic components and accessories).
[b] 30% of Col. 4, as in the 1963 input/output table.
[c] Assumes that imported products were about 20% cheaper than domestic products.

Michael Boretsky

tended compass. Readers interested in reviewing proposals to avert the continuation of the outlined trends might wish to consult another paper of mine (*15*) which discusses such policy needs and some of the available options in considerable detail.

Notes

1. Promotion of immigration of industrious and technologically skilled manpower, import of capital, and protection of "infant" industries. Of the three elements of Hamiltonian policy, economic historians usually emphasize the protection of infant industries. His *Report on the Subject of Manufactures*, submitted to Congress on Dec. 5, 1791, in his capacity as Secretary of the Treasury, does not emphasize it any more than the other two. In fact, all three would seem to have been to him equal and indispensable elements of overall policy. (Cf. Alexander Hamilton, *Report on Manufactures* (Boston: A reprint by the Home Market Club, 1892).

2. Cf., e.g., W. E. G. Salter, *Productivity and Technical Change*, Monograph 6 (Cambridge, England: Cambridge University Press, 1960), passim; and Edward F. Denison, *The Sources of Economic Growth in the United States and the Alternatives Before Us* (CED, 1962), p. 265; and *Why Growth Rates Differ* (The Brookings Institution, 1967), p. 298.

3. Cf. William D. Nordhaus, "The Recent Productivity Slowdown," *Brookings Papers on Economic Activity*, no. 3 (1972), 493–536.

4. This is true in terms of all commonly used indicators of inflation, such as GNP deflator, wholesale price index, and consumer price index, except in terms of changes in unit export values. As the U.S. Bureau of Labor Statistics has recently demonstrated, however, the index of unit export values is totally meaningless. Cf. *Department of Labor Bulletin*, no. 71-363 (June 30, 1971); and no. 72-16 (Jan. 14, 1972).

5. At the end of the year OECD estimated that in the whole year of 1973 real economic growth in the EEC countries would average about 6%, compared with an annual average of about 4.8% in the 1960–71 period; in the United Kingdom, about 6.7% compared with an average of 2.9% per year in the 1960–71 period; in Canada, 7.3% compared with 4.9% per year in the 1960–71 period; and in Japan, 11%, about the same as in the 1960–71 period. See OECD, *Economic Outlook*, no. 14 (December 1973), 23–24.

6. In 1973 the U.S. GNP deflator increased by 5.3% compared with an average of about 3% per year in 1960–72, but in West Germany it increased by about 6.3% compared with an average annual increase of 3.6% in the 1960–71 period; in France, by 7% compared with an average of 4.4% in the 1960–71 period; in the United Kingdom, by 5.8% compared with an average of 4.2% in 1960–71 period; in Japan, by 10.5% compared with 4.8% in 1960–71; and in Canada, by 6.5% compared with an average of 3.0% in the 1960–71 period. See *ibid.*, p. 35.

7. Office of the Prime Minister, Statistics Bureau, *Survey on Science and Technology* (1972), p. 333. (The source is in Japanese. Translation of the headings of relevant statistical tables was kindly provided by Professor Ozawa of Colorado State University.)

8. The data on German R & D expenditures in current prices are from Der

Trends in U.S. Technology

Bundesminister für Bildung und Wissenschaft, *Forschungsbericht (IV) der Bundesregierung* (Bonn, 1972), p. 202. The estimates of "real" rates of growth represent the respective percentage increases in prices deflated with percentage increases in the GNP deflator.

9. Cf., e.g., National Science Foundation, *The Effects of International Technology Transfers on U.S. Economy* (July 1974), passim.

10. As Paul B. Simpson of Oregon University demonstrated more than ten years ago, the same is largely also true with the export of capital. See his "Foreign Investment and the National Economic Advantage: A Theoretical Analysis," in Raymond F. Mikesell, ed., *U.S. Private and Government Investment Abroad* (Eugene, Ore.: University of Oregon Books, 1962), pp. 503–538.

11. From 1950 through 1972, Japan, for example, was spending on purchases of foreign technology from all countries about $158 million per year, merely two-tenths of one percent of GNP; and the aggregate Japanese outlays for that purpose amounted in that period to about $3.6 billion, not quite 3% of what the U.S. industry (own funds) spent on R & D at the same time and only slightly better than 1% of U.S. total expenditures on R & D in the same period. (Data on Japan's purchases of foreign technology for 1950 to 1962 are from Daniel L. Spencer and Alexander Woroniak, "The Feasibility of Developing Transfer of Technology Functions," *Kyklos*, 20 (1967); and for subsequent years from the *Japan Statistical Yearbook,* various issues, balance of payments series.)

12. The twelve innovations developed exclusively in the U.S. started with Bell Lab's discovery of transistor effect in 1947, which was followed by point-contact transistor (Bell Lab and Raytheon Co.), alloyed-junction transistor (GE, Bell, RCA, Raytheon), grown-junction transistor (Bell), zone refining (Bell), chemical etch process (Philco Corp.), epitaxy (Bell), planar process (Fairchild Corp.), gunn effect (IBM), beam-lead technology (Bell), metal-oxide silicon (MOS), transistor and integrated circuit (General Instrument, General Microelectronics, and Fairchild), and large-scale integration (Bell, Texas Instruments, and Fairchild). Of the other 4 pivotal innovations, diffusion process originated in the United States, United Kingdom, and the Netherlands (Bell, Texas Instruments, and Mullard Co.), integrated circuit in Great Britain and the United States (British Plessey Co., and Texas Instruments, Fairchild, and Westinghouse), III-V compounds in Germany only (Siemens Corp.). Cf. OECD, *Gaps in Technology, Electronic Components* (Paris, 1968), p. 44.

13. See, e.g., *Forbes*, May 15, 1973, p. 24; *Aviation Week and Space Technology*, May 28, 1973, pp. 27–28; *Aviation Daily*, May 16, 1973, p. 94, and June 5, 1973, p. 194; and *Boeing News,* May 17, 1973, p. 1.

14. In making this statement, I am, of course, aware that a great many economists see devaluation merely as a change in accounting procedure and a virtue of policy, since, in 1973, after the two devaluations, the United States achieved a trade surplus of $700 million (on balance-of-payments basis) compared with a $6.9 billion deficit in 1972. There is no denial that at times devaluation might be a necessary and useful tool of trade policy, but envisaging it as a standard tool of foreign economic policy is sheer nonsense. Moreover, I calculate that even if we assume that the $7.6 billion improvement in the U.S. trade balance in 1973 was totally attributable to devaluation, society's cost of this improvement amounted to over $10 billion worth of sweat. (In

Michael Boretsky

1973 U.S. imports of merchandise only amounted to $69.6 billion; in pre-devaluation dollars these imports would have cost about $56.8 billion, or $9.8 billion less. However, in addition to merchandise, quite a few other imports also became more costly, and thus the total cost to society was well in excess of $10 billion.)

15. Program of Policy Studies in Science and Technology, *U.S. Technology: Trends and Policy Issues*, Monograph No. 17 (Washington, D.C.: The George Washington University, October 1973) obtainable from National Technical Information Service, U.S. Department of Commerce, Springfield, Virginia 22151, publication number PB 227-930, $5.50 (microfiche $1.45).

16. U.N.; OECD; Angus Maddison, *Economic Growth in the West* (The Twentieth Century Fund, 1964); idem, "Comparative Productivity Levels in the Developed Countries," *Banca Nazionale del Lavoro Quarterly Review*, No. 83 (December 1967); Edward F. Denison, *Why Growth Rates Differ* (The Brookings Institution, 1967); M. Bornstein, *A Comparison of Soviet and the U.S. National Product* (JEC, 1959); Peter G. Peterson, *U.S.-Soviet Commercial Relationships in a New Era* (August 1972); Murray Feshbach, "Manpower in the USSR: A Survey of Recent Trends and Prospects," *New Directions in the Soviet Economy*, Part 3 (Joint Economic Committee, 1966), and subsequent communications; and individual country data.

17. Source of data on the individual country's expenditures in national currencies and full-time-equivalent employment of professional manpower: OECD, International Statistical Year for R & D, *A Study of Resources Devoted to R & D in OECD Member Countries in 1963/64;* Vol. 1, *The Overall Level and Structure of R & D Efforts* (Paris, 1967); Vol. 2, *Statistical Tables and Notes* (Paris, 1968); and *International Survey of the Resources Devoted to R & D in 1967 by OECD Member Countries;* Vol. 5, *Statistical Tables and Notes* (Paris, 1970). Data on relative cost of R & D: U.S. Department of Commerce and C. Freeman and A. Young, *The Research and Development Effort in Western Europe, North America, and the Soviet Union. An Experimental Comparison of Research Expenditures and Manpower in 1962* (Paris: OECD, 1965).

D. World Affairs

This section looks at technology on the international level to see how it affects American technology policy. We begin with an examination of our military posture with respect to new technology. The following chapters extend the perspective to consider needs of the world community as influenced by our technological actions. Chapter 9 addresses the needs of the underdeveloped nations; Chapter 10 considers the implications of technological change for international institutions. In all of these facets of world affairs, technology stands out as a driving force behind policy development.

In Chapter 8 Herbert F. York and G. Allen Greb present a history of military research and development since World War II. They detail the institutional apparatus developed for the direction of military technology, with emphasis on the roles of scientists and engineers. They stress the need for some organizational check on the military R & D system.

The notion of special assistance to underdeveloped countries is new, according to Gunnar Myrdal (Chapter 9). But it is no simple matter. Technology is not readily transferred to underdeveloped nations; it must be adapted with respect to climate, industrial institutions, available skills, and domestic markets. In order to assist these nations we must work closely with indigenous institutions such as universities to provide technical training and assistance consistent with local opportunities and values.

In Chapter 10, Eugene B. Skolnikoff examines a spectrum of technology-rooted issues with implications not confined to individual nations—environmental pollution, management of the oceans, population control, etc. Skolnikoff foresees increasing tension as these issues grow in political urgency and a resulting need not only for international cooperation, but also for greater national sensitivity to any loss of control. Such issues may well emerge as the "high politics" issues, like national security is today, becoming more central concerns of national governments.

8. Military Research and Development: A Postwar History
HERBERT F. YORK AND G. ALLEN GREB

A brief overview of the history of the American half of the Technological Arms Race that has dominated the research and development scene in the United States since World War II will be presented here. In doing so, the focus is mainly on the historic development of the apparatus that has been created for managing and directing our military technology programs. And within that context the roles and functions of the scientists and engineers, who either were in positions of authority themselves, or who were the direct and intimate advisors of those who were, will be specifically examined. In order to avoid spreading too thinly over too many topics, two readily separable special subjects are omitted whose histories have been relatively thoroughly covered elsewhere: nuclear weapons and basic research.

Such a review clearly demonstrates that the so-called "action-reaction cycle" has, on the largest scale at least, played a very fundamental role in determining the course of events. Specifically, four major events which took place within the Soviet bloc elicited major long lasting programmatic and organizational responses on the U.S. side. In three of these cases—the Soviet A-bomb, the Korean War, and Sputnik—the response directly followed the triggering event. In the fourth case—the roughly concurrent development in the USSR of an H-bomb and a long-range missile—the response of the United States, in fact, anticipated the events.

In each case, organizational changes led to improved capabilities for conducting a greater number and a larger variety of more sophisticated development programs than had been possible before. With

Reprinted by permission of the *Bulletin of the Atomic Scientists,* January 1977, pp. 13–26. Copyright © 1977 by the Educational Foundation for Nuclear Science.

Military Research and Development

only one important exception, all the new organizational mechanisms either became permanent or were later replaced by other stronger organizations which eventually became permanent. All of these organizational mechanisms continued to control and influence the course of technological affairs long after the events which precipitated them had faded and long after the original set of problems they had been created to cope with had been solved. In this way, these new mechanisms contributed not only to solving important problems, but also to producing that special property of the arms race which Secretary of Defense Robert McNamara had in mind when he said it had "a mad momentum of its own." Ironically and significantly, the one important exception was the organization that consistently played a moderating role, the President's Science Advisory Committee. An overview of these events and the people who participated in them also brings out another most important characteristic of the postwar scene, the interconnectedness of it all. Nearly every one of the principals mentioned in this article knew most of the others on a personal basis for a very long time, and a substantial majority of them served in more than one—often many more—of the various positions and committee posts described here. Most of them also had participated in the dramatic technological events of World War II, and most of the rest were students—in the most fundamental sense of the word—of those who had. As a result, information and ideas easily and quickly penetrated both vertical and horizontal organizational barriers (for instance, within and between services) that otherwise would have seriously hampered such interchanges and greatly slowed the whole process.

During World War II, American programs for the development, deployment and use of advanced military technology embraced many different branches of science and technology, and a large number and great variety of institutions were involved in their execution. Some of these institutions dated from long before the war; others were created during the war to exploit those new ideas and new technologies which were judged to be beyond the capabilities of the pre-existing institutions.

Like the programs themselves, the organizational mechanisms for managing them and for making decisions with respect to them were also very complex and involved many independent elements. This inherent organizational complexity on the one hand, and the need for maximum speed during wartime on the other, made it necessary to create special mechanisms for coordinating these R & D programs with each other and for appropriately relating them to the other elements of the total military effort.

During the war, this special mechanism was in the hands of a relatively small group of men working on a few interlocking boards and committees. Chief among this group was the triumvirate of Vannevar Bush, James B. Conant and Karl T. Compton. Bush, a professor at MIT before the war, was simultaneously director of the Office of Scientific Research and Development (OSRD), which reported to the President; chairman of the Joint New Weapons and Equipment Board (JNWEB), which reported to the Joint Chiefs of Staff; and chairman of the Military Policy Committee, which acted as a board of directors for the Manhattan Project. Conant and Compton (along with Bush) were members of the National Defense Research Committee (NDRC), which was the parent and later an integral part of the OSRD.

When the war ended, the apparatus for managing and conducting military R & D was, like the rest of the great war machine, largely demobilized and dismembered. Since then, it has passed through the following four successive phases of remobilization and reorganization.

Phase I: 1945-1950

The first phase, extending from 1945 to 1950 and coincident with the formative stage of the Cold War, was a period characterized by demobilization, uncertainty and marking time. After the war ended, the duties and authorities of the Military Policy Committee and the Manhattan District were taken over by the various mechanisms created by the Atomic Energy Act of 1946. The history and output of this set of institutions has been well and thoroughly examined elsewhere, and will not be further pursued here.[1]

The Joint Research and Development Board (JRDB). Those functions of the Office of Scientific Research and Development (OSRD) and the Joint New Weapons and Equipment Board (JNWEB) that remained after the war ended were taken over by a new Joint Research and Development Board. The board had five members: Bush, who was named as chairman, plus two high-ranking officials from each of the two services. All of the members of the board itself served part-time, and consequently a full-time secretariat was created to conduct its day-to-day business.[2]

The main function of the board was to "coordinate all research and

Military Research and Development

development activities of joint interest to the War and Navy Departments so that the War and Navy Departments will establish and carry out a strong, unified, integrated, and complete research and development program in the field of national defense."[3] The emphasis was on such management concepts as coordination, integration and avoidance of unnecessary duplication; the board had no budgetary authority or any other forms of direct control over programs.

Most of the work of the board was conducted by specialized committees dealing with particular technologies. The committees, like the board itself, consisted of a mix of part-time "civilian experts who are specialists in each of these fields" and high ranking military officers "who put in the hard-boiled 'military requirement' kind of thinking."[4] Most of the committees, like the parent board, had a full-time executive director.[5]

The Research and Development Board (RDB). The passage of the National Security Act of 1947 created a number of new high level bodies including the National Military Establishment, the Departments of the Air Force and the Army, the Munitions Board, the National Security Council, the National Security Resources Board, the Central Intelligence Agency and the Research and Development Board.

The Research and Development Board was organized Sept. 30, 1947, and replaced the Joint Research and Development Board with only minor modifications in structure and functions. It was the highest level R & D management unit in the Pentagon. Vannevar Bush continued to serve as chairman of the Research and Development Board until October 1948. At that time, he resigned and was replaced by Karl T. Compton, who had been president of MIT since 1930, and who had also served as a member of the National Advisory Committee for Aeronautics (NACA) and as chief of the OSRD's Office of Field Services.

The board did not accomplish nearly as much as had been hoped, and many of those personally involved, including Bush himself, later expressed their disappointment in this regard. The reasons were the complete lack of any kind of real authority and the extreme decentralization of such responsibilities as the board did have to a very large and complex system of committees, each of which was made up of intelligent and capable but part-time people.[6] In addition, the very cordial and supportive personal relations which existed between Bush and Secretary of Defense Henry Stimson did not long continue after Stimson was replaced by James Forrestal. In situations such as this, where the formal statement of authority is weak and unclear, the role of personal relationships is extremely important.

The board did, however, have an important direct impact on the future course of military R & D through its functioning as a mechanism for the exchange of information and ideas among those with authority and influence in some other capacity. Its military members generally had real authority over R & D in their respective services, and many of the civilians served on other bodies which often had more direct influence over events than the Research and Development Board itself had.[7] Perhaps even more important, the board served as a school where a number of the future leaders, managers and explicators of defense R & D received an education in national security affairs and kept active in the public policy arena during the period that spanned the time from the end of World War II until the Soviet atomic bomb and the onset of the Korean War.[8]

Project RAND and the U.S. Air Force Science Advisory Board. Also dating from this earliest postwar period is the oldest and most influential of the so-called "think tanks"—RAND—and the most influential of the many military advisory committees—the Air Force Science Advisory Board (AFSAB). The idea of creating a permanent specialized body to study the application of modern technology to national defense arose in a wartime group which conducted operations analyses for the Secretary of War. The group was headed by Edward Bowles of MIT, and included Frank Collbohm, Arthur Raymond, William Shockley and David Griggs.[9]

Believing that the kind of service they provided would continue to be needed after the war, Bowles and his colleagues pondered various possibilities for doing so. They eventually settled on a plan for creating what we now know as Project RAND as an autonomous division of the Douglas Aircraft Company.[10] (Raymond and Collbohm were on leave from Douglas at the time.) The original letter contract to Douglas asked them to study "the broad subject of intercontinental warfare, other than surface, with the object of recommending to the Army Air Forces preferred techniques and instrumentalities for this purpose."[11] Soon after, the scope was broadened to include all aspects of "aerospace power."

The first two tasks specifically assigned by the Air Force to the new Project RAND were to (1) study the feasibility, design and military utility of an earth-circling satellite, and (2) make a comparison of ramjets and rockets (including ballistic missiles) as prospective strategic systems.[12] These two remarkably prescient studies continued for many years and resulted in a great many reports and briefings which very strongly influenced subsequent Air Force programs. Since its beginnings, RAND has continued to evolve (it became an indepen-

Military Research and Development

dent corporation in 1948) and to expand its horizons beyond technology and into areas outside those of interest to just the Air Force or even the Defense Department. All through its history it has had a strong influence on certain Air Force programs, though perhaps never again quite so great as that which it exercised in respect to the first military satellites and the first ICBMs.

Institute for Defense Analyses (IDA). In 1948, the Secretary of Defense and the Joint Chiefs concluded that they, too, needed an operation analysis group working closely with them. To that end, they set up the Weapons System Evaluation Group (WSEG). The group was commanded by a senior flag rank officer, it was physically located in the Pentagon, and it consisted of a mixed group of military and civilian experts of various kinds. Within just a few years, it was found that the kind of civilians needed could not be recruited in sufficient numbers under such an arrangement. So the Institute for Defense Analyses (IDA) was created in 1956 to provide a separate civilian-oriented administration for the group; initially it was a consortium of universities put together by James Killian, but now it is a free-standing not-for-profit corporation.

In 1958, IDA's scope was expanded to include similar work for the top civilian offices in the Department of Defense. Like RAND, it continues today to do the same sort of work. The Navy and the Army have had similar institutions working for them, namely, the Center for Naval Analysis (CNA) and Research Analysis Corporation (RAC).

The Air Force Science Advisory Board was the direct outgrowth of the wartime collaboration of two exceptional men—General H. H. Arnold and Professor Theodore von Karman. General Arnold was Chief of Staff of the U.S. Air Force during the war; von Karman was a very distinguished professor of aeronautics at Cal Tech. In September 1944, shortly before the Battle of the Bulge, Arnold contacted von Karman and asked him to organize a long-range postwar planning effort. The formal charge was set forth in a letter dated Nov. 7, 1944. It said in part:

> I believe the security of the United States of America will continue to rest in part in developments instituted by our educational and professional scientists. I am anxious that all Air Force postwar and next-war research and development programs be placed on a sound and continuing basis. In addition, I am desirous that these programs be in such form and contain such well thought out long-range thinking that in addition to guaranteeing the security of our nation and serving as a guide for the next 10 to 20 year period, that the recommended programs can be used as a basis for adequate congressional appropriations.[13]

Herbert F. York and G. Allen Greb

One result of this request was a very thorough and comprehensive report entitled *Toward New Horizons*. The report consisted of 33 volumes, each one covering a topic of major interest. It was especially influential over the next decade because of its comprehensiveness, because of the way its recommendations and ideas paralleled the ideas and prejudices of many of the top Air Force officials, and because, as in a number of other crucially important instances, the people who wrote it continued to have special influence over Air Force and other high technology programs for many years to come.[14]

Another result of Arnold's request was the establishment of a permanent new organization known as the Science Advisory Board (SAB). Von Karman was named its first chairman. This board continues to this day, and provides an especially effective mechanism for keeping the lines of communication open between the top levels of the Air Force and those elements of the U.S. intellectual and industrial communities which do work of interest to them.[15]

The other services and the Defense Department itself have created roughly parallel arrangements, the most important of which is the Defense Science Board which today assists both the Secretary of Defense and the Director of Defense Research and Engineering.

Phase II: 1950-1953

The second phase, extending from 1950 to 1953, was a period of some reorganization and partial remobilization brought on in part by the first Soviet atomic bomb, but mainly by the sudden onset of the Korean War and all the fears for national security it engendered.

Neither the first Soviet nuclear test nor the final triumph of the Chinese communists, both of which took place in 1949, had any important direct influence on defense programs other than those of the Atomic Energy Commission. The Defense budget for fiscal 1951, the details of which were worked out between the Pentagon and the White House during the first several months following the Soviet test, was the lowest since World War II and it faithfully reflected the tight fiscal policies of President Harry Truman and Secretary of Defense Louis Johnson. In the following spring (1950), Congress approved this budget with only minor modifications. (The AEC, however, responded with an overall expansion of its program and with the initiation of a special high priority effort to develop the H-bomb.)

The sudden onset of the Korean War on June 25, 1950, followed five months later by the massive intervention of so-called volunteers from the Chinese Peoples Liberation Army, radically changed this

Military Research and Development

situation. These events were seen by American leadership as signaling a shift in communist policy from one of expansion through subversion to one which included direct military action as well. The Sino-Soviet bloc was viewed as a monolith and the intervention of the Chinese "volunteers" strongly reconfirmed this idea of a shift to direct action. This change in American perception about the form of the Soviet threat in turn made the technological threat seem much more dangerous and greatly reinforced the urgency to solve the problems it posed.

As a result, many of the civilian veterans of the great World War II military R & D programs—the OSRD, the Manhattan Project, the MIT Radar Project, rocketry at Cal Tech—regarded it as time to remobilize their efforts in the defense of freedom. The main root of their inspiration was their perception of the military threat posed by monolithic Communism but other, more personal, factors played a role as well. As is often the case with veterans, many of them recalled their wartime activities with pleasure and pride. They remembered the camaraderie, the adventure and the feeling of fulfillment and accomplishment they derived from their contribution to the Great Victory. When they met, they often reminisced about the mud at Oak Ridge, operating new types of radar in flight over Germany, meetings with generals and admirals, and sitting at the right hand of presidents and prime ministers.

A quick series of supplemental budget acts brought the fiscal 1951 budget up from $13 billion to over $48 billion. At the same time, military policy was completely revised and four basic goals were set:

—Fighting a war in Korea

—Strengthening the mobilization base for whatever might come next

—Creating stronger forces to act as a deterrent to the enlarged threat

—Strengthening U.S. alliances throughout the world

These general changes in military policies stimulated corresponding changes in defense-related research and development. New organizations were created at all levels. A Science Advisory Committee was established in the Executive Office of the President, a "Missile Czar" was appointed in the Pentagon, and civilian scientists were brought in and given positions of authority and influence at the top of the Air Force. Special ad hoc studies were funded to determine how American science and technology could best be used in solving both the short-term and long-term national security problems. These included Projects Hartwell, Beacon Hill, Vista, Charles, East River, and

Lincoln, and the Long Range Objectives Panel of the Research Development Board. There was a revival of interest in long-range ballistic missiles, new programs for the development of tactical missiles and warheads were established and old ones were expanded, programs designed to produce an effective continental air defense were initiated, and the development of all types of aircraft and avionics was stimulated.

The Science Advisory Committee to the Office of Defense Mobilization (SAC/ODM). The twin questions of what should be done and how to mobilize to do it were raised and discussed at various levels of government. Consideration of these questions at the highest levels confirmed the already widespread conviction that the existing structures—the Research Development Board plus certain other coordinating groups—were far from adequate for the job. The board was seen as being too complex, and as having too little (or no) authority and insufficient information.

As I. I. Rabi put it "[the Research Development Board's] very organization makes it impossible for it to do a good job because the committees and panels consist largely of people who serve part time and are not in the position to analyze the problems and see the whole picture."[16]

It was widely believed by scientists that they could not be effectively mobilized if it involved working directly for the military. It was therefore usually suggested that any new structure should have its apex either in the office of the Secretary of Defense or, as in the World War II case, in the White House. As part of the solution finally adopted, a Science Advisory Committee reporting to the White House Office of Defense Mobilization (SAC/ODM) was created.

The origin of the idea for establishing this committee is rather complicated. In August 1948 (after the Czech Coup and Berlin Crisis) Vannevar Bush, chairman of the Research Development Board, asked Irvin Stewart to chair a committee to study the general question of "mobilization of the civilian scientific effort in the event of an emergency."[17] In December 1949 (after the explosion of an atomic bomb by the Soviet Union in August), the Stewart committee submitted its recommendation to Compton, who had by then replaced Bush. Their recommendation, in effect, called for a new Office of Scientific Research and Development which, like the original, would report directly to the President. This involved matters obviously beyond the scope of the Pentagon, and so it was referred to the White House.

After the Korean War began, which presumably constituted something like the "emergency" Bush had had in mind two years earlier,

Military Research and Development

the problem was passed on to William T. Golden, a special consultant to the Bureau of the Budget. On Dec. 18, 1950, Golden submitted a report to the President which recommended "the prompt appointment of an outstanding scientific leader as Scientific Adviser to the President." Among other things, this Scientific Adviser was "to plan for and stand ready promptly to initiate a civilian Scientific Research Agency, roughly comparable to the Office of Scientific Research and Development of World War II."[18]

The final result was the establishment of a Science Advisory Committee reporting, not to the President, but to one of his principal assistants, the director of the Office of Defense Mobilization. Apparently one of the reasons for inserting the committee at this relatively low level was the desire on the part of the Bureau of the Budget to maintain a direct line to the President on all policy matters, including science.[19] Oliver E. Buckley, who had recently retired as president of the Bell Laboratories, was appointed at the first SAC/ODM chairman by President Truman on April 19, 1951.[20]

Despite its august membership and central bureaucratic location, the committee at first did little to fulfill the hopes of those who had made the recommendations that led to its creation. It met only a few times a year and took no initiative of its own. As a result, the Science Advisory Committee played very little part in the remobilization and redirection of American science that was then going on.[21] Later, after the inauguration of Eisenhower, both the work and the influence of the committee steadily expanded, reaching its apex just after the launching of Sputnik in 1957.

A "Czar" for guided missiles. During the late 1940s, the variety, number and complexity of missiles being proposed, studied and developed steadily proliferated. Neither the available funds nor the available talent, it was thought, was enough to go around, and demands arose to straighten out the growing confusion and to eliminate the "unnecessary duplication."[22]

In response, the Research Development Board in January 1949 set up a Committee on Guided Missiles chaired by Clark Millikan, like von Karman before him, a Cal Tech aerodynamicist and director of the Guggenheim Aeronautical Laboratory.[23] The functions of the committee, like those of the board itself, were limited to recommending the allocation of resources (including money), establishing liaison and cooperation where useful, and somehow avoiding unnecessary duplication. It did help higher level administrators to understand problems better, but its authority was not adequate for straightening out the confusion and producing order in an increasingly chaotic situation. Despite growing expressions of malaise by many in the military-technological community,[24] no new organizations were created having the ability to bring order to the missile program, irrelevant projects were not cancelled, and no new programs were initiated during late 1949 and early 1950.

It was only after the sudden onset of the Korean War in June 1950 that the level of alarm reached a point that officials were moved to do something about the situation. John McCone, then Undersecretary of the Air Force, Army General Kenneth D. Nichols, and others recommended, in effect, that a so-called "Manhattan Project" for missiles be established and that it be headed by a single individual with access to and a direct chain of command from the President in order to avoid military bureaucracy, and to cope effectively with red tape and service rivalry.[25] These proposals were passed to Defense Secretary Johnson and President Truman.

The President, like nearly all other interested persons, agreed that the current system was inadequate and that a strong centralized form of control was needed. Truman turned to K. T. Keller,[26] an old friend, to organize and direct the new project. After some hesitation, Keller accepted the assignment, and his appointment as Director of Guided Missiles, reporting to the Secretary of Defense but having access to the President, was announced on Oct. 26, 1950. The position and its occupant became well (and favorably) known by its sobriquet "The Missile Czar." It was widely believed that there was some sort of a "mess" which had to be cleaned up, some sort of impasse that had to be overcome, and the press and Congress welcomed the idea of someone with real power to "set things straight."[27]

In the course of attempting to draft appropriate directives, Keller and Nichols (who became Keller's deputy and who had been deputy chief of the real Manhattan Project five years earlier) soon discovered that setting up something analogous to the Manhattan Project was not feasible at that time. The Navy Department especially opposed the idea of a new and very special authority taking over its missile projects. Bowing to such pressures, Keller and Nichols determined to go to the other extreme. They decided to ask for no authority at all for themselves; instead they proposed to set up a very small office which would make clear and simple recommendations to the Secretary of Defense who by statute already had all the authority necessary.[28]

While these new proposals were being developed, George Marshall and Robert Lovett became respectively the new Secretary and Deputy Secretary of Defense. They quickly approved this new approach.

Military Research and Development

They, too, believed that sufficient authority already existed in their own offices to handle the missile situation, and they agreed that what had mainly been lacking were clear and simple decisions concerning which missiles were ready for production, which should remain in R & D and which should be abandoned altogether.

The brevity and conciseness of the resulting "charter" establishing Keller's office are as unique as the arrangement it describes. He and a small staff of technically trained officers from all four services reviewed one missile at a time in depth.[29] It very quickly became known that a recommendation by Keller had the force of a decision by the Secretary, so both the services and the contractors did their best to be responsive to questions put by him and his staff.

Among the installations Keller visited was the Redstone Arsenal at Huntsville, Alabama. Wernher von Braun and his team of German scientists had recently been transferred there from Fort Bliss, Texas, and were beginning work on a tactical Army missile that later became known as the Redstone. Keller was very favorably impressed by von Braun. He felt von Braun was a "doer" and that he had in his organization a number of other people who "know production." As a result, he authorized preparations for pilot production of the Redstone missile. For von Braun this was an important break. Prior to that, he and his "team" had largely been limited to making paper studies and giving advice, and this support by Keller helped to bring about their first opportunity to build real rockets since coming to the United States five years before.

After Dwight Eisenhower took office in January 1953, Keller and his staff were invited to stay on in the new Administration.[30] Keller's style of operation, however, turned out to be incompatible with that of the new people, and he soon resigned.[31] Even so, the system of setting missiles apart from the rest of military technology, and putting some special person with special access to higher authority in charge of them persisted for approximately the next decade.

Phase III: 1953–1958

The third phase, extending from 1953 to 1958, saw further remobilization and reorganization growing out of the so-called "New Look" at military strategy that took place at the end of the Korean War and in the wake of the successful test of the first superbomb, the hydrogen bomb.

Herbert F. York and G. Allen Greb

Three closely related events in late 1952 and early 1953 led to major changes in both the content and management of military R & D:

—The explosion of the first huge superbomb at Eniwetok in November 1952

—The inauguration of Eisenhower and the consequent complete change in top level defense leadership for the first time in a dozen years

—The cessation of hostilities in Korea.

These events led to a thorough reappraisal of U.S. defense policy and strategy, known as the "New Look." This reappraisal in turn led to a reduction in troop levels coupled with a greater emphasis on military technology, a plan for increasing the size and power of the Air Force and the proclamation of the doctrine of "Massive Retaliation." This last was not so much a new strategic doctrine as it was simply a warning to the Russians and the Chinese that in the next round the United States would not allow the Sino-Soviet bloc to choose a battlefield in some remote corner of the world favorable to it. Instead, no matter where or through whatever proxy they might strike the first blow, the United States intended immediately to shift the battle to the arena best suited to itself, to an arena where its more advanced technology and enormous strategic air power would make an overwhelming difference.

These events and reappraisals led to three major changes in the way R & D was managed. *First,* the Research Development Board and its system of part-time committees was abolished and replaced by two separate full-time staffs, each headed by an Assistant Secretary of Defense. *Second,* the practice of handling the increasingly large and important missile program separately from the rest of military R & D was continued but greatly strengthened and expanded. And *third,* the human resources available to SAC/ODM were effectively mobilized in the form of two special study panels: the Technological Capabilities Panel (Killian, 1954) and the Security Resources Panel (Gaither, 1957).

The first two of these changes were not directed at R & D management alone but had the broader objective of confirming for the first time the absolute authority of the Secretary of Defense over the operations of the individual military departments. President Truman had desired a similar unification of the armed services long before but lacked both the military experience and knowhow necessary to overcome service objections, particularly those of the Navy. President Eisenhower, on the other hand, as an acknowledged former national

Military Research and Development

military leader, could advocate such measures with little fear of opposition from the military establishment.

Assistant Secretaries of Defense for Research and Development and for Applications Engineering (ASD/R & D and ASD/AE). The Defense Reorganization Act of 1953 resulted in the appointment of two new Assistant Secretaries of Defense: Donald Quarles for Research and Development (ASD/R & D), and Frank Newberry, for Applications Engineering (ASD/AE). The motivation behind this aspect of the reorganization was the persistent notion that modern technology was not being exploited to its full potential. It was believed the diffuse and weak coordination of research and development by the Research Development Board was inadequate to the task and beyond hope of any simple minor reform, and that the "Missile Czar" approach might be good for certain specific programs but it was not suitable for research and development as a whole.

Quarles had been part of the R & D operation from its beginning, being first a member and then chairman of the Research Development Board's Committee on Electronics. From March 1952 until he took on the new R & D post in September 1953, he served as president of Sandia Corporation, a subsidiary of Western Electric and, in effect, the AEC's laboratory responsible for all the ordnance design aspects of nuclear weapons. In 1955, Quarles became Secretary of the Air Force and in 1957 he became Deputy Secretary of Defense, a post he held until his death in 1959. During his six years in the Pentagon, and despite an unassuming, somewhat reticent personal character, he exercised very great influence over both the content and the style of defense R & D during a period of exceptionally rapid change. Quarles was succeeded as ASD/R & D first by Clifford C. Furnas (1955–1957) and then by Paul D. Foote (1957–1958).

The other new post—Assistant Secretary of Defense for Applications Engineering—had a very different history. The exact role of the new position was never adequately worked out. Its first (and only) incumbent, Frank D. Newberry, was 73 at the time of his appointment, having retired from Westinghouse six years before. After four years, he retired from this new post, which was then abolished and the most important of its purported functions assigned to the Assistant Secretary of Defense for Research and Development.

Special arrangements for missiles. In June 1953, at a time when planning for these new assistant secretaryships was well advanced but before they had been established and filled, and while K. T. Keller was still in office as the Director of Guided Missiles, Secretary of Defense Charles Wilson decided to establish yet another top level to review the missile programs of the Department of Defense. Wilson appointed Trevor Gardner, who had recently joined the Air Force staff as Special Assistant for Research and Development, as chairman of the new study. Gardner, like von Karman and Millikan, had also been involved in the rocket development program at Cal Tech during the war.

To the extent that the purpose of the exercise was to save money by eliminating unnecessary programs, the Defense Department management picked the wrong man. Gardner had entered the Pentagon only a few months earlier with the firm conviction that not enough was being done to adapt modern technology to the needs of national security, and everything he learned after arriving on the Washington scene evidently reinforced this view.

The selection of Gardner for this task, and Gardner's selection in turn of a very special group of men—including John von Neumann, George Kistiakowsky, Simon Ramo, and Jerome Wiesner—to sit as a special committee, known as the Strategic Missiles Evaluation Committee (SMEC), to assist him had several far-reaching consequences. Among the more important were the initiation of several high priority programs for building long-range rockets for missile and space applications, and the further education of several persons, including those named above, who later held high positions in government.[32]

Also under Gardner's aegis special new mechanisms were created for providing what is now called "general systems engineering and technical direction" (GSETD) for the missile and space programs. The first of these, formed in 1954, was the Guided Missiles Research Division (GMRD) of the newly founded Ramo-Wooldridge Corporation. In 1960, largely because of persistent complaints by other private firms about unfair competition, many of the programs handled by GMRD/RW were transferred to the Aerospace Corporation, a new not-for-profit corporation set up by the Air Force just for that purpose.

Ivan Getting, out of MIT by way of the Air Force Headquarters and Raytheon and one of the principal authors of *Toward New Horizons*, has been the president of Aerospace Corporation since its founding. Both of these organizations still exist and still provide the Air Force with the general systems engineering and technical direction it needs for its programs. Earlier, in 1958, the Air Force had established the Mitre Corporation to provide similar services in connection with various electronics and communications programs.[33] This particular management tool was not adopted by the other services.

Technological Capabilities Panel (TCP). Another important element

Military Research and Development

in the general ferment of reorganization and review that characterized military research and development during the first Eisenhower years was the Technological Capabilities Panel (TCP). The Science Advisory Committee, reporting to the White House Office of Defense Mobilization created this panel at the direct request of President Eisenhower.

The President's request grew in part from his own personal concerns about a possible surprise attack on the United States by modern bombers carrying nuclear weapons, and in part by the urgent suggestion of several members of the Science Advisory Committee who had similar concerns.[34] As a result, the Technological Capabilities Panel was formed in March 1954 and worked throughout the rest of the year. The director of the panel was James R. Killian, Jr. of MIT and the deputy director was James B. Fisk of the Bell Telephone Laboratories, both of whom were members of the parent Science Advisory Committee. Three subpanels were created to perform the detailed work of the Technological Capabilities Panel, and each one reviewed a great many proposals for applications of technology to a variety of national security problems.[35]

Most of the ideas had been present in the U.S. research and development system for some time; but because of uncertainties connected with the Korean War and the change of Administration, and because of confusion over Russian technological capabilities and plans, they had either been blocked entirely or had been supported only at relatively low levels of funding. The Technological Capabilities Panel came along at just the right time to help untangle this web of confusion. By clearly and firmly endorsing certain of the many already existing ideas and proposals, it was able to exercise an unusual degree of influence over the technological future.

The panel's support of the ICBM programs, by then already underway, aided in securing the "highest national priority" for them. Its recommendation that both land-based and sea-based intermediate-range missiles be developed led directly to the initiation of the Thor, Jupiter and Polaris programs. The panel's conclusions and recommendations concerning technological intelligence, both the need for it and the means to get it, led directly to the development and use of the U-2 spy-plane, and served to give essential support to the Air Force's plans for building reconnaissance satellites.[36]

The Security Resources Panel of 1957, also known as the Gaither Panel, had a similarly stellar membership (including, in fact, many of the same people), but it did not lead to similarly significant results. By 1957, most of what we now recognize as the important strategic ideas had already been sorted out, and the appropriate developments were underway. The only area in which the Gaither report had much novelty involved civil defense. There, it urged a program that would have been extremely costly, and probably even socially disruptive, if it had been adopted.[37] For that reason, the President did not like either the report's tone or its details, and he did his best to keep its conclusions and recommendations from being circulated outside his immediate staff. Their general nature did, of course, become known, and the Gaither report subsequently became a political football in the national debate over the missile gap that followed in the wake of Sputnik.

Phase IV: 1957–1976

The launch of Sputnik on Oct. 4, 1957, triggered the fourth and final phase of the postwar reorganization and remobilization of military R & D. Sputnik did not come as an intellectual surprise to the American missile and space community—there had in fact been much advance information about it—but it did come as a visceral surprise both to the experts and the American people. Since it was in the main a technological shock, the President and the rest of the American body politic demanded a technological reply, and this final set of reorganizations was designed to provide just that.

The changes included the elevation and reconstitution of the science advisory apparatus in the White House, an approximately parallel elevation and rearrangement of the technology-management apparatus in the Pentagon and the transformation and expansion of the old National Advisory Committee for Aeronautics into the National Aeronautics and Space Administration (NASA). Despite some earlier uncertainties, the last of these changes only peripherally affected military R & D, and will not be further discussed here.

The President's Science Advisory Committee and the Special Assistant for Science and Technology. The changes in the White House apparatus consisted of three parts:

—The elevation of the Science Advisory Committee so that it reported directly to the President rather than one of his chief assistants (hence a new name: the President's Science Advisory Committee, or PSAC), and the consequent increase in the frequency and intensity of its meetings

—The establishment by the President of a new full-time position, the Special Assistant to the President for Science and Technology, whose incumbent by custom came to be elected the chairman of PSAC

Military Research and Development

—The appointment of the above Special Assistant to the chair of the Federal Council for Science and Technology, a coordinating mechanism that acquired somewhat more usefulness as a result

A few years later, during the Kennedy administration, the White House Office of Science and Technology was established by Presidential Reorganization Plan 2 of 1962, and the chairman of PSAC also came by custom to fill this job. This last change, however, was more form than substance and the power and influence of PSAC and its chairman continued to derive mainly from the first two elements described above.

The first Special Assistant for Science and Technology in this new era was James R. Killian, Jr., president of MIT and a long time high-level consultant on research and development and its management in both the White House and the Pentagon. Although (or perhaps because) he was not a research scientist himself, Killian was extraordinary in his ability to understand both the technological and political aspects of the problems involved, and to discuss these problems and their solutions in ways that both statesmen and technological experts could appreciate.

As a result, Killian's and PSAC'S influence were substantially greater than even the political situation and the august arrangements would indicate. George B. Kistiakowsky, a Harvard chemist with long experience on various scientific advisory boards, succeeded Killian in the summer of 1959. Though his background, personality, and style of operation differed from Killian's, Kistiakowsky continued to maintain the same close relations and special degree of influence with the President and other high officials.

During the last years of the Eisenhower administration, Killian, Kistiakowsky and PSAC reviewed virtually every important high technology program of the Department of Defense, and many of those of the AEC and the CIA as well. Few programs or ideas that did not meet their approval got very far. They were also the major factor within the government in generating real movement toward both the nuclear test ban in particular and strategic arms control in general.

Much of the actual work on these problems was conducted through special panels created for the purpose. These panels, however, whose chairmen were almost always members of PSAC itself, acted as integral parts of PSAC rather than as nearly autonomous groups, as had been the case of the earlier Technological Capabilities Panel and the Gaither panel. Thus, although they produced many excellent studies, none of them ever again achieved the special, separate identity of those earlier two groups.

President Kennedy continued the practices set in motion by Eisenhower, and appointed Jerome B. Wiesner as his Special Assistant for Science and Technology. Wiesner had excellent relations with the same variety of people as his two predecessors, but he had more competition as well. Another very capable academic, McGeorge Bundy, became the President's Special Assistant for National Security Affairs, and he involved himself in a forceful way in some of the problems that had been the virtually exclusive province of Killian and Kistiakowsky. In addition, the continued development of the Pentagon staff made it possible for McNamara and his assistants to present increasingly strong cases for the technological systems they wanted.

As a result of both of these factors, the influence of PSAC began to wane. During the Johnson and Nixon administrations (Donald Hornig, Lee DuBridge and Edward David were the special assistants) this trend was substantially accelerated by the failure of PSAC to provide much help with what Johnson and Nixon regarded as their paramount military problem: winning or otherwise "honorably" ending the war in Indochina. PSAC'S influence over military technology steadily decreased and, finally, in 1973 the committee and virtually all of the related White House level organizations were abolished.

The demise of the President's Science Advisory Committee involved additional, non-military factors as well. These included the failure of the committee as viewed by Johnson to give adequate advice about implementing his "Great Society" programs; the public opposition of several former, and one active, members of PSAC to certain favored programs, particularly the ABM and SST; and the growing clash of personalities between Johnson and Nixon on the one hand and their respective scientific advisors on the other.

Currently, there are plans and actions to reestablish something closely akin to the former arrangement. Guyford Stever, previously the director of the National Science Foundation and one time chief scientist of the Air Force, has been appointed to a post superficially resembling the one held by Killian and Kistiakowsky just after Sputnik. However, the basic situation is so completely different from what it was when PSAC flourished (1957 to roughly 1963 or 1964) that it seems unlikely to us that anything similar to the type of influence that PSAC had over military or nuclear affairs can be reestablished.

Pentagon Changes

In the Pentagon, the main changes engendered by Sputnik were incorporated in the Defense Reorganization Act of 1958.[38] They in-

Military Research and Development

cluded abolishing the position and the office of the Assistant Secretary for R & D, creating the office of the Director of Defense Research and Engineering (DDR&E), and giving the latter both much more authority and greater scope. In brief, the Assistant Secretary had been largely limited to advisory and coordinating program actions' whereas the DDR&E had direct authority to approve, disapprove or modify all R & D programs of the Defense Department. This last meant that such key programs as ballistic missiles and satellites ceased to be separately managed, and were combined with all other research and development programs. The formal status of the DDR&E in the Defense Department was also legally modified so as to give him precedence equal to that of the service secretaries and ahead of all the other assistant secretaries.

A special new action unit, the Advanced Research Projects Agency (ARPA) was also created. This agency became in effect an operating arm in the office of the Secretary of Defense by means of which various special projects (most originally in the space and missile defense areas) could be accomplished without having to cope with service red tape. It took almost a year to straighten out the conflicting relationships between DDR&E and ARPA; but since late 1959 the authority of the DDR&E over ARPA'S programs has never been less than that which the director exercised over the programs of the military departments.[39]

These changes originated in Eisenhower's desire that "an arrangement be devised which would serve to avoid duplication and reduce interservice rivalries."[40] The details for creating the new arrangement were worked out primarily by Charles Coolidge, a former legal counsel in the Defense Department, and James Killian.

The first Director of Defense Research and Engineering was Herbert F. York, who previously had been the first chief scientist of Advanced Research Projects Agency and before that, director of the AEC's Lawrence Livermore Laboratory. More important in this context, York was one of the new people brought into the President's Science Advisory Committee by Killian, and he continued to have a close working relationship with both Killian and Kistiakowsky after moving over to the Pentagon. This personal element probably played an important positive role in consolidating the authority and influence of all three men and their respective offices.

During the first year of the Kennedy administration, Harold Brown assumed the DDR&E post. Like York, he also had been a director of the Lawrence Livermore Laboratory and a member of the President's Science Advisory Committee. He therefore brought with him a similar orientation and he had a similarly close working relationship with the President's Special Assistant, then Jerome Wiesner. Moreover, although McNamara's style of operation was in general quite different from that of his predecessors, the basic concept of the office of the DDR&E fit well into his modus operandi, and so the role of the office remained unchanged in all important respects. Indeed, by continuing and expanding upon the practice of bringing in from the outside a strong group of deputies and assistants, Brown was able to further strengthen and consolidate the authority and influence of his office.

In 1965, President Johnson appointed Brown Secretary of the Air Force, and John S. Foster, Jr., yet another former director of the Livermore Laboratory, became the DDR&E. Foster continued for an unusually long time in the post leaving only in June 1973, well into the Nixon administration. He was succeeded by Malcolm Currie, who still holds the position. Like all three of his immediate predecessors Currie is a physicist; but he came to the Pentagon not via the Livermore Lab but from Hughes Aircraft, where he had held numerous important R & D positions over a 15-year period, and the Beckman Corporation, where he had been vice-president for research and development.

At about the same time the current system of R & D management was being established in the Department of Defense, similar structures were created in each of the three military departments. In each case, the highest civilian R & D official is called the Assistant Secretary (of the Army, Air Force, or Navy) for R & D,[41] and the highest uniformed official usually has the title Deputy Chief for R & D. Since the 1950s, the top-level civilians have all been professional scientists or engineers with substantial management experience, and the officers usually have had both advanced academic training in technology as well as considerable R & D management experience. Generally, the authority of these officials is not great; their main function is to serve as staff assistants to the service secretary and chief of staff, respectively, and in effect to the DDR&E as well.

The military R & D management systems have been remarkably stable since Sputnik. External events comparable in substance and scale to those which in the pre-Sputnik era produced major reorganizations have occurred since 1957–1958, but they have not led to further organizational changes. Such events include the Bay of Pigs, the orbiting of the first Soviet cosmonaut, the Cuban missile crisis, the acceleration of the Indochina war in 1965, and the U.S. withdrawal from Vietnam and final defeat in the 1970s. Internal adjustments were made in response to some of these events and, as is normal, the

Military Research and Development

whole process became more bureaucratized with time; but no overall changes occurred like those following the Soviet atomic bomb, the Korean War, or Sputnik.

Indeed, in 1970, following the first Nixon election, a so-called President's Blue Ribbon Panel on Defense Reorganization recommended substantial changes in R & D management, including the dissolution of the DDR&E post. These were not taken up by the then Secretary of Defense Melvin Laird, except for one which called for strengthening the test and evaluation function under a new assistant secretary. In the end, however, even this function was added to those of the DDR&E, thus strengthening that office.[42]

Since 1960, two countervailing influences, each of which might have resulted in substantial changes in the authority and status of the office of the Director of Defense Research and Engineering, have, in effect, cancelled each other out. On the one hand, the relative importance of research and development within the defense establishment has diminished somewhat. As an example of this, we note the R & D budget measured as a percentage of the total defense budget rose rapidly from 1945 until 1961, then levelled off for a few years, and since 1965 has substantially declined.[43] In addition, other technical experts have in recent years frequently filled other authoritative Pentagon posts (for example, Secretary of the Air Force Brown, Deputy Secretary of Defense Packard, Secretary of Defense Schlesinger) and have served within the offices of other assistant secretaries of defense. As a result, the authority of the DDR&E within the Department has been diluted. On the other hand PSAC, which provided the only external competition for authority over military R & D by a group possessing high technical competence, has been abolished and no comparable body has taken its place. The net result of these events and trends is that both the functions and the authority of the Director of Defense Research and Engineering have remained constant over almost two decades.

Of all the trends and events in the administration of military R & D during the past 10 years, surely the decline and demise of the President's Science Advisory Committee is the most important. The only effective "check and balance" operating in the military R & D system has been removed by eliminating the only inside group which had access both to information and power and which, at the same time, was composed in the main of persons not having a vested interest in some major element of the program.

How long this state of affairs will continue is an open question. The current plan for the creation of a science advisory apparatus in the

White House could restore an effective check and balance on military R & D; but the underlying situation is so different that such would not be an automatic result. Only time will tell.

Notes

1. Richard G. Hewlett and Francis Duncan, *A History of the United States Atomic Energy Commission,* vol. II, *Atomic Shield, 1947–1952* (University Park, Pa.: Pennsylvania State University Press, 1969).

2. The first executive secretary of the Joint Research and Development Board (JRDB) was Lloyd Berkner. He was aided by Major General A. C. McAuliffe of the Army and Captain J. H. Thack of the Navy.

3. From the *Charter of the JRDB,* June 6, 1946, amended July 3, 1946, as jointly promulgated by the Secretaries of War and the Navy.

4. U.S. Congress, House, Subcommittee of the Committee on Appropriations, "Hearings: Military Functions, National Military Establishment Appropriation Bill for 1949," Part I, 80th Cong., 2d Sess. (1948), p. 112.

5. In 1948, according to Lawrence R. Hafstad, executive secretary of the Research and Development Board, the board had a total staff of 202 civilians and 57 military personnel, a budget of $1,483,000, and reviewed about 15,000 projects each year. "Hearings" (n.4), pp. 112, 116, 136.

6. U.S. Congress, House, "Hearings" passim; personal interviews with Vannevar Bush and Eric Walker, former executive secretary of the board, 1972.

7. Such persons included several members of the General Advisory Committee of the Atomic Energy Commission (J. Robert Oppenheimer, James B. Conant, and Hartley Rowe); several members of the Air Force Science Advisory Board (Detlev Bronk, Hugh L. Dryden, and William Pickering); and of the National Advisory Committee on Aeronautics (Vannevar Bush, Karl T. Compton, and Dryden).

8. Such persons included Lawrence Hafstad, later chief of reactor development for the AEC; Ralph Lapp, later one of the most prolific writers on the arms race; David Beckler, who served for many years as the executive secretary of the President's Science Advisory Committee; Donald Quarles, later president of Sandia Corporation, Assistant Secretary of Defense for R & D, Secretary of the Air Force, and Deputy Secretary of Defense; Clark Millikan, a more or less permanent member of all major committees advising the government on ballistic missiles until his death on Jan. 2, 1966; W. M. Holaday, "missile czar" in the late 1950s; Mervin Kelly, chairman of several important ad hoc committees dealing with various defense questions; Ivan Getting, later chief scientist of the Air Force (1950) and the long-time president of the Aerospace Corporation (1961–present); Clifford C. Fumas, later Assistant Secretary of Defense for R & D (1955–1957) and a member of the board of trustees of the Aerospace Corp.; William Pickering, long-time director of the Jet Propulsion Laboratory; Robert R. Gilruth, later director of the NASA Space Center at Houston; and many others who continued to perform as members of key advisory groups for as many as 20 years after the Research and Development Board ceased to exist.

9. James Phinney Baxter III, *Scientists Against Time* (Cambridge, Mass.: MIT Press, 1968), chap. 26.

Military Research and Development

10. The origins, practices, and programs of RAND are discussed in a number of books. One of the most thorough is Bruce L. R. Smith, *The RAND Corporation: Case Study of a Nonprofit Advisory Corporation* (Cambridge, Mass.: Harvard University Press, 1966). RAND itself has published several pamphlet-length histories.

11. RAND, *The RAND Corporation: Its Origin, Evolution, and Plans for the Future* (Santa Monica, Calif.: RAND, Feb. 1971), p. 5.

12. RAND (n. 11), p. 10.

13. General H. H. Arnold to Theodore von Karman, Nov. 7, 1944, Mixed Files, Air Force Space and Missiles System Office, Los Angeles.

14. Among such persons were von Karman himself; Major Teddy F. Walkowicz, who held various key positions at the top of the Air Force until his retirement, and who then worked on most of the Rockefeller panels that have since studied defense programs; Hugh Dryden, who directed the NACA Laboratories until they were absorbed into NASA in 1959; and Lee A. DuBridge, who served as chief science advisor to President Eisenhower (1953-1956) and Nixon (1969-1970). Others occupied important positions in defense industry and in certain non-profit federal contract research centers where they influenced the programs of those institutions to conform to Air Force thinking and requirements.

15. See Thomas A. Sturm, *The USAF Scientific Advisory Board: Its First Twenty Years, 1944-1964* (Washington, D.C.: USAF Historical Division Liaison Office, 1967).

16. I. I. Rabi, "The Organization of Scientific Research for Defense," Academy of Political Science *Proceedings,* 24 (May 1951), 360. Rabi had been a leader in the MIT Radiation Laboratory and later served as member and chairman of the Science Advisory Committee to the Office of Defense Mobilization.

17. Vannevar Bush, Chairman, Research and Development Board to Irvin Stewart, Aug. 13, 1948, U.S. President's Science Advisory Committee (PSAC) Records.

18. William T. Golden Memorandum for the President, "Mobilizing Science for War: A Scientific Adviser to the President," Washington Bureau of the Budget, Dec. 18, 1950, PSAC Records.

19. Daniel J. Kevles to York and Greb, Sept. 21, 1976.

20. The original membership included Oppenheimer, Conant, and DuBridge, all of whom (along with Buckley himself) were members of the General Advisory Committee of the Atomic Energy Commission. Other public members were James R. Killian, Jr., Robert F. Loeb, and Charles A. Thomas.

21. Oliver E. Buckley to President Truman, May 1, 1952, OF 3000D, Truman Papers, Truman Library for general information on the activities of the SAC/ODM, its mode of operation, and accomplishments to that date.

22. A useful estimate of the number of missile programs being supported directly by the government in the period 1949-1950 is 35. Many other smaller efforts were being supported indirectly through overhead, deductible R & D, and similar means.

23. Other members included C. C. Furnas, later an Assistant Secretary of Defense, and Laurence A. ("Pat") Hyland, later chief executive officer of Hughes Aircraft.

24. An *Aviation Week* editorial, for example, complained that there was not yet one service-tested or service-accepted missile of any type or range. It

went on to declare, "somebody should start cleaning up the woeful condition; set up a directing agency without the preconceived prejudices and theories of any other group, and set to work with open minds and the realistic goal to produce and produce and produce" Robert H. Wood, "Overselling the Missile," *Aviation Week*, 53 (July 31, 1950), 46. Although this was written just after the Korean War started, it gives an accurate picture of the thinking during the prior period as well.

25. This and much of what follows is derived from 1973 personal interviews and correspondence with Nichols.

26. The initials stand for Kaufman Thuma, but he was never identified other than as "K.T."

27. According to Roswell L. Gilpatric, who became Assistant Secretary of the Air Force in 1951, Keller "just took the bit in his teeth and . . . proceeded to assign out a lot of programs so tht there wasn't much opportunity for in-fighting and byplay while I [Gilpatric] was there [in the Air Force office]." Transcript, Gilpatric Oral History Interview (January 1972), 37-38, Truman Library.

28. See K. T. Keller papers, *passim*, Truman Library; for example, Keller to Oliver Fraser, Oct. 31, 1950; Keller to Lt. General H. R. Bull, Feb. 12, 1952; "Final Report of the Director of Guided Missiles," Sept. 17, 1953.

29. Among them were Admiral John H. Sides, who later was director of the Weapons System Evaluation Group; and Colonel Charles H. Terhune, who was deputy chief of the Air Force long-range missile program during the mid and late 1950s, and after that was chief of Electronic Systems Command.

30. Nichols believes Eisenhower originally invited Keller to be his Secretary of Defense. Personal interview, 1973.

31. Keller himself explained his resignation in slightly different terms. "I believe the thing [the missile program] is getting more into the administrative stage. . . . Having reached the age where I don't want to take on an administrative job, I think it is fitting that this work be passed over to the Defense Department, and plans are being made to that end." Keller to Brigadier General G. G. Eddy, Aug. 7, 1953, Keller Papers.

32. After only 2½ years, in February 1956, Gardner resigned his post in protest over what he regarded as insufficient funding for the Air Force's missile programs (New York Times, Feb. 8, 9, 10, 1956).

33. William Leavitt, "MITRE: USAF's 'Think-Tank' Partner for Space-Age Command and Control," *Air Force and Space Digest*, 50 (July 1967), 58-62, 65.

34. These growing concerns of the President and members of the Science Advisory Committee, in turn, had been in considerable part stimulated by Trevor Gardner, who, in addition to the missile program review described above, undertook a review of the overall posture of the Air Force's strategic forces.

35. Subpanel One, chaired by Marshall Holloway of Los Alamos, dealt with strategic weapons. Subpanel Two chaired by Lee J. Haworth, dealt with air defense. Subpanel Three, chaired by Edwin H. Land of the Polaroid Corporation, dealt with Intelligence. *The Report to the President by the Technological Capabilities Panel of the Science Advisory Committee*, 2 vols. (Feb. 14, 1955), vol. 2, pp. 185-90, Gordon Gray Papers, Eisenhower Library.

36. TCP report, Gray Papers, vol. I, pp. 16, 24-25, 38, 44; vol. 2, pp. 63-66.

37. ODM Science Advisory Committee, Security Resource Panel, Deterrence and Survival in the Nuclear Age. NSC 5724 (Washington, D. C., 1957), pp. 7-13. Also published recently as a joint committee print: U.S. Congress, Joint Committee on Defense Production, 94th Cong., 2d Sess. (1976).

38. Then Secretary of Defense Neil McElroy (1957-1959) has referred to this act as "the most significant improvement of the operation of the Department of Defense that took place during my time in the Eisenhower Cabinet.... The changes that were made could be considered somewhat technical, but were very real in their accomplishments." Transcript, McElroy Oral History interview (May 1967), 71-72, Eisenhower Library.

39. Actually ARPA was created ten months before the position of DDR&E was filled. ARPA could be and was established by fiat of the Secretary of Defense but setting up the DDR&E required new congressional legislation. The arrangements described above, however, came into effect very soon after the first DDR&E was installed in December 1958.

40. Killian to York and Greb, Sept. 9, 1976.

41. The first person to hold this title in the Department of the Air Force was Trevor Gardner (1955); in the Navy, James H. Wakelin, Jr. (1960); and in the Army, Richard Morse (1961).

42. *Blue Ribbon Defense Panel Report,* July 1, 1970, p. 1. Gilpatric, Deputy Secretary of Defense from 1961 to 1964, reports that "much of what we were able to do in the three years of the Kennedy administration had its roots and beginnings in what we had started during the Korean War. There was a definite connection there and a relationship that I was very well aware of having been there both times." Gilpatric Oral Interview, 40-43.

43. U.S. Bureau of the Census, *Statistical Abstract of the United States,* for various years, 1945-1975.

9. The Transfer of Technology to Underdeveloped Countries

GUNNAR MYRDAL

The proposition that the developed countries, in their dealings with the underdeveloped countries, should show a special concern for their welfare and economic development and should even undertake a collective responsibility for aiding them is an entirely new concept. It began to be articulated only at the end of World War II, with the dissolution of the colonial power structure. From the outset, perhaps

Reprinted with permission. Copyright © 1974 by Scientific American, Inc. All rights reserved. From *Scientific American,* September 1974, pp. 173-182.

Gunnar Myrdal

inevitably, this novel and humane idea was suffused with expectations that were, as we can now see, excessively hopeful. Certainly this was true with regard to the prospects for the transfer of technology. In the developed countries, it was observed, there now existed a highly productive technology; the underdeveloped countries could simply take over that technology and so could be spared the slow process of inventing it themselves. The difficulties confronting the underdeveloped countries here as in other quarters have only gradually become understood. Hard lessons have been learned, partly from what has happened to those countries over the past 25 years and partly from closer study of their problems.

In the first place it had to be learned that scientific-industrial technology, to be maximally useful in the underdeveloped countries, cannot simply be transferred but must be adapted to the conditions prevailing there. The tropical and subtropical zones where these countries are mostly located have a different climate, and the importance of climate among all the problems besetting their economic development has been, in my view, grossly underestimated. The factors of production, capital and labor, are locally available in quite different proportions. Educated, experienced and skilled managers, engineers and workers are relatively scarce. Domestic markets are small, depriving new industries of economies of scale unless they can rapidly find large export markets. The external economies provided by a diverse surrounding industrial system are, of course, also absent and take time to develop.

Technology transfer imposes its own by no means insignificant costs in royalty payments for licenses and know-how. According to the estimate of a United Nations agency, these costs amounted to $1.5 billion in 1968 and have been increasing steeply. The transfer of technology through joint ventures is often accompanied by extraordinary costs imposed by the overpricing of imported intermediate goods, the underpricing of final goods for export, tax subsidies, various forms of tax avoidance and so on, the cumulative cost of all of which it is impossible to estimate. Even when the transfer is effected on long-term credits from individual developed countries, the almost universal practice of tying such "aid" to imports from those same countries implies the payment of prices higher—by 20 to 40 percent —than if the choice of imports from abroad were altogether free.

Against all these costs it has nonetheless been argued that the transfer of technology carries a net advantage. Edward S. Mason of Harvard University pointed out in the 1950's that the underdeveloped countries "could hardly be at an absolute disadvantage as compared

Transfer of Technology to Underdeveloped Countries

with the initiators of industrial development, since they always have the alternative of devising techniques themselves as did their predecessors in development." Yet is it realistic to assume that the underdeveloped countries have such an option?

In the 1950's it was also frequently pointed out—and one hears this hopeful thought voiced even today—that in the history of economic development in the West it was often advantageous to be a latecomer. Britain has continually paid a price for being the first country to undergo industrial revolution. Time and time again, in one industry after another, other Western countries have succeeded in catching up with and then surpassing her. Taking over and advancing a technology first developed in Britain was a recurring historical event.

The countries that accomplished this feat, however, were very different from the underdeveloped countries of today. They were all, like Britain, located in the Temperate Zone. They had access to a competitive capital market where they could borrow for as little as 3 percent and sometimes even less. And they did not have exploding populations.

More directly pertinent to their ability to absorb new technology from abroad, the western European and North American countries could boast almost general literacy in their populations. In regard to elementary education they had passed Britain, which lagged in this respect until fairly recently. The technology they already had was much higher and more diversified than that now possessed by underdeveloped countries. They had universities and technical colleges and were generally part of Western civilization. In all these respects they were much more favorably placed to benefit from the transfer of technology.

Moreover, the technology they acquired was very different from that of today. Most of the technical innovations of the late 18th and early 19th centuries were of a rather simple mechanical kind, whereas today they grow out of scientific discoveries concerning the nature of matter and energy. The entrepreneurs themselves, in textiles and other industries, took an active part in improving the machines in their factories. Charles H. Wilson has observed that technology then "involved no principles that an intelligent merchant could not grasp."

To this it should be added that technology in those days was much less capital-intensive and in general did not confer so much advantage of scale. Private enterprises, often small at least to start with, had their field day, and they effectively carried out the transfer of technology. The state often subsidized and protected them but seldom found reason to direct them.

Gunnar Myrdal

Technology today is much more intimately related to science. From this difference stem the many other differences that constrain its transferability compared with the simple technology of the earlier years of the scientific-industrial revolution. And the connection between scientific discovery and technological innovation becomes closer every day.

This brings us to what can be called the dynamic problem. The transfer of technology must now take place in a setting of more rapid advance in science and technology than ever before in history. As the American historian Henry Adams and others since have observed, this advance is accelerating, proceeding on an exponential curve; the pace of history is accelerating in parallel, and the driving force is scientific and technological advance.

It is not only that the underdeveloped countries must attempt to approach, by adoption and adaptation, a much higher level of science and technology but also that they must reach for science and technology that are continually and rapidly moving to a higher level. The developed countries are the site of practically all this advance. It is financed by their governments, philanthropic foundations, universities and industries. It is almost exclusively directed to the interests of the developed countries, although it produces occasional spin-offs useful to the underdeveloped countries.

All the elements cited here in the relation between the dynamics of scientific and technological advance and the prospects of the underdeveloped countries have been analyzed and accounted for separately by others in topic-by-topic studies. The conclusion that they add up to a general disadvantage for these countries is not, however, commonly drawn.

Thus the contribution of technological advance to the deterioration of the trading position of South Asian countries is recognized in studies of trends in their traditional exports. Synthetic fibers have largely displaced hemp, for example, in ship cordage and other technically demanding uses, and plastic film has been displacing jute in sacking. According to a recent U.S. Congressional committee report, coffee, tea and cocoa may soon have competition from synthetics. Rich countries will continue to develop manufactured substitutes for imported natural commodities. Not many of the commodities exported by the poor countries are as difficult to replace as the petroleum for which the few exporting countries in the Middle East, Africa and South America have recently succeeded in exacting a much higher price.

The dynamic advance of industrial technology also militates against

Transfer of Technology to Underdeveloped Countries

the successful development of manufactured exports by underdeveloped countries. Industries in the developed countries hold their markets firmly by dint of product and market research that brings rapid response to changing market preferences and by the internal and external economies that are provided by a broad and diversified industrial foundation. For contest with such competitors, enterprises in underdeveloped countries are ill-equipped. Typically their managers are weak in business acumen and their production workers are weak in skills; they lack capital, and they have little experience in standardized, high-quality mass production. The inefficiency of both management and labor and the absence of a supporting industrial environment will often, in spite of lower wages, tend to raise unit costs and so negate, in part at least, the international comparative advantage of low wage scales.

To all these well-investigated disadvantages facing new industrial enterprises in underdeveloped countries, industries in the developed countries have added research departments devoted to the improvement of their products and production processes. The resources of any one of these laboratories are amplified by the country's entire scientific establishment. What a new industry in an underdeveloped country must face is not only that the technological competence of its established competitors is high but also that it is constantly rising. Even if a license to use a particular technology is available, the contract does not often license future improvements; when it does, the obstacles to catching up are great.

With these difficulties in the way of reaching out with exports into world markets, underdeveloped countries have commonly adopted import substitution as the strategy for pushing their industrialization. By import restriction they can create markets at home. By the same action they should also be able to save foreign exchange. This advantage, if it is won at all, is secured only in the fairly long run. New industries, even those set up for the home market, need capital equipment from abroad and a continuous supply of spare parts, intermediate products and sometimes raw materials. Efforts to set up supporting industries that can provide substitutes for these imports raise again the same type of new import needs.

Seen from a planning point of view, import restrictions are more often a necessity than a policy. Shortage of foreign exchange usually comes first, and the protection provided by import restrictions for nursling industries is incidental. How the restrictions are handled determines whether or not they contribute to planned development. Import restrictions must either stop or make very expensive the importation of less necessary goods. That is rational. But domestic production of these goods then enjoys the very highest protection, often sky-high, which is not rational. Because the governments have not usually proved themselves able or willing to soak up "too high" profits by excess-profit taxes, they are compelled to resort to a large variety of direct controls. These controls have not often succeeded in reducing such profits, much less in redirecting investments according to plan.

Underdeveloped countries, quite apart from what their constitutions proclaim, are commonly ruled by upper-class elites. These people hanker for luxury goods, are glad to see them produced at home when restrictions bar their importation from abroad and usually have nothing against high profits. Protection thus becomes either unplanned or badly planned. It tends to create noncompetitive, high-cost and high-profit industries. This result in turn negates the planners' efforts to encourage entrepreneurs to enter more primary industries or to tackle the difficulties of the export market described above.

Keeping all of this in mind, it is surprising to learn that a few countries, in Latin America and Southeast Asia, have broken through the market barrier in certain fields and come out as exporters of manufactured goods. This is a sign that in these cases the difficulties of the transfer of technology have been overcome. Many of these cases reflect active cooperation from industries in developed countries. An important role has often been played by multinational corporations—an item to their credit that must not be forgotten in assessing their role in the world economy.

From the early 1950's there has been considerable discussion of how technology transferred from the developed to the underdeveloped countries could be adapted to be more labor-intensive. Underdeveloped countries are regularly short of capital, whereas their labor forces are grossly underutilized and become ever more so owing to the rapid, and increasingly rapid, growth of the populations in the labor-force age range. Some such adaptation of imported technology normally takes place without much difficulty or planning. The handling, packaging and shipping of raw materials and finished products can easily be, and for the most part are, conducted by labor-intensive methods. Lack of skill and sometimes of physical stamina, combined with low wages, makes it a rather normal thing that machines are run at a slower pace or with more workers tending them—except when an enterprise has found it advantageous to hire only literate workers and pay them salaries that enable them to improve their nutrition and general health.

Transfer of Technology to Underdeveloped Countries

When it comes, however, to the technology implied in the industrial processes and embodied in the machines, which is what is usually meant by the transfer of technology, there is little choice but to take it all without much change. For one thing, a different technology in this more fundamental sense would be likely to call for more technological creativity than underdeveloped countries command. The only capital equipment available, in any case, is what the developed countries are currently employing. That, along with the problem of finding spare parts, explains why underdeveloped countries have not equipped themselves with secondhand machinery from earlier years of technological advance, as some well-wishers have suggested they should. What is more, entrepreneurs are reluctant to accept "second best" equipment. That is often a rational attitude and not merely an oversensitive one. The most modern technology may in fact be optimal for promoting economic development. In many branches of industry the most up-to-date capital-intensive equipment is capital-saving in the important sense of consuming less capital per unit of output.

Careful studies have shown that the adoption on a major—and otherwise totally impractical—scale of more labor-intensive technologies in India would not have secured a substantial increase in the employment of labor. Modern industry forms, and for a long time to come will form, a very small part of the total economy there, and a still smaller part in other countries of South Asia.

When industrialization was first advanced as the road to development, it was thought that the growth of modern industry would draw off the underutilized labor force bottled up in agriculture and the loosely organized nonagricultural pursuits. A U.N. report in the early 1960's stated: "The reason for emphasizing industrialization is that industrial development would absorb rural underemployed persons." The plans of underdeveloped countries echoed this theme; throughout the 1950's and into the early 1960's and sometimes even now economists in both kinds of country have endorsed it. In the long view it is also entirely rational. If India and most other underdeveloped countries do not by the end of this century have a much larger portion of their labor force—which by then will have doubled in size—employed in industry, there will then be no hope of preserving even their present low average levels of living, no matter what improvements have meanwhile been accomplished in agriculture. The underdeveloped countries have serious reason to press on with their industrialization.

For several decades ahead, however, the employment effects of industrialization cannot be expected to be very large. The impact on employment is a function not only of the rate of industrialization but also of the absolute size of industry. For some time the effects may in fact be negative, owing to "backlash" disemployment of people from traditional industries and crafts that are either competed out or modernized and so made less labor-intensive. This was the experience in the Soviet Central Asian republics of Uzbekistan, Kirghizia, Tadzhikistan and Turkmenia. In the throes of a rapid industrial expansion that began in the late 1920's, the proportion of the labor force employed in manufacturing (including crafts and traditional industry) actually decreased until fairly recently, when industrialization progressed to a much higher level. The statistics on "employment" and "underemployment" in underdeveloped countries—usually inadequate—do not contradict the judgment that industrialization, at its present level, does not create much additional employment, even allowing for its side effects on employment in the building of the "infrastructure" and in services.

In the same overoptimistic view of the employment-generating effects of industrialization, the influx of population to the cities was regarded during the 1950's as a response to the demand for labor in industry. In reality it must be recognized as a flight from agriculture. The cities in all the underdeveloped countries now have a much bigger labor force than can possibly be employed in industry, however fast it is growing. They are simply displaced rural people.

The transfer of technology to the agriculture of the underdeveloped countries presents a totally different problem. Apart from the highly commercialized plantations in some of these countries, which should more properly be reckoned with industry, yields are almost everywhere low. To the low yields correspond serious nutritional deficiencies, which at present are threatening to become worse. In most countries there is not much uncultivated land to put to the plow. The only hope for increased production is intensified cultivation aided by advanced technology.

Contrary to a common misconception agriculture in these countries is not labor-intensive; it is labor-extensive, even though an exceedingly large portion of the labor force is confined to agriculture. In many countries a part of that labor force does no work at all; the nonworkers cannot all be classified as unemployed—members of the Brahman caste in Bengal do not work but hire workers. Most of the workers who do work do so only for short periods—per day, week, month or year—and not very intensively or efficiently. This is what, in false analogy to conditions in developed countries, is called "unemployment" or "underemployment."

Transfer of Technology to Underdeveloped Countries

The low yields per acre reflect the underutilization of the labor force. An increase in the input of labor and in labor efficiency would raise yields even without technological innovation or any additional investment except work. Much of any such increase in input should be directed to better use of existing technology—to improving the land, to constructing works for water conservation and distribution, to building more and better roads and to generally improving the condition of the villages.

Technological innovations can improve yields still more. The new technology must, however, be highly labor-intensive. Otherwise it will swell the stream of migrants to the city slums. The agricultural technology of the developed countries—aimed almost from the beginning at the improvement of yields while the agricultural labor force was declining, first relatively and then absolutely—is not on the whole adaptable to conditions in the underdeveloped countries. Certain important exceptions can nonetheless make all the difference. These include the artificial insemination of livestock, new methods of preventing plant diseases and, of course, high-yielding seeds.

Properly adapted to local conditions, such techniques should be applied by labor-intensive methods for the double purpose of securing increases in both yields and employment. Whereas tractors have their place and may in some situations increase employment, mechanization in general disemploys human workers and would not contribute much to increased yields per hectare, which are better secured by labor-intensive methods. It is often maintained that the agriculture of the underdeveloped countries requires some "intermediary technology." This should not be understood to imply a technology that has become obsolete in developed countries. What should be transferred is a technology that engages the latest results of scientific research, adjusted to the highest possible utilization of the labor force.

There is thus a fundamental difference in the conditions that should rule the transfer of technology to the industrial and to the agricultural sectors in underdeveloped economies. In the former the use of modern technology makes it possible to establish new industries with minimal disturbance of the institutional framework and minimal diffusion of skills throughout the labor force. Industrial expansion may thus evade direct confrontation with the social and institutional obstacles that have for so long inhibited economic development. The price paid for this strategy, however, is the perpetuation of the enclave industrial economy familiar from colonial experience. The "spread effects" of industrialization, that is, the generation of employment in ancillary industries and services and the percolation of productive skills and of the rational analytical attitude to the surrounding society, have proved very much smaller in many countries than was expected in earlier years and is still often assumed.

In contrast, the task of raising yields in agriculture through highly labor-intensive technology must confront the attitudes and institutions that have so long held these rural societies stagnant. Most important is to change "the relation between man and the land," creating the possibilities and the incentives for a man to work more, to work harder and more efficiently and to invest whatever he can lay his hands on to improve the land, in the first instance his own labor. This is the essence of the demand for a well-planned land and tenancy reform. Such reform has been on the agenda in all underdeveloped countries; with only a few exceptions it has been botched. As a result the auxiliary reforms often attempted in market organization, extension work and credits have been less effective. Hence, whereas industrialization can go forward without involving the masses, the successful transfer of technology to agriculture must change the people and all the social and economic relations there.

Once, on a visit to a famous agricultural school in a university in the American Middle West, I found myself talking with a large group of Indian and Pakistani students enrolled there. "What are you doing here?" I asked them. "What can you learn that can in any way help your people at home, toiling in fields around their mud huts?" Their answers showed they shared my misgivings.

Students so trained might just as well stay on in this country and swell the "brain drain." If they go home, they will either be unemployed or, thanks to their "connections," get administrative positions for which their education in the U.S. has not fitted them. For effective technology transfer it would be better to send teachers to the universities and schools in the underdeveloped countries. Some of them would at least get to understand what the practical problems there really are. Those who come to the developed countries for training should not be students, least of all undergraduate students, but people who are already established at home, for example physicians who want to learn a particular new technique or engineers who are out to master a new industrial process.

We in the developed countries should do whatever we can to enlarge and raise the standards of the universities and research institutions in the underdeveloped countries. Since many of them are bent on producing results that will impress us rather than solve their own practical problems, we have need of more knowledge about conditions in their countries and more imagination if we are to be of help to them. A very important part of our "aid" can be the redirection of our own research efforts to their problems. That should be done in collaboration with research institutions either already established or newly created in the underdeveloped countries. A model for such enterprise was provided by the work of the Rockefeller Foundation and later also by the Ford Foundation in the development of high-yielding strains of cereal grains. Another is the research conducted by the Population Council in the improvement of birth-control techniques and their adaptation to conditions in the underdeveloped countries, initiated at a time when the U.S. Government and public were not prepared to support such work. These are outstanding examples of what we should do on a large scale in many other fields.

It is evident that the transfer of technology must proceed on a much larger scale if increase in the production of material necessities is to overtake the increase in the population of the underdeveloped countries. The need for aid in the larger part of the underdeveloped world must therefore be recognized as substantial. Moreover, this need cannot be thought of as a short-term exigency; it will be long-lasting.

If my striving toward realism leads me to a darker view of the prospects for economic development than that still current among most of my fellow economists, this does not lead me to defeatism. On the contrary, my conclusion is that development requires increased and, in many respects, more radical efforts: speedier and more effective reforms in the underdeveloped countries and aid appropriations in the budgets of the developed countries that approach the amount those countries spend for other important national purposes, such as social security or education, not to mention armaments.

10. The International Functional Implications of Future Technology

EUGENE B. SKOLNIKOFF

Technological advance and the increasing application of technology have had profound effects on the international issues confronting nations, effects often quite different from those originally anticipated on common sense grounds. Technology, instead of minimizing consciousness of national sovereignty, *seems* to have exaggerated it; instead of discouraging the emergence of weak, small states, has made proliferation of states possible; instead of bringing about sharp changes in the attitudes and assumptions of governments toward foreign affairs, has allowed continuation of "traditional" approaches to the workings of the international system.

This essay will outline some of the future international implications of continued rapid developments in technology, with emphasis on the impact of technology on international intergovernmental machinery. Likely technological developments, in a time frame of 10-20 years, are going to generate important performance demands on the international system going well beyond, in scale and intensity, the requirements placed on the system today. These demands will result from the increasing constraints on independent national action coupled with much more intensive requirements for international

Copyright © 1975 by Columbia University. Reprinted from the *Journal of International Affairs*, vol. 25, no. 2, 1975, pp. 266-286, with permission.

Eugene B. Skolnikoff is Professor of Political Science and Chairman of the Political Science Department at MIT. He has long been interested in the interaction of technology and foreign affairs, both from policy and scholarly perspectives. From 1958 to 1963 he was in the White House on the staff of the President's Special Assistant for Science and Technology. He is the author of *Science, Technology, and American Foreign Policy* (MIT Press, 1967) and many articles in related fields. He is also chairman of the Science and Public Policy Studies Group, a national organization of universities working in the field.

Prepared for delivery at the Sixty-sixth Annual Meeting of the American Political Science Association, Biltmore Hotel, Los Angeles, California, September 8-12, 1970.

International Implications of Future Technology

activities carried out by international machinery. Moreover, because of the diffusion of decision-making with regard to many technologies, these demands have a certain "inevitability" which means they cannot be prevented or ignored by governments. The conclusions drawn will raise important questions about the viability of the prevailing model of the international system: sovereign states vying for security and advantage, with the primary locus of decision-making within the states. They will also raise questions about the ability of the existing international machinery and the existing attitudes of governments to cope with the changed situation. The actual technical situation justifies a position somewhere between the apocalyptic and the complacent. There are in a variety of areas, particularly pollution, the seeds of unparalleled disaster. More information could show that the world now faces irreversible environmental changes that require drastic and urgent international political action, though available information does not support such a conclusion.[1]

The specific functions of international organizations can be represented in the following typology:

A. Service
   1. Information exchange
   2. Data-gathering, analysis, and monitoring of physical phenomena
   3. Consultation and advice
   4. Facilitation of national and international programs
   5. Coordination of programs
   6. Joint planning
   7. Small-scale funding
B. Norm Creation and Allocation
   1. Data-gathering and analysis for establishment of norms
   2. Establishment of standards and regulations
   3. Allocation of costs and benefits
C. Rule Observance and Settlement of Disputes
   1. Monitoring adherence to standards and regulations
   2. Enforcement of standards and regulations
   3. Mediation, conciliation, and arbitration
   4. Appeal of standards and regulations
   5. Adjudication
D. Operation
   1. Resource and technology operation and exploitation
   2. Technical assistance
   3. Conduct of research, analysis, and development
   4. Financing of projects

Eugene B. Skolnikoff

I. Environmental Alteration

This is perhaps the newest major technological subject to receive critical public and political attention. Suddenly, governments find themselves under growing pressure to protect the "environment." Every international organization is also directly involved in some way; for a few, it is becoming a significant part of their activities. Developments of technology for deliberate manipulation of the weather, already extant or likely, can be thought of conveniently in three categories: micro-modification, modification of storms, and large-scale climatic modification. The inadvertent modification of weather or climate is also possible.

The evidence associated with cloud-seeding experiments in micro-modification suggests that rainfall can be increased locally from 5 to 20%. T. F. Malone anticipates that by 1980 "naturally occurring rainfall can be either augmented or diminished locally by proven techniques" and that by the end of the 1980's "the probability is high that rainfall several hundred miles downwind from the site of the operations can be increased or decreased at will."[2] Problems are sure to arise within countries and between nearby countries with regard to water distribution. Second, a need for international cooperation in operational matters is likely also to emerge, for at times the actual seeding may have to be done over foreign territory or international waters. Eventually, a different kind of international issue will emerge: how to allocate what is, in effect, the finite resource of atmosphere-borne fresh water. Clearly, a major problem of international resource allocation will exist.

The prospect of being able to divert or suppress hurricanes is now with us. Experiments in the U.S. (August 1969) using silver iodide appear to have been successful in reducing the maximum velocity of Hurricane Debbie by 31 and 15% on alternate days.[3] If modification activities could deflect the course of a hurricane, the possibility of claims for damage or water deprivation could result. Operational activities with regard to hurricane modification must necessarily be carried out in an international environment, even at times over several different national territories. Also on the technological horizon are bold schemes for melting the Arctic ice cap by means of chemicals (carbon black), a dam across the Bering Straits (often proposed by Russian scientists), dispersing the Arctic cloud cover, or pumping warm Atlantic water into the Arctic basin.[4] The purpose would be to alter radically the climate of the Northern Hemisphere, in particular to warm and provide increased moisture for the vast Siberian region.

International Implications of Future Technology

The probable effects of removing the Arctic ice cover vary from forecasts of entirely benign and beneficial effects (more temperate climate with increased rainfall in Siberia, Europe, and as far south as the Sahara), to predictions of the onset of a new ice age and substantial increases in the ocean level as the Greenland ice cap melts. Obviously much more information of the likely effects is required before any such project can be allowed to proceed. But, it must be realized that the resources required for such a project may *not* be beyond the capabilities of a single large state (a Bering Straits dam is entirely feasible at costs comparable to large continental dams). Moreover, if the economic payoff is as fantastic as warming Siberia, the incentive to proceed would be enormous. What international machinery does the world have to govern such a project?

Weather modification technology can also lead to strategic or tactical military capabilities, thereby affecting the balance of power. It can be used as a weapon in economic warfare, diverting water needed for agriculture or hydroelectric power. Conversely, it can be used as a tool for enhancing economic development and welfare. Such uses of this technology may depend critically on the international means developed for controlling it, for allocating its benefits, and preventing its misuse.

This technology depends directly on increased scientific understanding of the atmosphere. The underlying research and data-gathering is being carried out partially under national auspices, but the major components are two international programs associated with the World Meteorological Organization (WMO) and the International Council of Scientific Unions (ICSU). One is called GARP—the Global Atmospheric Research Program—and the other the World Weather Watch (WWW).[5] These two related programs may themselves create important requirements for new or revised international machinery to operate the programs when they are farther along. The potential implications of the knowledge gained through the international programs also establish a requirement for, in some sense, "controlling" the application of the knowledge in accordance with internationally agreed purposes.

Inadvertent modification of weather and climate is occurring as a natural concomitant of this century's accelerating application of technology, the growth of the world's population, and the urbanization being experienced in all countries. The world *could* be heading for a major catastrophe. From 1880 to 1940, the average temperature of the earth increased by 0.4°C, while in the last 25 years, the temperature has decreased by 0.2°C.[6] Are these climatic fluctuations of the last

Eugene B. Skolnikoff

80 years natural variations, or a result of man's activities? There are a number of ways in which man's activities could disturb the atmospheric heat balance.

One is the accretion of carbon dioxide in the atmosphere, as would be expected from an increased consumption of fossil fuel. The result of a CO_2 "blanket" is to trap solar energy reflected from the earth's surface, creating a hothouse effect. The accretion of CO_2 would correlate with the increase in the earth's temperature until 1940, but not with the subsequent decrease. A possible explanation of this phenomenon is that the CO_2 concentration is being overwhelmed by atmospheric particle pollution. That is, dust from urban, industrial, or agricultural activities can affect the thermal balance primarily in the opposite direction to the CO_2. The dust forms a barrier to solar radiation, thus reducing the energy reaching the earth. In addition, the dust forms nuclei for low-level cloud formation which serve to reflect some additional solar radiation back to space. "At present, on the average, about 31% of the earth's surface is covered by low cloud; increasing this to 36% would drop the temperature about 4°C, a drop close to that required for a return to an ice age."[7]

There are also other possibilities of considerable concern. The earth's albedo (the percentage of incoming solar radiation directly reflected outward) is being changed by man-made alteration of the earth's surface. Dense urban areas and highways reflect more radiation than forest or agricultural land. These changes could also lower the surface temperature. Additionally, there are unknown factors such as rocket exhausts in the upper atmosphere which may affect the transfer of radiation, or water vapor exhaust from jets and SST's which spread ice crystals at very high altitudes producing a haze and cloud cover with undetermined, but potentially significant, consequences.[8] The list of possible calamities could be considerably extended. But the simple point is that there is a good probability that the changes in the atmosphere brought about by "multiplying man's" multiplying use of technology are leading the planet to significant climatic changes. This danger may arise in critical form within the next twenty years.

The world has already seen several dramatic illustrations of premeditated large-scale actions with potentially substantial global environmental effects. The fallout from atmospheric atomic tests is one example. Others have included high-altitude U.S. and Soviet nuclear tests and the orbiting by the U.S. of a belt of copper filaments for a military communications experiment. The latter two illustrate the interesting kinds of problems these capabilities raise.[9]

International Implications of Future Technology

In the case of both the U.S. 1962 high-altitude nuclear shots—called Project Starfish—and the copper filament experiment—Project Westford—advance notice of the experiments was given by the U.S. Government. In both cases, extensive analysis was made within the U.S. of the predicted long- and short-term effects of the experiments. In the Westford case, in fact, the government made impressive efforts to publicize the analysis in advance, and to encourage scientists in other countries to make their own analyses. This did not prevent a substantial negative reaction from the world scientific community. The Westford experiment followed the predictions exactly: the filaments have fallen out of orbit as expected.[10] The same cannot be said for Project Starfish. Rather, there were substantial effects that did *not* accord with the predictions: some of the released electrons became trapped in the earth's magnetic field, with some long-lasting effects on scientific experimentation.[11]

Two points must be noted in these examples. One is that the scientific analysis prior to the experiments was not infallible. The other is that the U.S. Government, even though it demonstrated substantial responsibility in allowing prior publication and analysis of security-related experiments, was still prepared to proceed unilaterally in the face of doubts raised by the worldwide scientific community, and in the face of the knowledge that miscalculations would affect the entire globe, not just the U.S.

Several lessons for the future requirements for international machinery emerge from this listing of only a few phenomena associated with man's large-scale tampering with the environment. It is essential that we know more about the processes of the environment to plan for the full range of effects that will follow specific large-scale actions. Machinery to control technology could take many forms, varying from independent national capabilities to some kind of "impartial" international body. Machinery with a genuine international capability would necessitate development of ancillary mechanisms for appeal, adjudication, monitoring, enforcement, and assessment of damages and claims.

It may be necessary to contemplate international responsibility for, or even international operation of, some large-scale technology as a way of guaranteeing equitable distribution of benefits and genuine concern for possible harmful side-effects. The high cost of some projects may also force them into the international arena.

With a rapidly growing population and with rapid increases in industrialization and urbanization, both the needs and the wastes of society grow at exponential rates. A world population expected to

Eugene B. Skolnikoff

grow to nearly 5.0 billion by 1985 from a little over 3.5 billion in 1970 would lead to more than a 40% increase in requirements for food, energy, and natural resources just to maintain the present unsatisfactory economic levels.[12] These requirements will, in fact, be substantially increased by the economic growth levels in all countries and the corresponding increase in industrialization. But to meet these requirements it will be necessary to use massive quantities of fertilizer and insecticide, transport and burn growing quantities of fuel, dispose of more agricultural and industrial waste, transform more agricultural land into houses and highways, cut down more forests, find more fresh water supplies, etc. All this will substantially alter our present environment and add substantially to the environmental pollution problem. It is not a single problem, but an enormously complicated interrelationship. The implications of a change in one aspect cannot be approached adequately without consideration of the total system.

The use of DDT and related organochlorine compounds to control insect pests has been a major factor in making it possible to feed and improve the health of the present world population. But now we realize that the accumulation of these persistent pesticides in the food chain is toxic to some forms of animal life. Many animal species have been endangered, and the background concentration for man has in some cases exceeded presently accepted limits. Perhaps the most disturbing recent development is a report that DDT interferes with photosynthesis of marine phytoplankton, a phenomenon that could have catastrophic effects on the living resources of the sea.[13] Several countries have now banned, or severely curtailed, the use of DDT-related compounds; the U.S. acted in mid-October, 1969. There is controversy as to whether effective substitutes for DDT are available. When they are, the costs are likely to rise. Thus, we quickly come to hard choices between starvation or disease on one hand, and gradual accretion of DDT levels on the other, between economic growth and stagnation. Who is to make these choices? Are there alternatives? Ultimately, who pays? It is instructive that in present debates in the U.N. and other international bodies, the developing countries are understandably much less concerned with pollution problems than the developed countries.

The industrial sulphur effluents of one country are claimed to come down as sulphuric acid in another.[14] The problem can be solved most easily by using sulphur-free fuels, but that would cut off the market for oil from the Middle East and Venezuela. Oil pollution at sea, with attendant serious effects on bird and marine life and on recreation, becomes a grave menace as the size of tankers increases in order to satisfy growing energy-demands of civilization. A single accident can become a disaster; and the exploitation of seabed oil resources, with the likelihood of occasional accidents, is sure to increase sharply in the near future. Whatever measures are taken to reduce these risks, they will, at the least, raise the price of oil—with political consequences.

The general problem of solid waste disposal has astonishing dimensions. In the U.S. alone, the magnitude of solid wastes today is estimated to be 140 million tons of smoke and noxious fumes, 7 million automobiles, 20 million tons of paper, 48,000 million cans, 26,000 million bottles and jars, 3,000 million tons of waste rock and mill tailings, and 50 trillion gallons of hot water along with a variety of other waste products.[15] The waste problem is aggravated by the conscious development of plastic and other containers that are not degradable in the environment.

The expected growth in the use of nuclear power as a major source of energy will greatly aggravate the problem of safe disposal of atomic waste, which in turn may change the economics of the nuclear power industry. Even the relatively low-level release of radioactivity from "normally" operating nuclear power plants may be a severe problem.[16]

Implications of environmental alteration for international functions. Research, analysis, and information about the changing environment are needed rather desperately. The goals are several:

a. Simply to know on a continuing basis what, in fact, is going on

b. To determine the likely effects of present trends, and to establish tolerances

c. To develop alternatives to, or modification of, current practices when necessary

d. To establish hard data on the costs and benefits of alternative courses of action for political decision

This is clearly the most immediate set of requirements, and one that has been recognized now by many international organizations.[17] Developing the information is the (politically) easy part of the job. The implications of the information for further action will prove more difficult.

The primary need is for establishing international norms for effluents, for solid waste disposal, for tanker routing, for actions in the event of ship accidents, etc. Free flow of trade, in fact, will also require uniform standards among nations. Where pollution has more subtle effects, and requires for its amelioration unaccustomed domestic limitations in certain fields, the political problems will be more serious. For example, if limitations are required on the amount of grassland covered over each year (as regarding the earth's albedo), or if limitations must be put on the total use made of specific technologies each year, then we will be seriously affecting areas never before subject to any form of international regulation (or in many countries even national regulation). Limitations on research and development itself may even become a serious political issue if a judgment can be made that the direction new technology may take would seriously exacerbate environmental problems. The prospective technologies of peaceful use of nuclear explosions and weather modification, and the actual technologies of supersonic aircraft might be limited.

Creation of norms implies other functions as well, especially the question of allocation. The costs of setting standards, or of banning the use of certain technologies, will not be evenly distributed. The costs of giving up or replacing some pesticides and herbicides will not only be measured in dollar terms, but also in terms of human life and health. Similarly, the costs of controlling industrial waste will fall unequally on certain nations because of their particular geographical positions, their dependence on certain resources, or their emphasis on certain industrial processes. Who should pay—the producer, the consumer, or the nations most offended by the particular pollutant? How is this to be determined? Moreover, a nation's competitive position in international trade may be substantially affected by the measures it must take. Regulations must, therefore, be international simply to maintain fair competition; universally-applied regulations may damage the competitive position of some countries, thereby raising issues of equity.

The allocation function requires an appeals and adjudicative mechanism. A means of developing technical analyses which are recognized as fair and impartial will be essential. A damage assessment and claim procedure will be necessary to settle violation of the international standards. Finally, a monitoring and enforcement procedure is implied to insure compliance. This could continue to be "passive," as is generally the case now, in the sense that nations tend to conform voluntarily to standards on the basis of a calculation of their own best interests. A more "active" means of enforcement may be necessary if the present pattern is not adequate.

II. The Oceans

The oceans have been the focus of some of man's earliest technological developments, and also his first attempts at codifying interna-

International Implications of Future Technology

tional law. Today, they are the scene of application of the latest technology, and as a result the cause of much re-thinking and re-shaping of international law. Rapidly advancing technology for underwater extraction of organic and mineral resources; improving knowledge of the likely extent of the resources, especially seabed oil; and continued exponential growth in energy demand—all combine to make the political and legal questions associated with ocean resources controversial. Technological developments, spurred by the promise of high return on investment, will make it possible to operate in deeper and deeper underwater environments. With the expectation that the "world's greatest supplies of fossil fuels" lie on the continental rise, one can quickly see both the motivation and potential payoff of technological developments.[18] Other resources of the seabottom are also of potential interest, particularly manganese, nickel, cooper, sulphur, and the "detrital" minerals.

The major international questions, of course, have to do with who owns the resources on or under the seabed; thus, who has rights beyond the continental shelf to their exploitation and to their benefits? Whatever institutional arrangement is established for regulation of the seabed, it would not only have to have a means for deciding who has access to the seabed, but also a procedure for distributing publicly whatever benefits accrue. In addition, a means for monitoring and enforcing decisions, regulations, and standards could well become necessary, whether it was carried out by an international mechanism, or by agreement among national entities.

There are several important aspects to the international issues surrounding living resources of the sea. One is the trend in present fishing practices and technology related to maximizing yields. It is a controversial issue, but some scientists estimate the potential harvest of the sea to be not very much greater than present annual harvests (about 60 million metric tons). The Food and Agriculture Organization (FAO) estimates less than twice the present catch size as the likely maximum that can be expected.[19] In fact, with improved technology and greater fishing effort, *depletion* of fish resources would result unless effective control methods are undertaken. A second aspect is the increasing knowledge of the migratory patterns of fish. This knowledge can create difficulties if it allows a nation to take fish that have been traditionally harvested by other nations at a different stage of the migratory cycle, or if the fish are taken at a stage of the cycle before they breed. A third aspect is the possibility of aquaculture, that is, artificially improving the nutritive value of the sea, and "raising" fish.

Eugene B. Skolnikoff

There are several other oceanic activities of great economic or political importance which will also be the focus of extensive technological development in the near future. Transportation will go in the direction of larger and faster vessels and probably true submersibles. Military applications will tend toward larger and quieter nuclear missile submarines for the deterrent force, more devices for submarine detection, and probably also more anti-submarine forces. With the increasing vulnerability of land weapons, the undersea missile launchers will steadily take on more importance, possibly replacing land-launched missiles entirely. The undersea environment will inevitably be a primary focus for arms control measures, as it already has.

In view of these increased uses, the problem of congestion of the ocean environment arises. Many applications are pertinent to the same areas of the ocean, potentially congregating ships for fishing and transport, submarines, unmanned buoys, and permanent drilling platforms in the same vicinity. Entirely new regulations and rules of the road may have to be developed.

Implications for international functions. Whatever regime is established for the deep seabed, it will have to be concerned with the exploitation of resources beyond the limits of national jurisdiction. Presumably, licensing authority would be involved, which means establishing criteria and making choices. If that regime is given genuine management responsibility for the seabed, or ownership of the resources, then the whole ramification of functions from licensing through norm creation, rule observance, appeal, adjudication, and operation all follow.

For fisheries regulation, one sees the same basic functions: denying benefits and rights to one country in favor of another; the existence of machinery for appeal, monitoring, and enforcement of decisions; and an ability to develop or obtain independent technical information. Similar steps will be necessary to prevent the serious depletion of protein-supplying species endangered by overfishing. With rather less impact, the need to handle the growing congestion resulting from uses of the seas will also require international machinery.

III. Outer Space

The space-launching powers will undoubtedly continue space activities, though with less political urgency.[20] The interest in receiving economic returns from space, and in minimizing the costs by sharing, will undoubtedly grow. Satellite applications that can assist in the planning and management of forestry and agriculture activities are

International Implications of Future Technology

almost certain developments during the time period of interest. Based on the ability to scan large areas with a variety of sensor devices, satellite systems will provide invaluable information for allocating water use, anticipating crop size, and many other land management activities. Such services are likely to become essential once they are adopted as a regular input to a nation's economic system.

Another exciting application of satellites is in the search for mineral and organic resources. It is quite likely that satellites can be used to prospect for minerals, or at least to identify promising locations for ground exploration. Such information may have considerable economic and strategic interest, raising troublesome questions about information control and access, control of the application of this technology, and international operation.

The number of satellites serving as relay communications systems will increase substantially in order to meet the expanding demand for communications channels. The systems themselves are likely to be increasingly sophisticated, with much larger channel capacity, well-defined beam (geographical) coverage, increased power, and other specialized capabilities. Whether there will be a single global system or multiple national systems is not clear at this time. The demand for frequencies will add enormously to the problems of allocation of the frequency spectrum. Direct broadcasting from satellites to augment home or community antennas is under development today. The U.N. Working Group on Direct Broadcast Satellites estimates that direct broadcast into community or augmented home receivers will be technically feasible by 1975.[21] However, the actual development of direct broadcast systems will depend on complex economic, political, and technical criteria.

The increased interest in observing the planet and its environment for research, exploration, exploitation, monitoring, and enforcement purposes will lead to the development of satellite-centered data-gathering and dissemination systems. These will be sophisticated satellites, with their own sensors, and tied to sensors and read-out stations throughout the globe. Such systems will also add to the pressure for frequency and orbital space allocations, and will raise questions of international versus national management, economic efficiency, security of information, joint planning, etc.

Expansion in air transportation, as well as introduction of high-speed marine transport, will require greatly improved navigation aids, a need very much in evidence today in transatlantic air routes. Satellite navigation systems are likely to offer the most attractive answer to such needs. Adequate allocations for frequency and orbital requirements, as well as the question of management, will again be central.

Functional implications of space. These stand out quite clearly, and, in fact, have already resulted in the creation of several new international institutions, most notably INTELSAT, the consortium for communications satellites, and the European cooperative space research and launcher development organizations (ESRO and ELDO). The first functional international requirement will be operating mechanisms for many of the systems mentioned above. In most cases, national ownership and operation will not be adequate as a permanent arrangement for political reasons. Some of the information generated by the space capabilities will have serious economic and even strategic consequences, most notably information on likely new resources, on fish movements, and on weather. Important questions about the ownership of the information and its availability to national or commercial interests will have to be dealt with by international means.

Whole fields of human activity will come to depend on some of the systems once they are in operation—for example, weather forecasting, navigation and air-traffic control, and crop-planning. For reasons of equity and political feasibility, therefore, there will be strong pressure that these be internationally-run systems. Many of the systems may involve differential application costs as well as differential benefits to some portions of the globe. Questions of allocation of benefits and costs, and of pricing policy will, therefore, have to be dealt with by international machinery. Some communication systems may have a profit potential, or at least seriously affect the profit potential of existing land-based systems in the same fields. Complex problems of equity will have to be settled through international action. The procurement of equipment will also raise questions of economic equity. Which countries will receive development and procurement contracts? INTELSAT is plagued with this problem now. The desire to share in the economic and technological benefits of development and production of communications satellites is clearly one of the motivations behind competing proposals.

Interest in spreading the costs of space development and exploration is likely to encourage more cooperative space efforts than are envisioned today, possibly involving international machinery in the process. Direct broadcast satellites will raise thorny problems of technical standards for the satellites themselves, television standards, regulation of broadcast coverage and interference (including jamming), and, of course, frequency and orbit allocation. Some form of international control of broadcast content cannot be excluded. Greatly increased activity in the near-earth environment will, in turn, require a space regime capable of establishing norms and making allocations for various contemplated uses. The existing and potential military uses of space (essentially surveillance, communications, and tracking) will greatly complicate that task.

IV. Natural Resources

The availability of natural resources—mineral, organic, and hydrological—is a perennial source of concern. Whatever the short-term situation, the world has a finite supply of mineral and fossil fuel resources. The world's demand for metals has been growing at a rate of more than 6% per annum for nearly a decade.[22] The entire metal production of the globe before World War II was about equal to what has been consumed since. Some materials are, or soon will be, in short supply: helium, mercury, tin, silver, manganese, chromium, titanium, tantalum, and tungsten.[23] Of these, helium and mercury have unique properties and no satisfactory substitutes are presently known. The nuclear-energy minerals have the same character. Thus, mineral conservation measures may be required and would have to be of international scope to be meaningful. The present situation seems to be that not as much is known as should be about the actual global resource situation.

The assumption that the market mechanism can be relied on to call forth new supplies from lower-grade ores is not valid for those metals in which sharp discontinuities of concentration occur. Lead, zinc, and mercury show this characteristic, as do some of the more common metals outside of their basic ore deposits.[24] The assumption that cheaper energy can make extremely low-grade ores profitable ignores the fact that low-grade ores would require enormous costs in the handling and processing of rock.[25] A persistent and vexing problem is the efficient use of existing resources in the light of future needs rather than of present technology and economics.

Functional implications of resource scarcity. The most immediate necessity is international development of a geochemical census of the earth's crust in order to determine the real resource situation. The uneven distribution of the world's resources, and the uncertain extent of those resources, may together create increasing pressure for an international approach to their management and efficient use. The U.S. and other developed countries are, by far, the largest consumers of raw materials today, and may be preempting the possibility of other countries using those resources for their own future development.

Eugene B. Skolnikoff

This could become a serious political issue producing pressures for machinery to manage, allocate, and regulate the use of resources. In addition, the growing use of resources has a direct relationship to the pollution problem. Regulations growing out of the need for pollution control may well be realized as limitations on national consumption of various kinds of raw materials.

V. Food

The Green Revolution of the late 1960's, which appears at least to have postponed a threatened worldwide famine, is a coincidental product of several factors: timely development of new technology in the form of new seed strains, good weather in critical areas, and the willingness of farmers to plant the new varieties. But this development is only a temporary reprieve, not a long-term solution, as long as population pressures continue to increase. The FAO's "Indicative World Plan for Agricultural Development" (IWP) designed to highlight food-related issues and requirements until 1985, points out that agricultural production on the average must show an annual average increase of between 3.2 and 3.8% as compared to the average 1962 figure of 2.8%, a substantial long-term change in the average.[26] Moreover, this assumes no change in already inadequate levels of food quality and caloric intake. If some increased demand is also postulated, the annual increase required for developing countries will be 3.9%.[27] In short, the danger of a catastrophic worldwide famine is likely to be with us for a long time.

Functional implications of food problems. The international implications largely depend on one's point of view, a different situation from some of the other topics discussed here. It is conceivable that the situation will remain roughly as it is: concern and limited attempts to help, some international machinery devoted wholly to the problem, but in the main a national approach. It is also possible that the food and larger development problems will receive increasing attention from the world community as a whole, and lead to substantially new or expanded functions on the international scene including management and allocation of resources—money, fertilizer, pesticides, water, machinery, etc.—to bring about adequate food production, distribution, and quality. The particular kind of event that could bring about such functional developments most rapidly would be the onset of a major crisis, whether it be widespread famine or some environmental crisis.

International Implications of Future Technology

VI. Population

Population control is without question the single most serious environmental problem the world will have to face in the next 50 years. From the economic, the cultural, and the sociological points of view, it appears essential that population growth be brought under control. But that clear imperative encompasses many problems. One is simply: how? Others involve judgments about optimum sizes of populations, distributions among races and nations, motivations, and other issues of considerable subtlety and sensitivity. It is clear that the finite resources of the earth, including open space, cannot support a continuously expanding population indefinitely. And whatever the optimum size of population may be, the preservation of at least some present Western values would be jeopardized by populations very much larger than those postulated by 1985 or 2000 (in excess of 6 billion). Of course, values may change along with population growth and new technology, as they have today in comparison with the past.

A time may come when enforced birth control is unavoidable. This social control may well be based on international, at least as much as national, considerations. In the immediate future, direct intervention in birth control will probably remain an individual, or at most a national, matter. But the international consequences of unchecked birth rates are likely to create substantial political pressure for restrictions which will have to be universally applied, or at least universally agreed upon. This, in turn, implies some kind of international negotiation and international machinery to help carry out the agreement. There remains the question of how birth rates can be controlled even if there is the personal or political will to do so. Many legal, sociological, technical, and psychological birth control techniques are now in use, most centered around the idea of family planning. The family planning approach is sound, but the decision is left in the hands of the family itself. It is entirely possible that individual preferences may keep the birth rate up too high on the average, even in developed, long-lived societies.

At the same time, the prospects are not good for development of chemical agents that might be adequate to the task, due to legal and political problems at least as much as to technical ones. One definition of adequacy implies a chemical that has long-term effects so that only one dose is required. The most desirable goal would seem to be an agent that produces indefinite sterility, but is reversible on a temporary basis. If state control is ever required, such a technology would be the easiest to adapt. However, present practices for developing and

Eugene B. Skolnikoff

testing new drugs make it most unlikely that any such drug could be developed.[28] The difficulties are that: (1) new drugs are tested by the FDA in the American environment, not the environment of other countries; (2) safety requirements are so stringent that development of new drugs are discouraged on cost grounds, and the prolonged test period impedes influencing the population problem soon enough; and (3) there exists no independent national or international scientific body of appeal to which FDA decisions can be presented and challenged on scientific grounds.[29] To the extent that the necessary technology for population control is yet to be developed—a controversial point since many believe that existing technology used in the family planning approach, coupled with health and development programs will do the job—some new international procedures will be required to change the ground rules for that development.

VII. General Observations

Stanley Hoffmann has observed succinctly that "The vessel of sovereignty is leaking." The consciousness of independence and license for nations to do as they please, mentioned originally, is illusory. Many technological developments today make it a partial truth at best. Self-interest in the use of many technologies makes it mandatory for nations to reach agreements that constrain their freedom of action, for to do otherwise would deny the use of the technology, or bring about various forms of retaliation. The constraints are usually freely entered into, and in that sense can be thought of as self-imposed, but they are nevertheless limitations on sovereignty. What is striking is how far this erosion of sovereignty has already gone. Essentially all of the functions which will be necessary by the 1980's in fact have their counterparts today. Some are rudimentary and *de facto*. But others, even in the politically difficult regulatory area, are surprisingly extensive and effective. The status of today's functionalism is critical in evaluation of the capability of the existing international system to evolve into what will be required in the 1980's. However, even if the entire complex of intergovernmental organizations were performing well today, the increase in the tasks and responsibilities of the system that will be called for in a very few years will not be realized unless conscious steps are taken now to prepare for that time. Moreover, the general performance of international machinery today, notwithstanding its effectiveness in some areas, leads to considerable skepticism that it provides an adequate base for expansion and increase in responsibility without substantial modification.

International Implications of Future Technology

The growing constraints on freedom of national action, and the increased responsibility flowing to international organizations, will mean that the locus of decision-making will increasingly be forced from the national to the international arena. Nations will not necessarily lose their voice in the control of specific issues, but an increasing number of issues will have to be settled in an international environment. It seems quite likely that there will be a tendency for interest groups within nations to look increasingly to the international organizations in their area of concern, accentuating the expansion of international decision-making.

A countervailing trend, however, will be the growth in political and economic saliency of the technological subjects discussed here. Of relatively marginal interest to governments in the past, their implications will force them closer to the center of the political stage. Inevitably, that will also imply greater national sensitivity to any loss of control, and thus increased resistance to any delegation of responsibility to international bodies over which governments exercise only limited control. That could also mean that these subjects will begin to look more like the "high politics" issues of national security and power relationships of today, and become more central concerns of governments.

But these issues are so pervasive in society, and cut across so many of the direct concerns of individuals and interest groups, that governments will never be able to speak with a single voice in these matters. In fact, the problem of integration of policy at the national level, often raised with regard to the effective operation of international organizations, will become even more difficult. It may be that we must look to the international level to provide the necessary integration, rather than primarily to the national level. If so, it is another example of the way in which the focus of decision-making is likely to move toward the international scene.

The nature of the issues emerging from advancing technology and its side-effects emphasizes the connectedness of things. Increasingly, issues can not be neatly divided into boxes labelled "oceans," "agriculture," "health," and so forth; rather, they interact with each other. This is no less true domestically than internationally and will be a major problem in the future, making current questions of jurisdiction and coordination of international organizations pale into simplicity. Additionally, the complexity and size of modern technological and organizational systems make the task of innovation exceedingly difficult.

We can point to what may be the hopeful beginning of increased

Eugene B. Skolnikoff

public interest in some countries in the substantive issues that have been raised in this analysis, particularly environmental control and pollution. It will surely take such public interest, expressed in political activity, to bring about the kind of controls on man and his works, whether national or international, that will be required for survival. Accompanying this increased public interest in protection of man's environment there seems to be a growing recognition that governments do not have the right to act unilaterally in technological areas when the effects may spread beyond national borders. It remains to be seen how this recognition will develop, but it will be a prerequisite for substantial movement in the direction of international decision-making.

There is also, at least in the U.S., a disturbing reaction to technology itself, a reaction which could take on dangerous "Luddite" aspects. To the extent that the solutions to technology-caused problems may lie in technology itself, this anti-technology reaction could seriously inhibit progress toward protecting the future. It may manifest itself first in pressure to keep budgets for science and technology to a minimum and to exercise tight control in those very areas that are essential for understanding and ameliorating the physical and social problems the world faces.

The political environment in which international organizations will have to function will also change. Developments in the nuclear arms race and political tension in general will substantially affect the possibility of meeting the issues raised here. Political developments between the super powers will obviously have a profound influence on the evolution of international machinery. China must be a part of the resolution of these technology-related issues. One can anticipate major conflicts of interest becoming serious sources of tension between North and South.

Whatever the political developments of the next two decades, "inevitable" technological developments will pose major new demands on the international system. Determination of the optimum course of system evolution must depend, in part, on an evaluation of existing international machinery in terms of requirements for the next 20 years. Whether "evolution" will be enough, or whether we will need a "revolution" in the existing international order, is a controversial judgment. Will governments and organizations recognize the extent of the problem in time, even if moderate change in the system would be adequate? Clearly the need for understanding the issues involved is urgent.

International Implications of Future Technology

Notes

1. *Man's Impact on the Global Environment,* Report of The Study of Critical Environmental Problems (SCEP) (Cambridge, Mass.: M.I.T. Press, 1970).

2. T. F. Malone, "Current Developments in the Atmospheric Sciences and Some of Their Implications for Foreign Policy," in *The Potential Impact of Science and Technology on Future U.S. Foreign Policy,* Papers presented at a Joint Meeting of the Policy Planning Council, Department of State, and a Special Panel of the Committee on Science and Public Policy, National Academy of Sciences, June 16–17, 1968, at Washington, D.C., pp. 82–97.

3. "Advances in the Eye of a Storm," *New Scientist,* 44, no. 681 (Dec. 25, 1969), 630.

4. P. M. Borisov, "Can We Control the Arctic Climate," *Bulletin of the Atomic Scientists,* 25, no. 3 (March 1969), 43–48.

5. Malone, p. 87.

6. Gordon J. F. MacDonald, "The Modification of Planet Earth by Man," *Technology Review,* 72, no. 1 (October-November 1969), 27–35. The information that follows is drawn from this article, except where noted.

7. Reid A. Bryson and Wayne M. Wendland, "Climatic Effects of Atmospheric Pollution," presented at the American Association for the Advancement of Science National Meeting, December 27, 1968 (mimeo), quote some of this evidence.

8. Bryson and Wendland.

9. See Eugene B. Skolnikoff, *Science, Technology and American Foreign Policy* (Cambridge: M.I.T. Press, 1967), pp. 84–92, for an extended discussion of the communications experiment.

10. I. I. Shapiro, "Last of the Westford Dipoles," *Science* 154, no. 3755 (December 16, 1966), 1445–1448.

11. Bernard Lovell, "The Pollution of Space," *Bulletin of the Atomic Scientists,* 24, no. 10 (December 1968), 42–45.

12. United Nations Economic and Social Council, "World Population Situation," Note by the Secretary-General, doc. E/CN.9/231, Sept. 23, 1969 (Population Commission, 15th Session, November 3–14, 1969).

13. Paul R. and Anne H. Ehrlich, "The Food-From-The-Sea Myth," *Saturday Review,* April 4, 1970, pp. 53–65.

14. Kenneth Mellanby, "Can Britain Afford to Be Clean?," *New Scientist,* 43, no. 668, (Sept. 25, 1969), 648–650.

15. United Nations Economic and Social Council, 47th Session, "Problems of the Human Environment," Report of the Secretary-General, doc. E/4667, May 26, 1969, p. 5.

16. Barry Commoner, "Attitudes Toward the Environment: A Nearly Fatal Illusion," address for presentation at Unanticipated Environmental Hazards Resulting from Technological Intrusions Symposium, Annual Meeting of the AAAS, Dallas, Texas, Dec. 28, 1968 (mimeo.).

17. "Problems of the Human Environment," Report of the Secretary-General, *op. cit.,* which lays out the general objectives of the U.N. Conference on the Human Environment to be held in Sweden in 1972.

18. P. M. Fye, A. E. Maxwell, K. O. Emery, and B. O. Ketchum, "Ocean Science and Marine Resource," in Edmund A. Gullion, ed., *Uses of the Seas,*

Eugene B. Skolnikoff

The American Assembly, Columbia University (Englewood Cliffs, N.J.: Prentice-Hall, 1968), pp. 17–68.

19. Food and Agriculture Organization, Committee on Fisheries, Fifth Session, Rome, April 9–15, 1970, "Fishery Aspects of the Indicative World Plan and Proposed Follow-up," doc. COFI, 70, 3, Jan. 26, 1970.

20. The information for this section comes from many sources, but in particular the report of the U.S. National Academy of Sciences, National Research Council, "Useful Applications of Earth-Oriented Satellites," Washington, D. C., 1969, and "Selected Space Goals and Objectives and Their Relation to National Goals," Battelle Memorial Institute Report no. EM1-NLVP-TR-69-2 to NASA, July 15, 1969.

21. United Nations, Committee on the Peaceful Uses of Outer Space, Report of the Working Group on Direct Broadcast Satellites, doc. A/Ac.105/51, Feb. 26, 1969.

22. T. S. Lovering, "Mineral Resources from the Emerged Lands," in *Potential Impact, op. cit.,* p. 43.

23. Preston E. Cloud, Jr., "Approach to Assessment of the World's Mineral Resources," in *Potential Impact, op. cit.,* p. 30.

24. Lovering, *op. cit.,* p. 39.

25. *Ibid.*

26. G. Cheld, "Famine or Sanity," *New Scientist,* 44, no. 672, (Oct. 23, 1969), 178–181.

27. *Ibid.*

28. Carl Djerassi, "Prognosis for the Development of New Chemical Birth Control Agents," *Science,* 166, no. 3904 (Oct. 24, 1969), 468–473.

29. *Ibid.*

PART II

Science, Technology, and American Government

E. The Federal Executive

The executive branch of government is involved with a broad range of technological issues from presidential policy setting to detailed agency development work. We begin with an overview of the role of technology in executive affairs that puts forth issues and potential actions. Next we consider the scientific advisory system, which serves all levels of the executive. Finally attention is directed to one of the key nerve centers in the management of technology, the Office of Management and Budget (OMB), and how it operates.

The position paper of the American Association for the Advancement of Science (AAAS) presented in Chapter 11 begins with a statement of the essential technology-intensive issues. It notes the increased governmental emphasis on the application of technology to civilian problems today in contrast to the 1950s and early 1960s. Many of the problems addressed are long-range and not amenable to quick solutions. Appropriate executive branch responses to these issues are addressed, and specific roles are defined for science-technology advocacy, advice, and management. AAAS suggests that scientific and technological considerations need to be integrated into the policy-making process in much the same way as the economic, legislative, and political factors.

In Chapter 12, Martin L. Perl outlines the nature and extent of scientific advice in the federal government. He evaluates the advisory system as quite effective in providing useful information and recommendations on limited technical questions, but not on broad technical policies because of its reliance on the scientific establishment, confidentiality provisions, and the prestige of participation, which tends to socialize the advisers into conformity. Furthermore, Perl notes, the system as constituted hinders more open, comprehensive input from the scientific and technological community to the policy process.

In Chapter 13, John Walsh and Barbara Culliton portray the way OMB operates. This executive office agency plays a highly influential role in the federal executive. Understanding how it deals with technological issues is critically important. The authors point out that OMB's treatment of these matters has changed over time and has been influenced by the administration personalities. Technical issues in defense are typically treated more respectfully than those in biomedicine (although the latter had their day too). The budget is the key tool in pursuing policies, and the relative role of the president, OMB, and the agencies varies greatly with respect to it.

11. Organization for Science and Technology in the Executive Branch

AMERICAN ASSOCIATION FOR
THE ADVANCEMENT OF SCIENCE

In the last two years, since Presidential Reorganization Plan No. 1 of 1973, much has been said about the science advisory needs and arrangements of the White House or the Executive Branch (1).

This discussion has been largely preoccupied with organization and structure, and too little has been said about the requirements of complete presidential staff work, anticipatory planning, and the formulation of public policies toward science and technology.

Scientific advice in the White House must be viewed primarily in a substantive rather than an organizational context. A disembodied decision on structure is unlikely to serve well either the needs of science policy formulation or the quality of scientific research and development. Little would be gained by a political gesture restoring a seat at the table to the scientific community. It is a much deeper matter, which goes to the effectiveness of the national policy machinery in the decade ahead.

An examination of federal policy-making relative to science and technology must begin with the recognition that:

—National goals of every description depend significantly, although in differing degrees, on scientific and technological progress.

—The federal government's approach to R & D has been more tactical than strategic. (The most notable exception to this rule has been the National Science Foundation, which has consistently emphasized the long-term aspects of science.)

—Reliance on a tactical, crisis-type of scientific and technical response is wasteful and disruptive of human and material resources.

—Within the federal government, the financing of R & D is not yet viewed as investment, but only as discretionary expenditure.

Copyright © 1975 by the American Association for the Advancement of Science. Reprinted from *Science*, Vol. 187, March 7, 1975, pp. 810–814, by permission.

AAAS

—The institutional interactions among economic planning, international policy-making, national security planning, domestic social objectives, and science and technology are ad hoc rather than systematic, with predictable malfunctions of policies and outcomes.

Here, then, is the issue: *As we assess the position of the United States at home and in the world against the emerging concerns of the 1970's and 1980's, is there a clear need and opportunity to improve our national policy machinery by strengthening the role of science and technology in defining and meeting national goals? If so, how can we do this most effectively?*

The Array of Policy Problems

Science and technology policy in the mid-1970's operates in an environment which is different from that of the 1950's and 1960's. During the late 1950's to mid-1960's, the White House apparatus for science and technology policy evaluation and advice was largely directed toward space and military matters, and concerned with strengthening academic science and its infrastructure. The agenda is now tilted strongly toward consumer and public-oriented technologies in, for example, energy, transportation, health, education, natural resources, ecology and environment, and social systems.

This shift brings with it a powerful new set of issues, which must be addressed by the methods and results of the natural and the human sciences together with economics, law, politics, and public opinion research and analysis. Answers to questions of critical importance, such as, "Can federally supported R & D aimed at public technologies be responsive to marketplace needs and consumer preferences?" involve a frame of analysis very different from that used in military and space programs, where the federal government is both the producer and the consumer of the R & D work.

It is clear that most of the driving issues of the last half of the 1970's and into the 1980's will be related to major social problems of which scientific knowledge and technological development are pervasive and critical aspects. For the remainder of this decade, and into the 1980's and beyond, policy-makers will confront a formidable and changing array of problems, some examples of which are given below.

Social and economic problems
  —Food requirements and supply
  —Widening disparities in living standards within different nations
  —Access to efficient, humane, and cost-effective health services
  —Development strategies for have-not nations

Science and Technology in the Executive Branch

— Maintenance of labor productivity and price-wage stability
— Creation of jobs for an expanding and educated labor force
— Rationalization of all, and especially higher, education
— Aging, changing population distribution, and humane and effective social services programs
— Ethical and social impacts of applications of biomedical engineering
— Transformation of human settlements
— Impacts of population and resource imbalances on political and other institutional adaptations
— Economic, political, and social implications of changes in the global climate, in the context of population growth and resource constraints

Environmental, technological, and institutional problems
— Land use planning and management
— Environmental regulation
— Air and water management and regulation
— Oceans policies
— Consumer product safety and performance

Technological innovation
— Energy supply and utilization
— Reduced dependence on foreign imports of basic materials
— Technology and balance of trade

National security
— The limits of détente relative to safeguarding national security and scientific and technological leadership
— Weapons assessment and arms control
— Nuclear proliferation
— International implications of technological changes

The national security issues in policy-making obviously require independent and critical staff work. Choices among military hardware systems aside, the real dimensions of "national security" have assumed a wholly new scale and character. They concern the uses of the sea, the environment, and the resources of the planet. They address the equities of resource allocation among developed and developing societies. They confront choices as to population stabilization, and the uses of science and technology in creating alternative social and economic structures which can help to reduce dissatisfactions leading to conflict. These are presidential issues, and our national security is tied to them. For these reasons, we believe that any realignment of the science advisory process must provide for direct involvement, within that process, in the staff work on the *trans*national problems on which global security will depend in the future.

This indicative enumeration of policy problems comprises a formidable menu for policy-making. Many are long lead time problems rather than quick response issues. To deal with them, policy analysis cannot rely on ad hoc improvisation. The national policy machinery must be equal to them.

While science and technology alone will not have all the answers, they certainly will have important roles in illuminating questions of choice, feasibility, and alternatives. For that reason, we believe that the national policy machinery must have an effective science and technology component which functions in concert with other policy support staffs.

The emphasis, accordingly, is on ensuring complete staff work for policy analysis and decision-making. Of course, the mere existence of policy staffs does not guarantee results or head off bad choices, and it certainly does not substitute for political accountability. Staff work may be used well or badly. Its quality may be excellent or third-rate. If it is to count, policy-makers must demand and obtain the full measure of its capabilities.

The Structural Problem

Within the context of the preceding assessment of the federal government's needs relative to science and technology, we see three distinct but complementary Executive Branch staff support roles:

1. The science and technology policy advice role
2. The R & D management and coordination role
3. The science and engineering advocacy role

We believe that the first two functions can best be performed within the Executive Office of the President, and the third, while very much needed, is best kept out of the White House or the Executive Office and quite distinguishable from the others.

The science and technology policy advice role. There are two dimensions of science policy advice. The first involves frequent inputs to the traffic of *short-term* policy-making. The second involves *strategic planning* for the contribution of science and technology to national goals and objectives. Together, they require involvement in dealing with the problems of allocation of resources for science and technology, determining priorities among multi-agency programs, evaluating the quality of agency R & D programs, and fostering long-range planning for technology assessment, health of the scientific community, and other matters bearing on science and technology.

Short-term policy-making should focus on (i) budget allocations, (ii) evaluation of proposed legislation, and (iii) major program decisions or choices, such as the amount and distribution of R & D for energy, appropriate levels of expenditure, and best mix of civilian and defense R & D, or the treatment of multinational corporate R & D expenditures under proposed tax rules of the Internal Revenue Service. These are critical functions which require timely and informed science policy advice. If the policy advice role does not involve participation in them, it will have no clout or impact, and will be merely ad hoc.

The second dimension is distinct from the short-term policy-making discussed above. It is addressed to a long-standing gap in our national policy machinery.

The strategic planning dimension requires deliberate attempts to develop assessments of the quality and productivity of science and technology and to develop long-range goals for them in relation to the position and needs of the United States at home and in the world. The importance of this role is obvious if science and technology are to be approached in investment terms rather than simply as year-to-year work programs. Establishment of this role implies that government recognizes the character of the discovery process, accepts its long lead times, and means to create multiyear perspectives which will help to define and forecast the policy environment within which science and technology can be carried on.

An important dividend which should emerge from the planning and assessment roles is an annual guidance statement of the Office of Management and Budget (OMB) defining the ranges of new budget authority and outlays for the federal R & D effort, to be used as planning benchmarks in the preparation of the Executive budget. These guidelines should be consistent with the medium- and long-term economic, social, and international policy objectives of the Administration, and they should reflect the context of the real world in which budgetary choices have to be made. The development of the annual guidance statement does not preempt the role of the OMB in rationing resources among rival needs, but instead provides a rational framework which can help to extricate budgeting for science and technology from the constraints of incrementalism and inequities of "crash" R & D funding.

Whether both of these activities—short-term staff support and longer-range strategic planning—can be handled effectively by the same group in the Executive Office is arguable. However, it should be tried, because the realism of strategic policy planning will be fortified by immersion of the advisory staff in current decision-making with OMB and agency heads on the touchy issues which confront political executives from one day to the next. The danger to be guarded against is that long-range policy planning may be driven out by demands for quick response staff work for the White House. This problem can be met, in part, by using the National Science Foundation as a major support arm for science policy studies.

Both of these advisory functions must be situated in the Executive Office to have the necessary ad hoc policy input and the leverage required to set long-term science and technology goals. They must be accepted and tied into the delivery of staff work, and they must be headed by presidential appointees. Finally, they must be accessible to the outside world, where science and technology are largely initiated and performed, and not screened from "real life." The science and technology policy advisory staff would establish working relationships with the Domestic Council, the National Security Council, the Office of Management and Budget, the Council on Environmental Quality, and the operating departments and agencies.

It is not necessary to dramatize the personal involvement of the President as the chief client of the science advisory staff. There will be times when a President will need and want advice. He has to be the judge of this. The main objective, however, is to deliver sound, timely, and informed advice to the centers of the national policy machinery in and around the Executive Office—to make it an arm of complete staff work, so that facts, judgments, arguments, and alternatives work their way up the line. This is the best way to help the President. In organizational terms, this means that the restoration of a science advisory system need not necessarily require a personal science adviser to the President; it may be enough to provide the staff capability to work on even terms with the White House and Executive Office staffs and the heads of agencies.

Finally, a degree of institutional tension is one of the risks that a staff activity must run. Science advice includes the responsibility to criticize or oppose policy trends, *within the White House staff system,* when the grounds for doing so are within the competence of science and technology.

The R & D management and coordination role. While the administration of federal R & D programs must remain the responsibility of the mission-oriented departments and agencies, there is great need for "cross-cutting" coordination and oversight. This problem has never been handled well in the past. Interagency committees are poorly suited to the task. The OMB is necessarily concerned with issues of program content, cost-effectiveness, and dollar cost and, while some measures of oversight can be exercised through budgetary reviews, the process is selective, targeted, and may lack a balanced perspective.

What is needed, we believe, is assurance as to the priority, quality, balance, and end utilization of R & D. Crash programs, especially, call for objective evaluation and quality assurances, as in the case of cancer research and energy R & D. This evaluation requirement is particularly needed when these programs are supported by high and growing budgets while other fields of science and technology are relatively constrained by a lower support level. To make this kind of assessment, it will be necessary to reach out for help from organizations and individuals whose insights, skills, and experience will inject freshness and objectivity to the evaluation, including groups which can communicate the values and preferences of a diverse society.

Serious and unresolved questions exist as well with regard to the efficiency with which the nation's R & D capabilities are being employed; examples are failures to define R & D objectives in concert with industrial and other users of the results, disarray in the arrangements for handling scientific and technical information, contradictory practices among federal agencies with regard to patent rights, policy barriers to joint or cooperative R & D by industry, the absence of system-wide oversight of valuable federal laboratories and research centers, and costly practices in competitive proposal solicitation. Priorities for R & D will emerge from agencies' missions and roles, but good science policy requires these priorities to be reviewed and coordinated to make them realistic in terms of feasibility, manpower requirements, timing of expected results, available funds, and well-defined objectives.

Equally important is the need to manage the federal government's discordant impacts on technological vitality in the United States. The attitude prevails at all levels of government that technology is the result of market forces and the decisions made in the private sector. What is not recognized adequately is that the government's policies and activities have a tremendous influence on the rate of technological investment, innovation, and risk-taking. Nor is it clear that government understands the importance of lively technology in maintaining a positive international trade balance, in improving productivity, and in generating jobs.

Yet, the federal government influences the rate of technological enterprise in ways that are critical: through its regulatory and standard-setting activities (which need a sound scientific base), through its massive procurement operations, through its R & D expenditures, through its tax policies, through its trade and monetary policies, through its economic policies, through its personnel policies, and through its educational and research policies. The aggregate effects of these disparate interactions on the directions and the scale of technological enterprise are unseen but great. No focus now exists in the public policy structure for coordinating policies and decisions relative to technological thrust. No analytic focus exists for considering the impacts of changing policies or regulatory actions on technological risk-taking, or for evaluating the impact of government on the marketplace in which decisions that affect technological risk-taking must be made.

An Executive Office focus is needed to deal systematically with these problems of management policy, to carry out special projects and R & D management audits, and to deal with issues which today go by default. The National Science Foundation has tried heroically to address most of these needs, but its resources are inadequate and its position in the Executive Branch bureaucracy too low to be fully effective.

Managerial oversight of any multibillion-dollar diversified enterprise, including federal R & D, is no part-time affair. Federal agencies need a great deal of help to improve their programs. Large industrial corporations with substantial R & D activities have developed management systems which get the job done without strangling R & D initiatives or the incentives for research creativity and technical innovation. Some "technology management transfer" of this kind from industry to government would not be out of place.

This role is distinct from the advocacy and advisory roles, and should be kept distinct.

The science and engineering advocacy role. Because scientific research is long-range in nature—it is a discovery process whose benefits and costs must be inferred rather than quantified—its claims on resource allocation are often difficult to establish. This difficulty means that scientific research—especially basic research in the physical, biological, and social sciences—cannot compete for support on equal terms with the short-run operational responsibilities of government agencies. Budget levels for these short-run operations are resolved by bargaining and level-of-effort compromises. Year-to-year changes in budget policy erode the continuity of research and induce chronic

Science and Technology in the Executive Branch

uncertainty, while inflation forces up the costs of research manpower and laboratory investigation.

While science, as a claimant for federal support, cannot be exempted from the "ends-means squeeze," neither can it be expected to maintain its vitality under conditions of open-ended uncertainty. As a comparatively weak claimant on limited resources, science needs to have responsible champions to help its case to be heard, to identify and argue for pursuit of emerging opportunities, and to press for the maintenance of a lively and productive scientific enterprise. What must be guarded against is the creation of a special interest lobby for science, or what might be perceived as a lobby. The advocacy role within government must not be a partisan or special interest one: it should be selective and well supported with analytic assessments of the nation's research enterprise, judgments on the balance of effort among fields of research, and evaluation of the scientific and social merits of new opportunities in science. This is the kind of information which should be brought effectively to the attention of the Executive and Legislative branches and the general public.

These roles, we believe, should be carried out within government primarily, but not exclusively, by the National Science Foundation, working with the National Academy of Sciences, the National Academy of Engineering, and the Institute of Medicine. However, ours is a broad and pluralistic society, and openings must be made for the advocacy of views from many groups and organizations which reflect crosscurrents of change and the articulation of emergent needs. The strength of our national science and engineering endeavors will be enhanced if strongly held views on health research, applied social science, environmental science needs, basic research, and technological innovation and development, to name but a few, can find a ready, but critical and evaluative hearing within the federal government.

Furthermore, there is need for an Executive Office annual report on science, technology, and national policy addressed to the Congress. Such a report, prepared by elements within the Executive Office, but with inputs from several parts of the Executive Branch, should assess the health of science and technology applicable to the needs of the United States. Such a report could provide a degree of guidance to the scientific and engineering communities and be the focus for a continuing appraisal by the National Science Board and the director of the National Science Foundation, as well as the Academies, and such broadly based organizations as the American Association for the

AAAS

Advancement of Science, of critical questions affecting science and technology and requiring governmental attention.

Recommendations

The essential point is that the Administration must decide its posture toward the function of science and technology in the total national policy picture. Our view of the future persuades us that the country's goals and objectives are linked closely to science and technology, and that the arrangements for policy analysis, planning, resource allocation, and management should reflect that linkage.

The aim is not to aggrandize the image of science and technology, but to improve the quality and performance of public policy. If the Administration shares this view, we assume that it will take steps to put new machinery in place. If the Administration believes that our view is overstated, it should not adopt recommendations just because we make them. Otherwise, the disorder will only be compounded.

In summary, the AAAS Board of Directors believes that the needs of public policy-making require enhanced staffing and better organizational structures than currently exist. Science and technology policy considerations need to be integrated in a regular and recognized manner into the decision-making process in much the same way as economic, legal, political, and social factors.

Thus, it seems clear that a variety of new arrangements is called for and not just a simple reversion to an earlier model or the creation of a single new body of which too much is expected. The focus of the examination of federal science and technology must not be on transient questions of organization. What matters is what we are doing with and to science and technology, and how they can best help to define the direction and quality of our public policies.

In the context of this discussion of needs and functions, the Board believes that some of the suggestions already made for organizational arrangements have genuine merit. The Board would welcome an organizational initiative comprising the following elements:

1. A Council of Science and Technology Advisers in the Executive Office, headed by a strong chairperson, to provide continuing staff advice on scientific and technical aspects of domestic and foreign policy-making together with long-range policy research, planning, and public investment for the uses of the nation's scientific and technological resources in achieving major goals and objectives. We believe that a council composed of knowledgeable individuals is a workable mechanism for accomplishing the planning, policy goal-

Science and Technology in the Executive Branch

setting, and assessment and monitoring functions essential to effective staff work. At the discretion of the President, the head of the council could also serve as science adviser to the President.

An alternative to a council would be a single presidential appointee, assisted by a carefully chosen staff. This alternative would be appropriate in circumstances where a President might find a council unwieldy and slow-moving, and would prefer a simpler arrangement.

To ensure a strong and in-depth capability for planning and assessment to support policy-making, the Executive Office elements should be able to look to the National Science Foundation to mount and carry out a substantial level of science policy research, analysis, and reporting.

What matters, to repeat, is not so much the organizational mechanics but rather the explicit provision for lively and complete presidential staff work—staff work which captures and gives weight to scientific and technical considerations in the examination and choice of policy alternatives and program strategies. This is what the current issue is primarily about. The purpose of the organizational decision is to focus—visibly, clearly, and effectively—the initiative and accountability for delivering such staff work with continuity and impact. The organizational answers should match the demands of the assignment, and should be seen as doing so.

2. An Office of Research and Development Management with the responsibility to evaluate programs, set priorities, provide quality assurance, see to policy coordination, and stimulate new initiatives. This office can be either a separate unit in the Executive Offices or an element in the OMB headed by a presidential appointee.

3. Principal reliance on the National Science Board and the director of the National Science Foundation, working closely with other federal scientific and technical agencies, for assessments of the nation's needs and opportunities for the advancement of science and education for science and engineering. Effective outreach should be maintained with the National Academy of Sciences and the National Research Council, as well as with scientific, professional, and public interest groups.

Closing Comments

Organizational inventions tend to lose vitality over time, and to become preoccupied with problems of the past rather than the future. Organizational lag is one of the afflictions of bureaucratic life. We believe that our suggestions are appropriate for as far ahead as we can look, but we strongly recommend that future administrations keep an open mind and open options as to the character and appropriateness of any set of science policy and managerial institutions. Events may call for different arrangements, and the national policy machinery must have the ability to recognize the need for change and revitalization.

Notes

1. A selected few of the large number of articles and statements which have appeared on the subject of science advice for the Executive Branch over the last two years are: U.S. House of Representatives, First and Second Hearings before the Committee on Science and Astronautics, *Federal Policy, Plans, and Organization for Science and Technology* (93rd Congress, 1st session, July 1973, and 2nd session, June and July 1974) (see also Interim Staff Report of the Committee bearing the same descriptive title and issued May 1974); Ad Hoc Committee on Science and Technology, *Science and Technology in Presidential Policy-making—A Proposal* (Washington, D.C.: National Academy of Sciences, 1974); G. B. Kistiakowsky, *Science*, 184 (1974), 38; D. W. Bronk, *Science*, 186 (1974), 116; E. B. Skolnikoff and H. Brooks, *Science*, 187 (1975), 35.

12. The Scientific Advisory System: Some Observations

MARTIN L. PERL

Since World War II, scientists and engineers have been going to Washington in increasing numbers to help the government make decisions on technical questions. These questions concern every aspect of our technological society—nuclear weapons, missiles, space travel, cancer research, pesticides, and mental health. Some scientists and engineers go for 1 or 2 days a month; others take a leave of absence from their institutions or corporations and spend several years in Washington. Some serve on committees attached to the executive branch of the government; others serve through semigovernmental institutions like the National Academy of Sciences. A few work with

Copyright © 1971 by the American Association for the Advancement of Science. Reprinted from *Science*, Vol. 173, September 24, 1971, pp. 1211–1215, by permission.

The Scientific Advisory System: Some Observations

the Congress. All of these scientists and engineers, the committees they serve on, and the positions they hold in Washington together constitute the scientific advisory system (1). This article is about that system, or more precisely, about a paradox connected with that system.

The paradox is easily presented. Most people will agree that the United States is besieged with perilous technological problems—how to stop the arms race and bring about nuclear disarmament, how to stop the technological destruction of the natural environment, how to raise the standard of living, or at least prevent mass starvation, in the poor countries. Most people will also agree that these problems have become much more severe in the last two decades. But in these same two decades, the United States has received enormous amounts of scientific and technical information and advice from the scientists and engineers of this country. This information is almost always technically correct and thorough; it is almost always given with the intention of solving or mitigating the problems sketched above. The paradox is simply this: How have we gotten into so much technological trouble while getting so much well-intentioned and correct technological advice?

A broad analysis of this paradox might require a study of the relationship between the scientific advisory system and the "technostructures" postulated by Galbraith (2). Or one might examine whether the advisory system is an example of the "techniques" that Jacques Ellul (3) believes are the essence of our technological society. However, I restrict my analysis to a discussion of the role played by the advisory system in the technical decision-making processes in Washington. In addition, I do not attempt to present a complete description and evaluation of the scientific advisory system, nor do I discuss the role of the scientific advisory system in the larger decisions on military technology.

Few people realize the size and complexity of the scientific advisory system, and I know of no complete study of the magnitude and structure of this system. Therefore, I refer here to a recent, but not exhaustive, study (4) that was carried out by a group of Stanford graduates and undergraduates, for whom I was faculty adviser. The study notes that the Executive Office of the President has advisory committees that involve several hundred prominent scientists and engineers. The best known of these committees is the President's Science Advisory Committee. Outside the Executive Office of the President, but inside the executive branch of the government, is a much larger advisory apparatus. This apparatus consists of thousands of scientists and engineers who serve on hundreds of committees, as well as in various temporary positions. Primarily, they advise the Department of Defense and other departments concerned with scientific, technical, or medical questions.

Semipublic institutions also provide a great deal of advice to the executive branch. For example, the National Academy of Sciences and the National Academy of Engineering, through the National Research Council, supervise the work of about 500 committees involving 7000 engineers and scientists. Other large sources of advice are the "think tanks." The Rand Corporation advises the Air Force, the Research Analysis Corporation advises the Army, the Center for Naval Analysis advises the Navy, and the Institute for Defense Analysis advises the entire Department of Defense. Taken together, these public and semipublic advisory groups involve more than 15,000 or 20,000 individual scientists and engineers.

On the other hand, very little scientific advice is given to Congress. Some technical information and advice is obtained through panels or committees attached to congressional legislative committees, and a few individual congressmen, particularly senators, receive some unofficial advice and information. Finally, the Science Policy Research Division and the Environmental Policy Division of the Legislative Reference Service provide reports and summaries on technical questions. But the total amount of scientific and technical information and advice given to Congress is very small compared to that given to the executive branch.

The Scientific Establishment

There is a large overlap between the scientists who lead the advisory system and the scientists who belong to what has been called by a sympathetic observer (5) the "scientific establishment." The scientific establishment comprises most of the prominent scientists and research engineers in the United States. Many of these individuals are deeply involved in science administration and in the making of science policy, both public and private. But usually their prominence has been attained through research rather than through administration or teaching. The scientific establishment has five functions or attributes.

1. Many members of the establishment are the heads of professional societies, the heads of university or industrial laboratories, and the chairmen of university science departments. Many are or have been university deans and presidents. Thus, the members of the establish-

The Scientific Advisory System: Some Observations

ment tend to be the administrators of the worlds of scientific and engineering research and education.

2. Members of the establishment represent their professions, institutions, and organizations before the federal government in requesting funds for research and education.

3. In the eyes of the press and the public, the establishment represents science and advanced technology. It is the members of the establishment who are most often interviewed and quoted. This comes about in part from their accomplishments and in part from their administrative positions.

4. Members of the establishment are the models for young scientists and engineers interested in research.

5. The establishment tends to guide the directions that research takes. This promotes the classification of a research subject as fashionable or unfashionable. This is a useful function in that it encourages researchers to leave unproductive fields, but it can also create difficulties for iconoclasts.

The scientific establishment is by no means a closed or fixed group. Not all eminent scientists and engineers are in the group, and individuals move in and out of the group as their attitudes and interests change. It should also be recognized that the establishment is not always united on issues—particularly on the allocations of funds for research.

Evaluating the Advisory System

I am mainly concerned with evaluating what I call the specific effectiveness of the scientific advisory system. Specific effectiveness is the measure of how well the system carries out its specific functions in the government. As I have already indicated, these functions are set almost entirely by the executive branch and are carried out almost entirely for the executive branch. One specific function is the gathering of information and the presentation of recommendations on limited, purely technical problems. Thus, an advisory committee might be instructed to determine if a newly discovered physical phenomenon could be used to detect submarines. Another specific function is an advisory committee's being asked to recommend a general governmental policy on a technical issue, for example pesticides.

I am also concerned with the general effectiveness of the scientific advisory system. By general effectiveness I mean the total and overall effectiveness of the advisory system in relation to the general processes of making technical decisions. In this country, technical decisions, like other governmental policy decisions, are arrived at through a complicated process. Formally, the process involves the executive branch and the Congress, but in reality much more is involved. Before a decision is made, the question may be argued in the press and by the public. The question may become an important issue in political campaigns for elective office. State and local governments may become involved and take the lead in making a decision, or they may impede a decision. Often the crucial decision is made in the courts, and only later does Congress extend it in the form of legislation. This is by no means a linear process, and most issues have to pass through it several times before they are resolved. This totality, then, comprises the processes by which decisions, including technical decisions, are made in this country. By examining the relationships of the scientific advisory system to these processes, one can determine the general effectiveness of the advisory system.

An evaluation of the scientific advisory system is greatly impeded by the confidentiality of the advising process. The advice given to a government official or to a governmental agency is almost always received under the condition that it may be kept confidential by the official or agency. That is, the advice need not be released to the press, to the public, to Congress, or even to other parts of the executive branch. Large numbers of advisory reports are made public; but, unfortunately, it is just those reports which concern the most controversial and the most important technical questions that are often never made public, or only after long delay. This is unfortunate, not only for those who wish to study the advisory system, but, more important, for the process of making technical decisions in a democracy.

The largest portion of the work of the scientific advisory system is devoted to limited technical questions. "How does method A for water desalination compare in energy requirements to method B?" "How does missile guidance system A compare in reliability to missile guidance system B?" It is with these limited technical questions that the advisory system is most successful. This success results from the competency of the advisers and from the great amount of effort that is applied to these problems. Thus, the advisory system ranks high in specific effectiveness, with respect to limited technical questions.

But suppose the questions are not limited and are not purely technical. Suppose that another specific function of the advisory system, the recommendation of general technical policies, is involved. Or suppose that the technical decision has public policy, economic, or

The Scientific Advisory System: Some Observations

ideological implications. Such questions I shall call broad technical questions. These broad, technical questions severely test the specific effectiveness of the scientific advisory system.

Environmental Questions

The Stanford Workshop (4) studied six broad, technical questions related to the environment and public health: the supersonic transport (SST), cyclamates, the safety of commercial nuclear power plants, the safety of underground nuclear tests, pesticide regulation, and herbicide use in Vietnam. On broad technical questions, the work of the advisory committees may be divided into three parts. First, the committee studies the technical and scientific aspects of the question. Here, as in limited technical questions, the committee generally exhibits high effectiveness.

The second part of the committee's work is usually the development of a program for further study and research. In this, the local effectiveness of the advisory system seems to be reasonable but not high. For example, the 1963 report of the President's Science Committee, entitled *Use of Pesticides* (6), recommended an extensive research program to study the safety of pesticides. Many of those research recommendations appear to have been carried out. On the other hand, the government rejected an advisory committee recommendation that additional study be devoted to the safety of some types of commercial nuclear reactors before those reactors were licensed for use (4).

The third part of the advisory committee's work on broad technical issues usually involves recommendations that certain technical policies be adopted by the executive branch. *Use of Pesticides* recommended that there be an "orderly reduction in the use of persistent pesticides" and that, as a "first step," the government "restrict wide-scale use of persistent pesticides [such as DDT] except for the necessary control of disease vectors." With respect to such policy recommendations, which I call action recommendations, the effectiveness of the advisory system is low. The executive branch will usually ignore the policy recommendation of the advisory committee if (i) the recommendation is contrary to existing policies of the executive branch, (ii) the adoption of the policy would expose the Administration to congressional or electoral difficulties, or (iii) there are strong pressures from special interest groups that are opposed to the new policy. These pressures may often be traced to industries, labor unions, or municipalities, which think their economic well-being depends upon the continuation of the existing policy. In some cases, such as those related to atomic energy, the recommendations of the advisory committee may also be opposed by strong technological interests within the government itself. As an illustration of the failure of an action recommendation, consider the 1963 recommendation that the widespread use of DDT be drastically reduced: this "first step" has yet to be completed in 1971. Its beginning is the result of 8 years of public pressure and of litigation by environmental and consumer groups.

As another illustration of the fate of action recommendations, consider the SST (7). In the beginning of 1969, as the controversy over the SST began to increase, President Nixon appointed an advisory committee to study the issue. This was a rather high-level committee, involving the undersecretaries of many federal departments. The committee and its subcommittees were charged with studying not only the technological and environmental aspects of the SST, but also the economic, balance of payment, and international aspects. The appointment of the committee was attended by much publicity that emphasized the Administration's concern with the problem. In March 1969, the committee presented a report that was almost entirely unfavorable to the SST. Lee DuBridge, a committee member and the President's science adviser, wrote (8):

> Granted that this [the SST] is an exciting technological development, it still seems best to me to avoid the serious environmental and nuisance problems and the Government should not be subsidizing a device which has neither commercial attractiveness nor public acceptance.

In spite of this strong disapproval, the President and his Administration continued to support the SST fully and enthusiastically. To prevent the report from being used by the opponents of the SST, it was kept confidential, even though there is nothing in it having to do with national security or military matters. Not even Congress, which had to decide on future SST appropriations, was allowed to see it. Only in October 1969 was Representative S. R. Yates (D-Ill.) able to obtain partial release of the report.

It is reasonable to require, as one of the tests of the specific effectiveness of the scientific advisory system, that the executive branch be fairly responsive to the policy recommendations of its advisory committees. Furthermore, the crucial test is its responsiveness to action recommendations. By this test, the advisory system has substantially failed on broad technical issues.

The Scientific Advisory System: Some Observations

Failure on Broad Technical Issues

While some observers will agree with me that the scientific advisory system has not done well on broad technical issues, they argue that the advisory system has accomplished all that could be done. These supporters point out that there are immense political, economic, and ideological pressures that prevent rational decisions on the environment and public health. However, other groups have made progress against these pressures. For example, there is a strong environmental and consumer protection movement in this country. The originators and leaders of this movement are people like Rachel Carson and Ralph Nader, not members of any strong self-interest group. While there are scientists and engineers in this movement, few of them are members of the scientific establishment. Thus, we are still faced with the question of why the advisory system, with its large membership, its great technical and scientific competence, and its prominent men, has not been more successful on the broad technical issues.

There are a number of reasons for the system's lack of specific effectiveness on these issues.

1. *The many functions of the scientific establishment.* The functions and attributes of the scientific establishment severely limit the influence of the advisory system. In a democratic country such as ours, important decisions are not made through a set procedure of debates and position papers, but through a long and messy process. The scientific establishment, because of its functions of representing and protecting research and technical education, is reluctant to take part in much of this process. Usually its members enter the decision-making process through the advisory system at only one point—when the Administration is considering a technical issue. For this reason, the influence of members of the scientific establishment is easily negated. The withholding of reports from the public is just one aspect of that process of negation.

2. *Confidentiality and legitimization.* I have emphasized that the information and advice provided by the advisory system can be declared confidential by the official or agency that receives it, and that it is up to the official or agency to release the information. Although every government official is certainly entitled to some completely private and permanently confidential advice, the problem is that the use of confidentiality is so widespread that very often the only technical reports available on the subject are declared confidential. In that case, the press, the public, and the Congress are left with very incomplete technical information. Thus, on technical issues, the decision-making

process is seriously impeded and, in many cases, the system of checks and balances nullified.

There is another aspect to the confidentiality of the advice given by the advisory system. The press, the well-informed citizen, and the Congress know that the executive branch obtains vast amounts of correct technical information and advice. They know that this advice comes from the best and most prominent scientists and engineers in the country. The final technical policy decisions made by the executive branch become associated with this knowledge. One thinks either that the technical advice has been followed or that it has been seriously considered and then overridden by other, more serious and more profound considerations. Thus the scientific advisory system, as presently constituted, provides a facade of prestige which tends to legitimize all technical decisions made by the President.

The executive branch is well aware of the legitimizing effect of the advisory system. For example, public concern about a technical issue can often be mollified by appointing a committee to study the issue in detail. There is often the hope that, by the time the report appears, public pressure will have decreased. Indeed, this technique extends far outside the sphere of technical issues. If the report appears and is favorable to the policies of the executive branch, it can be released with much publicity. Otherwise, the principle of confidentiality can be imposed. Even an unfavorable report can be used by releasing not the report itself, but a distorted summary of it. Just such a maneuver was used (4) with the unfavorable report on the SST.

The legitimizing aspect of the advisory system is eliminated only when some members of the system directly or indirectly disregard the principle of confidentiality, for example in testimony before Congress on the antiballistic missile and the SST. However, such actions are still rare.

3. *Socialization in Washington.* The basic way to get something done in the executive branch is to work from the inside. This means that one must be practical and hardheaded. One must work for small gains and progress in small steps. For the adviser it is a slow process, with respect to both his influence and his achievements. The adviser works first in less important committees on more restricted issues. As he demonstrates his ability, his reliability, and his reasonableness, he progresses to more important committees and to more important issues. But when he finally achieves a position of influence, his freedom to act is quite limited. This limitation comes not from any rules, but from the methods he learned while working with the executive branch. Thus, in order to retain his position of influence, he may not

The Scientific Advisory System: Some Observations

protest some decisions he intensely dislikes. He wants to reserve his influence for some other issue upon which he has concentrated his interest. Ultimately, the adviser may fall into the trap of considering, above all else, the technique of preserving his influence in Washington [I use the term "technique" here as it is used by Ellul (*3*)].

Socialization explains a number of things. It explains, for example, why the principle of confidentiality is so universally honored in the advisory system. The socialization also explains why the legitimizing effect is so strong. I note again that this socialization in Washington is something that happens to economists, accountants, labor leaders, and businessmen as well as to scientists and engineers. I only emphasize it here because we scientists tend to think that our objectivity and our scientific training constitute a magic cloak that protects us from socialization. It does not.

I have given some of the reasons that the scientific advisory system has a great deal of specific effectiveness on limited technical questions, yet little specific effectiveness on broad technical questions. Now what about the general effectiveness of the scientific advisory system? How does it enter into the decision-making process for general technical questions in this country? The answer is evident from my discussion: the advisory system does not usually enter into the decision-making processes for general technical questions. Thus its general effectiveness is very low, the only exceptions being when individual members of the advisory system testify before Congress or work with congressmen. But most members of the advisory system do not believe in working in the decision-making process outside of the executive branch. They believe that, if they increase their general effectiveness, they will decrease their specific effectiveness.

The Scientific Community

My colleagues in the advisory system have sometimes agreed with the analysis I have presented. But they then say, "All right, we in the advisory system work from the inside doing what we can. Perhaps we are not as effective as you wish us to be. Why don't you work from the outside? There are 10 or 20 thousand people in the advisory system, but there are several hundred thousand scientists and engineers who are not in the advisory system. They can all work from the outside." There are, unfortunately, a number of reasons that this division of labor does not work.

1. *The scientific establishment as a model for the scientific community.* Those members of the establishment who are in the advisory system

are models for the less well-known and younger scientists and engineers. An example of consciously setting a standard of behavior is the recruitment of young theoretical physicists into summer work with the Institute of Defense Analysis. Until recently, it was customary to ask the brightest and most promising young theorists to join in this summer work. Since the invitation was extended by some of the best of the older theoretical physicists, it was very flattering to receive one. Being invited to work with the Institute was, at least for a while, a mark of attainment in theoretical physics.

It is difficult for the scientific community to work on broad technical questions from the outside when the leaders are working from the inside. After all, only a few well-known scientists, men like Pauling, Lapp, and Commoner, work on the outside. Therefore, scientists who wish to serve the country in the technical decision-making process have tended to join the advisory system. In the last few years, there has been some opposition to this tendency, primarily from the environmental and consumer movements and from the various student movements.

2. *The "don't rock the boat" attitude.* I have pointed out that the multiple functions of the establishment and the overlap of the establishment and the advisory system cause a very cautious attitude among advisers. There is a widespread feeling that the advisers should not oppose the technical policies of the Administration too vehemently or too publicly. If they do, members of the establishment fear, federal or even public support for science research and education may be adversely affected. There is certainly some truth in this fear.

This "don't rock the boat" attitude extends into most of the scientific community. This is partly because of the model of behavior set by the establishment; but there is a more compelling reason for this attitude. The natural way for the scientific community to critically and publicly examine the government's technical policies is to use the independent scientific institutions—the professional and scientific societies and the engineering and science departments of universities. Yet these are just the institutions that are being protected by the "don't rock the boat" attitude. For this reason, the scientific community and the scientific establishment will not use independent institutions in the technical decision-making process. It is usually said that these institutions must be kept "neutral."

3. *Professional rewards for service in the advisory system.* There is a grave imbalance between the professional rewards (other than direct monetary rewards) for helping the government make technical decisions from the inside and the rewards for helping from the outside. Almost

The Scientific Advisory System: Some Observations

all universities encourage the public service activities of their faculties if these activities bring honor or influence to the university; teaching or administrative duties may be reduced to allow for them. But almost always, these must be official public service activities. Working within the scientific advisory system is official public service, but, except for a very few universities, working with unofficial neighborhood or consumer groups to reduce the pollution from a local factory is not considered public service. Thus, for the energetic, ambitious young faculty member who wishes to help in the making of technical decisions there are strong career pressures that push him into the advisory system.

Even for the senior scientist the advisory system has career rewards. To be in Washington, to work with other members of the establishment, and to get to know government officials can be of help in a number of ways. It is helpful when seeking funds for a department or for the research of younger people. It also makes a scientist more influential in his home institution.

4. *The "it's in good hands" attitude.* Consciously and unconsciously the members of the advisory system often present the attitude that the role of the scientist and engineer in the technical decision-making process is completely filled by the advisory system. This often takes the form of such statements as, "Don't worry about it, it's in good hands." It is often implied that the members of the advisory system are professional experts on this or that technical question. Other scientists or engineers who are outside the advisory system are regarded as amateurs. This attitude depresses attempts by the scientific community at large to enter the technical decision-making process. It also encourages government officials to ignore scientists and engineers who are not in the advisory system.

Summary

The scientific advisory system is effective on limited technical questions, and such questions provide much of its work. On broad technical questions, however, the scientific advisory system is not effective. Unfortunately this category includes most of the crucial environmental questions. Finally, the advisory system, as presently constituted, combined with the multiple functions of the scientific establishment, is detrimental in important ways to the process of technical decision-making in this country. This is because the combined effect of the advisory system and the establishment is to impede the development

John Walsh and Barbara Culliton

of a more effective and comprehensive role for the scientific community in the technical decision-making process.

Notes

1. The scientific advisory system is described in several articles in R. Gilpin and C. Wright, eds., *Scientists and National Policy Making* (New York: Columbia University Press, 1964); a description of other advisory systems is provided by T. E. Cronin and S. D. Greenberg, eds., *The Presidential Advisory System* (New York: Harper & Row, 1969).
2. J. K. Galbraith, *The New Industrial State* (Boston: Houghton Mifflin, 1967; reprinted, New York: Signet, New American Library, 1969).
3. J. Ellul, *The Technological Society* (New York: Knopf, 1964; reprinted, New York: Vintage, Random House, 1967); originally published in French as *La technique ou l'enjeu du siècle* (Paris: Librairie Armand Colin, 1954).
4. F. Von Hippel and J. Primack, "The Politics of Technology, 1970" (unpublished report available from SWOPSI, Stanford University, Stanford, California).
5. D. K. Price, in *Scientists and National Policy Making*, ed. R. Gilpin and C. Wright (New York: Columbia Univ. Press, 1964).
6. *Use of Pesticides* (a report of the President's Science Advisory Committee, May 15, 1963 (Washington, D.C.: Government Printing Office, 1963).
7. W. A. Shurcliff, *SST and Sonic Boom Handbook* (New York: Ballantine, 1970).
8. L. DuBridge, *Congr. Rec.,* October 31, 1969, p. 32609.

13. Office of Management and Budget: Skeptical View of Scientific Advice
JOHN WALSH AND BARBARA CULLITON

The View from the Executive Office

The phrase "men with a passion for anonymity" was coined in a 1937 report of the President's Commission on Administrative Management and was used, often with malice, to describe the presidential

Copyright © 1974 by the American Association for the Advancement of Science. Reprinted from *Science,* Vol. 183, by permission. This article was edited from three separate articles, two written by John Walsh (January 18, 1974, pp. 180–184; January 25, 1974, pp. 286–290) and the last by Barbara Culliton (February 1, 1974, pp. 392–396).

OMB: Skeptical View of Scientific Advice

advisers and top bureaucrats who manned the machinery of the New Deal and who accepted anonymity in exchange for power. Today, the phrase and its overtones might be applied to the officials of the Office of Management and Budget (OMB), who are as influential in the U.S. government as the gentlemen of the Treasury in Britain or the inspectors of finance in France.

Because the agency operates behind the scenes and because it appears to have the last word with the President on decisions which often result in cuts in program funds, OMB is blamed for goring a lot of congressional oxen, is a source of fear and anxiety to federal agencies, and is looked on by many members of the scientific community as the scourge of the science budget. The current struggle between Congress and the Administration over impoundment of funds is a notable demonstration of OMB's license and limits in wielding Executive power.

A budget agency has existed in the Executive branch for a little more than half a century; its influence has increased steadily as the government has grown in size and complexity, particularly in the last decade with the proliferation of social and welfare programs.

By the end of World War I, there was abundant evidence of the flaws in the traditional view that Congress set the budget through the appropriations process and the President simply administered the spending of the appropriated funds. The Budget and Accounting Act of 1921 established a Bureau of the Budget (BOB) in the Treasury Department with the main responsibility for preparing the budget. During the New Deal, BOB acquired "clearance" authority over new legislation to ensure that it conformed with presidential policy and budgetary requirements. In 1939, BOB was reorganized out of the Treasury and into the Executive Office as part of a bigger, institutionalized staff for the President. In the 1950's and 1960's, BOB took a much stronger hand in shaping federal programs so that they would be more likely to achieve the goals set for them, and in the later 1960's heavier stress was put on evaluation of existing programs and on tighter central management of the sprawling new social programs. The Nixon Administration's heavy emphasis on effective management was symbolized in a 1970 reorganization which changed the name of a restructured Bureau of the Budget to the Office of Management and Budget. Along the way the budget agency became involved in matters such as the training of federal executives, information gathering, and long-term policy planning and short-term troubleshooting for the President. In the Washington crisis over the energy crisis, for example, OMB provided much of the staff work which produced the Administration response and also sent key cadre to the new energy agency.

To many scientists, OMB has seemed a sort of reverse Scrooge who regressed from benefactor to skinflint. To put the ups and downs of science in perspective, it is necessary to remember that the budget agency, like every other government agency, has had to react to the primary factors that have conditioned American government in the past two decades; these have been the Cold War, Great Society legislation, and the Vietnam war. In an attempt to understand the budget agency's point of view in handling the science budget in this period, *Science* talked to a number of past and present members of the staff of the budget agency, as well as to close observers in other federal agencies and on Capitol Hill. Particularly helpful in discussing BOB perspective in the 1960's were Charles L. Schultze, BOB director under President Lyndon Johnson and now at the Brookings Institution, and William D. Carey, who served in the Budget Bureau from 1942 until 1969 and held key jobs dealing with science and education affairs. Carey is now an Arthur D. Little, Inc., vice president.

In the 1950's the real money in federal R & D was in defense, atomic energy, and medical research. Carey recalls that general-purpose, basic research was "tagged on" to the budgets in these main growth areas of R & D. Each of the main items represented different subgovernments and were handled differently in BOB.

"The bureau," as it still tends to be called by BOB veterans, had less leverage on military R & D than on civilian science. As technical investment became more critical under the conditions of the Cold War in the 1950's, the bureau realized it could not approach the review of the defense budget in the same way it did other budgets. Time was a factor. If defense "numbers" came into the bureau in October, as other agency figures did, it was physically impossible to get a massive military budget through the "screen," sort out issues, and get decisions by January. Therefore, a joint review of the budget was devised, with the bureau sharing the review carried out in the office of the Secretary of Defense. BOB, in effect, hitched on to the Defense Secretary's review of the budget. The review was essentially the Secretary's, not the budget director's, and the edge in decision-making on the military budget passed to the Pentagon.

With the formidable Robert S. McNamara as Secretary of Defense in the Kennedy Administration, the process went a step further. McNamara, by all accounts, was his own budget officer. Up to that time BOB had been instrumental in setting a dollar limit on the defense budget. McNamara succeeded in breaching BOB power to im-

OMB: Skeptical View of Scientific Advice

pose a total figure. It was the time of a shift from the doctrine of massive retaliation to a policy of more flexible response and a consequent major buildup of conventional forces. Those who witnessed the process year after year say the R & D component of the military budget was determined largely by the Secretary of Defense.

With the AEC in the 1960's, BOB was content to win some and lose some. From the bureau's standpoint, the experience with biomedical research was less satisfactory. The BOB view was that biomedical research took off largely through pressure-group tactics. As one veteran of the period put it, the growth of the biomedical research budget was "determined by the interests of a well-organized research lobby wired into the White House and the appropriations committees. The appropriations committees came to regard the budget power with open contempt."

Former Senator Lister Hill and the late Congressman John Fogarty, who managed the biomedical research boom in Congress, may be candidates for canonization in the biomedical community, but they are regarded as apostates in the bureau.

Carey says that "from our point of view it was a rationing problem. It's always a rationing problem. We were faced year after year with preemption: portions of the [R & D] budget were uncontrollable. The bureau wanted to provide for a reasonable rate of growth and expansion. Our feeling was that things got out of hand. A policy of growth for growth's sake seemed unacceptable."

The prevailing feeling in BOB was that biomedical science was riding for a fall, and, when budget troubles finally hit biomedical research, the BOB staff did appear to react with more than a tinge of *schadenfreude*. It seems to be a misreading to regard BOB as antiscience. Carey, for example, points out that BOB was the first and most consistent advocate of federal support of R & D in the post-World War II period. BOB interested itself in preventing the collapse of the R & D thrust developed during the war, says Carey. It is rare for the bureau to take an advocacy role, but it did as far as research goes. BOB was involved in working out the compromise that created the National Science Foundation (NSF). Through the 1950's and 1960's, in good years and bad, says Carey, the staff of the bureau sheltered basic science and research. "We felt that basic research could not survive in open competition for money."

The relationship between the bureau and the scientific community has been a complex one. Carey recalls "a funny blend of attitudes in the bureau toward R & D. On the one hand sympathy and protectiveness, on the other hand exasperation—exasperation with the smugness and political unsophistication of the world of basic science."

John Walsh and Barbara Culliton
Problems with Biomedicine

Carey acknowledges that there are special problems for the budget agency in handling the substance of biomedical research, which is harder to deal with, he says, than military R & D or atomic energy research. "We have problems of noncredentialed experts trying to deal with [experts]. All the competent researchers in the field are mortgaged to NIH by individual or institutional interests."

The central problem for the bureau in relation to R & D Carey put this way. "The managerial mind—of which the bureau is part and parcel—has an uncontrollable itch to know where [the money] is going, and the truth is we budgeted by the stars with no help when it came to the problem of choices, of setting priorities for public investment. And after a while you realize you're not going to get any help."

From BOB's perspective, the plateauing of the science budget in the late 1960's was produced by a convergence of circumstances. Not only had the R & D budget grown very rapidly, but Johnson's Great Society legislation put a severe and not fully anticipated squeeze on the budget. Initial funding for the new legislation had been relatively small, but the ultimate costs were badly underestimated. In the same way, the costs of the Vietnam war were consistently undercalculated. In addition, the space program had peaked, and the NASA budget, which accounted for a major chunk of R & D funds, was headed down. Significantly, a BOB attempt to use NSF as a balance wheel to protect the R & D budget amounted to little. A proposal to shift R & D projects cut from mission agency budgets to NSF was put forward but cold-shouldered by Congress.

Any description of how OMB functions is likely to be oversimplified, partial, and rather abstract. The essential thing to remember is that OMB exists to serve the President, and the scope and style of its operations change with successive presidents. OMB is not autonomous or omnipotent, the impressions of those who consider themselves its victims to the contrary. Although OMB is somewhat insulated from the hurly-burly, it is no more immune to Washington political realities than its patron, the President.

Schultze says that in very simplified terms there are two groups in town. The first might be called the "President's party," which is comprised of his staff, including OMB, and state and local chief executives sympathetic to him. The other is the coalition of federal agency heads, senior congressional figures and their staffs, and a supporting cast of Washington lawyers and lobbyists. Presidents come and go, but the coalition stays on. Cabinet members in the past have usually been

OMB: Skeptical View of Scientific Advice

chosen for their expertise or influence with a particular constituency and have usually been placed somewhere in the middle between the two power groups. Presidents sometimes drive their Cabinet officers into the other camp. An alternative available to a Chief Executive is to appoint Cabinet secretaries primarily on the basis of their loyalty to him. Most observers feel that Nixon, with exceptions, has done this. Under Nixon the responsibility for setting policy and fashioning legislation has been shifted largely to the White House, leaving the Cabinet officers the jobs of day-to-day running of their departments and of liaison with Congress.

This key change in the locus of policy and legislative authority began before Nixon took office. It was during President Lyndon Johnson's push for his Great Society that the Administration's legislative program was framed, not by the departments, but in the White House on the basis of the reports of special task forces. This approach had been tried during the Kennedy Administration, but the real initiative was taken by Johnson.

As one veteran of the Executive Office in that era put it, "Johnson felt he had to have a hundred bills a year. You can't even have a hundred bad bills a year, but every year he had those task forces working."

There is a consensus that during the Nixon Administration there has been a further strengthening of the powers of the President's personal staff and of Executive Office agencies like OMB at the expense of the departments. What the long-term effects of this will be are far from clear. Inevitably, the new arrangements have altered OMB's relationships with Congress, with the departments, and with other elements of the President's staff. What is most elusive is the change in the way that the budget agency has traditionally looked at its own functions.

Schultze suggests that this traditional view can be understood in terms of "role playing." On the one hand, for example, the President needs a Secretary of Health, Education, and Welfare who is loyal to him but is also compassionate and aware of the need to put more resources into health, education, and welfare. The President also needs a counterbalancing, relatively hard-nosed, analytical input which is not politically oriented. In the past, the President has relied for this kind of advice on the White House agencies—the Council of Economic Advisers (CEA), the Office of Science and Technology, but principally OMB.

In the case of an education bill, for example, OMB should ask, "Are you playing up to the National Education Association or the chairman of the Appropriations subcommittee on education?" OMB should look at substantive matters, but with a fiscally fishy eye, says Schultze. "For the system to work, a President needs a lot of personal contact with Cabinet officers; there should be a lot of head-to-head discussion between Cabinet officials and the analytical group. These should not be nay-sayers, but rather a group of professional skeptics. (Schultze believes that Cabinet officers need their own counterparts to OMB's analysts.) OMB should keep the President informed on alternatives, and the budget office should be resigned to being overturned—but not too often."

Schultze and others who argue the case for the traditionalist view emphasize that the budget process has to be salted heavily with adversary relationships played out in front of the President, and they also emphasize that the President must ultimately make the choices. This obviously loads the schedule of an already heavily burdened President, and those who believe that the budget affords the President his chief instrument in making effective policy tend to take a technocratically brusque view of the time a President spends on political or ceremonial pursuits. As one former senior official put it, "A President spends an awful lot of time on crap."

The advocates of competitive interplay feel that the President should not be screened from seeing policy people in government. But how completely has the ideal been achieved in the past? President Johnson had a voracious interest in the details of government, and a fund of information built up during three decades in Washington, but it appears that the adversary process was only partially carried out. Particularly as the Vietnam war preempted Johnson's attention, the range of subjects in which he participated personally in the final stages of budget review narrowed. In some areas of economic policy, for example, Johnson continued to meet with three or four of the chief actors—the Secretary of the Treasury, the chairman of CEA, the director of BOB, and the head of the Federal Reserve System. But when it came to education policy or science policy, it was "all done by staff," says one senior official.

Not So Much the Arbiter

During the first Nixon Administration, this trend accelerated, reinforced by the President's own lack of enthusiasm for give-and-take with relays of policy protagonists and by his immersion in foreign policy issues. Since his second term began, the distractions of

OMB: Skeptical View of Scientific Advice

Watergate have further reduced Nixon's time to act in the traditional role of budget arbiter.

In part because Congress has been controlled by the Democrats, the Nixon Administration has placed less emphasis on new legislation and more on reappraising existing programs and re-allocating resources. Because of this and because of a passion for making government more businesslike, there has been an accent on the management arts. The Nixon Administration, however, has pursued the old aim of using the budget to pursue its policy goals. A second article will discuss how, in the process, OMB has changed and how it has remained the same.

The Accent on Management

Every recent American president has sought to remold the machinery of the Executive branch to make it more responsive to his policies, and Richard Nixon, more than his predecessors, has put his faith in the efficacy of modern management techniques. In the process, he has used the Office of Management and Budget (OMB), not only as a model of management, but as a source of managers for other federal agencies.

George P. Schultz, now Secretary of the Treasury and Nixon's chief adviser on economic matters, proved himself as the first director of OMB, which was established under a 1970 reorganization plan. Caspar W. Weinberger was Shultz's deputy at OMB and succeeded him as director before going to his present post as Secretary of Health, Education, and Welfare (HEW). James R. Schlesinger, now serving as Secretary of Defense after relatively short stints as chairman of the Atomic Energy Commission and director of the Central Intelligence Agency, got his start in the Nixon Administration as an assistant director of the budget agency. OMB staff members colonized the new energy office, and OMB's associate director for natural resources, energy, and science, John C. Sawhill, was picked as deputy director of the new agency. Members of the OMB alumni are spotted in many other jobs throughout the federal hierarchy, rather like new branch managers sent out from the home office.

It is true that OMB has become a wellspring of men and ideas comparable to the Department of Defense under Robert S. McNamara during the Kennedy and Johnson administrations. But, despite the emphasis on the "M" in OMB, the management role of the agency has not blossomed quite as rapidly as was expected.

John Walsh and Barbara Culliton

This is attributable in part to resistances within the system, to which presidents Truman, Eisenhower, Kennedy, and Johnson all ruefully testified. Reforms of the federal system, even those firmly backed by presidential power, tend to get lost in the wastes of bureaucratic time. And the Nixon Administration may have had unreasonably high hopes of managerial miracles. One sarcastic semiepigram repeated by a former insider now on the sidelines makes the point. "The Kennedy Administration operated on the fallacy that, if you appoint a good man, the organization doesn't matter. The Nixon Administration operates on the fallacy that, if you get the organization right, you can appoint a third-rate tractor salesman and it doesn't matter."

More concretely, OMB has run into stronger resistance from Congress. There is a consensus on Capitol Hill that OMB has greater power than its predecessor agency, the Bureau of the Budget (BOB), and OMB has been cast in the role of direct antagonist to Congress on the issue of Administration impoundment of appropriated funds. Congressional attitudes were expressed in the action to require Senate confirmation of the director and deputy director of OMB. The Administration view was that this infringed the President's untrammeled right to appoint his own personal advisers, while the feeling in Congress was that the director of OMB and his deputy had acquired new policy powers and should be subject to the same congressional scrutiny as other top political appointees.

Not an "Open" Agency

OMB is not an "open" agency in discussions of its actions with Congress, the press, or the public, and it is particularly secretive during the months preceding release of the budget. Junior members of the OMB staff approached for interviews by *Science* during the preparation of these articles, for example, made the stock response that it would be better to talk to "policy-level" officials. The budget agency staff has always been a discreet and elusive group. But, in this reporter's experience, it should be noted that those policy-level officials in OMB today are, if anything, more accessible to the press than their counterparts were in previous administrations.

When Malek is asked to discuss the Administration's stress on management he does talk a little like a management casebook, but hardly like a zealot. The main thrust of the Ash commission, he says, going back to first principles, was "to give the President stronger executive assistance in managing the government." BOB played a key role in developing the federal budget and in giving analytical support to the

OMB: Skeptical View of Scientific Advice

President. BOB had also done some work in promoting effective management. Malek says that both the Ash commission and the President felt the budget agency "should play a more effective role in assisting the President to manage government—to set goals, to chart directions, to do the kind of things that the chief executive of any enterprise is concerned with."

When Ash and he took over last year, Malek says they felt there had been "considerable transition" but that a lot more could be done to improve management. The immediate target was "to change the focus from mechanical things, like data processing, purchasing, and so forth, to an orientation to broader management aspects to assist the President."

This meant meshing the management and budget functions of the agency. Under the 1970 reorganization, OMB had started out with a bifurcated structure with separate management and budget branches. "Our feeling is that management and budget are one and the same tool of management and that it was important to integrate them." OMB's organizational structure was changed by eliminating a dual chain of command and consolidating activities under four associate directors with direct "line" authority. "The change was not traumatic or even dramatic," Malek asserts. Each associate director presided over budget and management divisions in his field. Some routine management responsibilities were transferred to the General Services Administration, the government's housekeeping agency. And a number of new upper-level staff positions for "management associates" were created. Efforts were reportedly made to recruit "hot-shot management types" from outside government to fill these new posts.

The management associates do a variety of things, but they are expected to spend about a third of their time working with other agencies on so-called "MBO's." The Ash-Malek formula for achieving better management features the concept of "Management by Objectives (MBO)." Malek calls MBO "a tool for the President [to use in managing government], a way to delegate responsibility to the Cabinet without abdicating responsibility." Each agency submits objectives each year. These are carefully reviewed in OMB, discussed with the agency, and submitted to the President. If he decides they are acceptable, the agency is held accountable for achieving the objectives. Progress is to be monitored at monthly meetings at which Ash and Malek are expected to be present. Most issues that arise, however, are supposed to be settled at the staff level rather than at the meetings.

How is MBO going so far? "Objectives link up with the budget, but not to the degree they should," says Malek. "We're not satisfied yet." Pressure on the agencies in behalf of MBO continues to be applied. Each agency was asked to submit its objectives for fiscal 1975 along with its budget requests. A task force has been commissioned to determine how the linkup between objectives and the budget process can be made more effective. "If [MBO] is going to endure, we think it is necessary to link it to something as old and established as the budget process."

In the case of research and development Malek concedes that it is difficult to set objectives. He thinks that the National Science Foundation (NSF) "has done a very good job in this area" in trying "to get something quantifiable or at least measurable."

It is true, says Malek, that with basic research you can't measure too well, but that, "given [good] information, you can set priorities. The cancer program is a good example," he says. "HEW has an objective related to the effectiveness of implementation of the cancer program and subobjectives related to the direction research on cancer is to take. Cancer has a little higher priority than other types of research.

"NSF is in the process of identifying all areas of research it could enter into—for example, how to contribute to the assessment of ocean resources. Objectives should have real influence—provide a solid foundation on which to base budget decisions." When will this happen? "Remember, this has only been going on since spring," is Malek's reply. Reception of MBO in the agencies is now "a mixed picture," according to Malek. "But across the board at the top of the agencies there is enthusiasm and good cooperation. The real test is when you look down the line in an agency. Then it varies."

MBO is inevitably compared to the late, unlamented PPBS (Planning-Programming-Budgeting System) in the Johnson Administration. PPBS was based on the use of cost-benefit analyses to compare alternative ways to achieve policy goals. It was evolved in the Department of Defense, and its application in other agencies was made mandatory by presidential order. PPBS is remembered for being awkward to adapt to civilian programs and for inflicting masses of paperwork of dubious value.

"What we're trying to avoid that was present in PPBS are the rigid requirements and a lot of paperwork," says Malek. "We're not trying to establish a system. We're trying to get across that this is a way of life, a way of thinking, how you do business. In an agency like NASA, this is extremely well established. It's nothing new to them. In other agencies it's slower in developing."

OMB: Skeptical View of Scientific Advice

Budget Examiners

Inside OMB, the reorganization has effected a not so subtle change in the way the budget agency functions. The archetypal figure in the agency is the budget examiner, who has been responsible for a group of programs or a small agency. A budget examiner might be a relatively junior civil servant, but still wield decisive influence in the budgetary process. One budget agency veteran says, "Here, a Grade 12 (there are 18 grades in the civil service scale) deals with assistant secretaries, elsewhere they're nobodies."

This is still true, but the examiners have a diminished role. In the old days, the budget examiner, his division chief, and perhaps the director of BOB or his deputy might negotiate with an agency head and his budget officers on final budget items. According to those who have watched the process, the examiners now have less contact with the agencies during the year and less direct impact in the review. The division chiefs have apparently also lost clout; the associate directors are said to have absorbed power formerly exercised by those above and below them in the chain of command.

In terms of relations with the White House, the reorganization has also had a decided effect on OMB. Shultz, the first OMB director, was soon drawn into the White House, where he functioned as an adviser to Nixon on broad matters of economic policy. He spent the balance of his time in an office in the West Wing rather than in the old Executive Office Building where OMB was located. OMB deputy Caspar Weinberger became de facto budget director. The same pattern has been repeated with Ash and Malek. Malek presides over the director's review sessions and Ash functions as a White House adviser and is somewhat removed from the day-to-day process of fashioning the budget.

OMB regulars see a "fragmentation" of the agency occurring. Before the present review began, Malek was regarded as more interested in management problems and executive development than in the budget process, and the verdict on him will have to await completion of the budget cycle. Under the circumstances, however, it is suggested that now, with four highly independent associate directors dealing directly with the White House on their separate fields of responsibility, there are "four little OMB's" operating.

This fragmentation in turn is traced to the 1970 attempt to reorganize OMB in parallel with an augmented Domestic Council that never came off. Another of the "oddball things that have happened,"

John Walsh and Barbara Culliton

according to one observer, affects the route of decision-making on R & D issues. Shultz and his assistant, Kenneth W. Dam, were tapped to fill the vacuum left by the abolition of the post of President's science adviser early last year. But, as the observer said, "The Shultz science operation is the vestigial remains of the supercabinet which didn't work." The arrangement may have its consolations for scientists, however, since Shultz is thought to be better informed and more open-minded on science than anyone close to the President, including Ash, who is reputed to take an "industry view" of R & D. Dam, by no great coincidence, is a former assistant director of the budget agency.

Does OMB have any other problems? A too rapid turnover in staff, says one staff member. "If you're responsible for an education program, you can't just look at the education budget; you've got to know what's going on in the country, and you have to have some feeling of where the agency is moving." (Although there has been a steady expansion of the White House staff during recent administrations, especially during Nixon's first term, the growth of OMB itself has been relatively restrained. The regular staff now numbers about 630, roughly two-thirds of them professionals, compared with a total of about 450 during the early days of the Kennedy Administration.)

Some critics charge that OMB has been "politicized" and point to the interposition of the associate directors as proof. It is true that a number of top budget officials are Administration appointees who presumably were picked for their sympathy with the President and his policies. But what these critics see as a break with the past is that these appointees are from outside OMB and outside government. Three of the four associate directors came to OMB by way of successful careers in management consulting firms or industry. One of these is Sawhill, whose post as associate director for natural resources, energy, and science has been vacant since he moved to the energy office. The fourth associate director has a government background. He is Paul H. O'Neill, whose domain, human and community affairs, includes biomedical research. In his service in both the Veterans Administration and the budget agency, however, O'Neill distinguished himself as a new model manager.

Longevity No Bar

Long service in the budget agency is not a disqualification for key assignments. Hugh F. Loweth, for example, who was recently named to head the energy and science division under Sawhill, joined BOB in 1950 and has been dealing with science programs since the middle

1950's. His division handles most of the civilian R & D programs outside the health field.

Other critics insist that politization is less of a problem for OMB than is a lack of policy guidance. William A. Niskanen, a recent OMB insider as assistant director for evaluation and now a professor of economics at Berkeley, argued in a recent monograph[1] that the White House lacks an adequate formal process for communicating policy on major issues to OMB, and he makes suggestions for remedying the deficiency.

Old hands at OMB demur on the existence of a policy gap. "It's an iterative process," says one veteran staff member. [Policy] is not handed down on tablets. It's very fuzzy. We're told, for instance, that the President wants to hold down civilian employment. You rarely get signals clearly."

What is implied in these divergent views are differing general conceptions of how OMB should operate. Those imbued with the old "bureau" tradition seem to feel there is nothing wrong with OMB that a return to closer communications downward with client agencies and upward with the President and his chief aides would not remedy.

On the other hand, the new breed of managers clearly believes that OMB should continue to move in the direction of improving formal policy structures and increasing the active management of programs.

There is a third view based on the belief that OMB has acquired too much power by default. From this perspective, reform of the whole budget process is needed to restore authority to Congress. OMB is not a popular agency with Congress, and the budget which is about to appear is unlikely to make it more popular. What is different this year, however, is that Congress has taken the first faltering steps toward disciplining its appropriations process to keep spending within budgeted limits. Congress, however, has shown an almost feudal inflexibility toward the kinds of internal transfers of authority that such a major reform would require. So unless and until such radical reform occurs, OMB, under whatever name and organization chart, is likely to persevere, because the budget remains the most effective combination of carrot and stick available to a president.

A Skeptical View of Scientific Advice

The Office of Management and Budget is the agent of the President. Its officials are paid to make sure that legislation and budgets proposed by government agencies conform to Presidential wishes.

John Walsh and Barbara Culliton

They are also supposed to see to it that the agencies interpret and carry out presidential policy correctly. In fact, one important feature of the OMB's Reorganization Plan No. 1 of 1973 was "... the goal of reorienting the Office to focus on its original mission as a staff to the President for top-level policy formulation and for monitoring policy execution—where reliance could not appropriately be placed on individual departments and agencies." At least that's what John C. Sawhill told Congress.[2]

The OMB is set apart from the rest of the government. Its officials are not paid to make people in government agencies happy. Nor are they paid to please the people on Capitol Hill. Frequently, they don't. Staffers at OMB are not hired to be popular. By and large, they aren't.

Members of Congress have called OMB people "gnomes," and "faceless, anonymous forces." Aides of House and Senate appropriations committees complain that OMB staffers neglect to return their phone calls. Department and agency heads resent the fact that they frequently are relegated to dealing with young budget examiners who rank relatively low in the federal hierarchy, rather than with OMB directors. They resent the fact that these often inexperienced examiners make the decisions that count. And everybody resents the fact that, even if he does manage to take his case to someone at the top, he still is not likely to extract as much money as he firmly believes he needs for his program. Nobody likes the man who tells you that you cannot have what you want.

Most people at OMB are well aware that they can intimidate and anger lesser mortals. Some seem to enjoy the power. Others sincerely decry it and make honest attempts to establish mutually satisfactory relationships with the congressional and agency people with whom they deal. But it is not always easy.

The Nixon Administration has been perfectly clear in saying that it does not want agency leaders to be advocates of their own programs. Frank Carlucci, undersecretary of Health, Education, and Welfare and former deputy of OMB, has said that "a public agency must serve first and always the broad public interest and take its direction and policy from the duly elected leader of the executive branch of government—be he president, governor, mayor, or county supervisor.... To me, public advocacy by a public agency is outright chaos. Sooner or later it places that agency in an adversary position with the chief executive." Richard Nixon does not want that.

Nevertheless, the fact remains that people running government agencies *are* advocates of their own programs. Inevitably, officials of

OMB: Skeptical View of Scientific Advice

OMB and the agencies are adversaries. It is no wonder there are hostilities.

The OMB's activities with respect to the National Institutes of Health and biomedical research policy and funding are revealing of the way in which the office works and of its relationships with the scientific community generally.

On the whole, OMB and NIH are not on very good terms. One apparent reason is that their respective leaders do not spend enough time talking to each other. Former NIH director Robert Q. Marston complained about that last fall, when he debated OMB associate director Paul H. O'Neill at the Institute of Medicine of the National Academy of Sciences. Marston, alleging that "the source and scope of expert advice used by the executive branch is very limited," said, "I think Paul brought out one of the reasons for this when he said that he has been distracted by other things. I suspect, Paul, that we will probably spend as much time today together talking about major NIH problems as we did during the whole time that I was director of NIH."

It is no wonder that O'Neill becomes distracted from the problems of biomedical research. A glance at the list of agencies for which he is responsible puts things in perspective. The NIH is not even mentioned by name; its presence is subsumed under the listing for HEW.

If biomedical scientists are unhappy that O'Neill does not listen to them, O'Neill also is unhappy with the quality of advice he receives when he does listen. In his opinion, scientists often implicitly ask for special treatment but do nothing to deserve it. He says he hears too frequently the idea that what scientists say should be accepted just because they are scientists. "I don't think we can turn our world over to people who couch their reasoning in terms of their expertise or their degrees," he says.

When it comes to broad recommendations, O'Neill finds that panels of scientists, like panels of any special group, first propose that "a new organizational element be attached, either in the Office of the President or in the Office of the Secretary, to deal with the subject." Having done that, "They will recommend more money for their thing without looking out across the broader world."

O'Neill is also wary of much of the advice he gets from scientists because he believes he cannot always trust it. Simply put, the same people who stand up in public and declare that budget cuts are destroying American science say, in private, that there are excesses in the biomedical research budget. The national cancer program, in particular, has prompted individuals to follow this kind of double standard. Says O'Neill, "... while I have had people tell me quietly and privately, 'Look, we think you are doing too much in one area of biomedical research,' they have not been willing to stand up where it counts in public or in the Congress and say, 'You are doing too much.' They turn the argument around and say, 'You are not doing enough in other areas.'" He wishes, he says, that scientists could learn to think in terms of options or alternatives, especially in view of the fact that national resources are limited and that biomedical research is competing for them at the margin. "There is a fundamental notion in economics that says we live in a world of finite resources," he says. People who want them are going to have to be pretty good at justifying their claim. O'Neill thinks that research scientists are not doing a very good job of that.

Science looks different to different people. The OMB wants to know, first and foremost, what good a given scientific enterprise is, what will come of it. Even though many OMB people concede that the products of scientific research cannot be anticipated the way those of an automobile plant can, the prevailing philosophy at the OMB is that science, like everything else, should pay off if it is going to get public support.

Most scientists hold as an article of faith that science deserves public support because it contributes to the advancement of man's understanding of himself and his world. The unhappy truth is that not everyone shares this faith, or, in any case, not as completely. Some of those people are in influential positions in the Nixon Administration. It is not an easy matter to resolve, but the fact that the two sides enter budget negotiations with different first premises does not help.

With respect to the Administration's willingness to support biomedical research, it is worth noting that the NIH budget has been going up, not down. Granted, its increases have been selective, going only to the cancer and heart institutes, but from the Administration's point of view its support of research has not declined. The argument is over how those additional funds are allocated, which brings one back to the issue of whether research should be expected to pay off in order to justify receiving large amounts of federal funds.

Given this adversary relationship between the scientific community and the OMB, the form in which scientists usually submit their advice to OMB leaves something to be desired on both sides, although OMB specifies the form in which it wants its information to come. Putting the President's budget together is a year-round process which approaches its final stages around the end of September, when the agencies send their budget requests downtown to O'Neill or one of his counterparts in the big, gray executive office building next to the

OMB: Skeptical View of Scientific Advice

White House. With each request comes a justification that supposedly explains why this program or that deserves to have more money than it stands a prayer of getting. These documents, which generally do not make very lively reading, arrive at OMB by the carload. Reportedly, they are seldom persuasive; it is inconceivable that they are even very carefully read.

One thing that many government scientists involved in preparing these budget justifications resent is the confidentiality in which they must be held. The justification that the Institute of General Medical Sciences submitted along with its budget request for fiscal 1974 is representative. The document is an attempt to explain why certain research in molecular biology, genetics, and other basic sciences should be supported. Stamped conspicuously across the front and on the top of every page is the following OMB-decreed warning: "Administrative Confidential, *Exercise Caution in Handling This Document.*" One would think it were a letter bomb.

Last summer, these and other top-secret budget memos were made public thanks to the good offices of senators Mike Mansfield (D-Mont.) and Warren Magnuson (D-Wash.), who had decided that the Congress ought to know what the OMB knows. So, at Mansfield's request, the General Accounting Office, which is Congress's accounting bloodhound, went out and got from the health and education agencies of HEW the information that had gone to OMB on the potential impact of reductions in their budgets. Then, Mansfield and Magnuson, data in hand, blasted the Administration for neglecting research and opened their files to the press.

Among their treasure was a memo from Frank J. Rauscher, Jr., the director of the National Cancer Institute, explaining how the cancer program would fare in fiscal 1974 under various alternative budgets. That memo and the circumstances under which it was prepared are illustrative of how OMB gets advice still later in the budget process.

It was November and everybody's original justification was in but, of course, there was not enough money to go around and the men of the OMB had to decide whom to favor. They decided that, among other things, they needed to know more about the cancer program. So, someone from OMB picked up the phone and called Rauscher and asked him to determine what would happen to cancer research if the original budget request of $640 million were reduced to $456 million or $550 million. And, the caller wanted to know, could he have his answer by the next day. It was more a command than a question.

The next day, after many frantic hours of work with his staff,

Rauscher sent in his reply, the thrust of which was that to cut the cancer budget would be a crime. The memo told OMB the following sorts of things: At $550 million,

> —Expansion of research leads in the immunologic treatment of cancer into clinical trials will be restricted. Immunodiagnosis and immunotherapy offer the most immediate promising results in the early detection and treatment of major cancers.
> —Each year 12,000 women die needlessly of cervical cancer because no more than 25 percent get Pap tests. Industrial contracts to develop equipment to automate the cytological screening of Pap tests would have to be postponed for at least a year.
> —Evidence is rapidly accumulating that four newly discovered viruses cause cancer in man. If this is true, it can lead to the development of methods for preventing specific types of cancer. These leads cannot be fully developed or studies on new viruses which may induce malignancies cannot be mounted at the $550 million level.

The memo, which is like hundreds of similar defenses from other agencies, drew mixed reactions from the Administration officials and others who saw it. Some, perhaps many, of its points were valid, they felt; some were not. For instance, the fact that women are afraid to go to a doctor for a Pap smear has more to do with the high death rate from cervical cancer than does the fact that cytological screening is not automated. And, as far as those four human cancer viruses are concerned, virologists with the finest understanding of the field are still trying to figure out how Rauscher did his addition.

It is difficult to see that advice of this kind does anyone much good. The OMB apparently takes it with a grain of salt. The people who give it resent being forced to produce these documents overnight, and they do not believe that what they have to say will have any effect anyway.

When government scientists are not busy defending their programs in reports and memos to ranking OMB officials, they are justifying their activities to its budget examiner. There is one budget examiner for all of NIH. Her name is Ann Stone. She has been a budget examiner for 2 years. A fairly recent graduate of Duke University, where she studied social sciences, she worked for HEW before joining the staff of OMB. As part of an internship program at HEW, she took a few management courses, but she has never had any formal training in hard science. Stone recognizes that this is a sore point with many of the scientists with whom she deals at NIH, but she believes that her lack of an academic background in science is an asset to doing her job. Many officials at NIH would, of course, prefer to see a scientist in her position, but most of them are realistic enough to know that is not going to happen.

Stone is involved in the budget process from the very beginning of its yearly cycle, and as she goes from institute to institute to review its programs, she asks a lot of questions that make people uncomfortable. (O'Neill says that OMB is asking questions today that are a lot tougher than the ones it asked in 1967, when he first worked in the budget office; NIH scientists agree.)

As Ann Stone makes her rounds, she asks for a description of institute programs, demanding a definition of what they are and where they are supposed to be going. She asks institute officials how they measure progress. She asks what the alternatives to any given program are and, as part of that, suggests that scientists start asking whether the federal government should be supporting certain programs at all.

The latter question has become something of a refrain for this Administration. It was the question OMB asked when it decided to phase out NIH training grants. Scientists defended the program that supported young biologists, but OMB decided that there was no reason taxpayers should pick up the bill for individuals who will go on to earn good incomes, especially since they believe the nation is not facing a shortage of biomedical researchers. When the Administration was subsequently persuaded to relent a little and restore some of the training money, OMB was there to make sure NIH executed the "new" training program in accordance with policy. The Administration had decided that training money should go only to persons working in areas in which there is a shortage of researchers, and NIH was instructed to determine which ones they are.

Instead of trusting NIH to do that job, OMB stepped in and asked a slew of detailed questions about what the areas of shortage are and how anybody had determined that and what scientific accomplishments might be anticipated by training persons in one field rather than another. The questions offended many NIH scientists who believe that no one at OMB knows enough to evaluate such scientific judgments. Officials at OMB say that the people at NIH have it all wrong. It was never their intention to ask NIH to make substantive changes as far as its scientific assessment goes. Rather, OMB asked those questions to force the scientists to think about what they were doing, to force them to consider alternatives and to set goals. The OMB intends to keep asking such questions until everybody learns.

Although Stone's questions are offensive, at times, to scientists who are put off by her cross-examination, what galls them more is their conviction that she and her immediate supervisor, Victor Zafra, are ultimately making decisions about what NIH is going to do. At the very least, they think that O'Neill should be directly involved every step of the way, and they are not happy about having their point of view filtered through Stone and Zafra, who also lacks academic credentials in science. In short, senior NIH scientists resent the fact that individuals whom they sometimes refer to as "just a couple of young kids" can tell them what to do.

Staffers at OMB like to play it down, but the truth is that their influence on the activities of anyone who works for the government is tremendous. The OMB's control of the budget is only one aspect of its power. It also controls legislation that the Administration proposes in order to be sure it is in line with what the President wants. If, for example, HEW were to draw up a bill for more work in population studies, and the OMB were to decide that it did not fit Presidential policy, there would be no bill.

In the same way, OMB controls what government officials say when they go before Congress to testify on pending legislation. For instance, Charles Edwards, the assistant secretary of HEW for health, is asked to appear to testify on a bill dealing with federal support of research. He prepares a statement and sends it over to OMB. There, it is read and circulated to the heads of any agencies other than HEW that might be affected by the legislation. If there are disagreements, O'Neill might call everyone together to iron them out. If the problems cannot be solved, the testimony might be sent over to the White House for a decision. But whatever happens, neither Edwards, nor the director of NIH, or anyone else can testify unless OMB OK's his statement. It is not uncommon for someone to find himself publicly saying the opposite of what he thinks because he lost a battle with OMB.

It is fine for O'Neill to say, as he did at the Institute of Medicine meeting, "I'm disappointed in scientists for not standing up for their point of view when it differs from the party line." It is fine, that is, as long as they do not work for the government. Government scientists who swerve from the party line laid down by the OMB can find themselves in trouble. Such is the power of the Office of Management and Budget.

Notes

1. W. A. Niskanen, *Structural Reform of the Federal Budget Process* (Washington, D.C.: American Enterprise Institute, 1973).

2. John C. Sawhill, former associate director of OMB for natural resources, energy, and science, testified on July 23, 1973, on the reorganization of the office before the House Committee on Science and Astronautics. He is now the number-two man in the Federal Energy Office.

F. The Congress

The style of the legislative branch in setting policy for technological matters is quite different from that of the executive. In Chapter 14, Herbert Roback tells how Congress operates with respect to science and technology. The actors shift over time—new members of Congress replace old stalwarts, OTA and the Budget Office enter the scene, the Executive Office of Science and Technology Policy (OSTP) replaces the Office of Science and Technology (OST), congressional largess toward NIH shrinks, etc. But the basic process that Roback describes changes only slowly. Congressional committee chairpersons work closely with agency heads to guide the budget authorizations and appropriations and to oversee agency functioning. He describes how the fate of projects and programs depends on the economic climate and personal politics.

Next in Chapter 15, Barry M. Casper discusses the role of congressional science fellows in enhancing in-house technical analysis capabilities and dealing with technical issues. However, he notes that there are disturbing patterns of potential abuse of the fellowship programs and that the fellows have not overturned the "cozy triangles" among congressional committees, executive agencies, and corporate interests.

In his analysis of OTA in Chapter 16, Casper finds serious shortcomings, but high potential. High-stakes Congressional politics have influenced personnel appointments and the selection of assessment topics and have created a difficult setting in which to produce far-sighted, hard-hitting assessments. Committee interests in protecting "turf" are reflected in gaping holes in OTA topical coverage, notably in military technology. Casper suggests that OTA needs to take on the technology policy issues that really matter in ways that cannot be ignored.

14. Congress and the Science Budget
HERBERT ROBACK

Congress makes the laws of the land and in this sense gives final form to national policies and their organizational underpinnings. Whether the Legislative or Executive branch takes the initiative in developing any given policy or organization is less important than the adequacy of its response to national need. If the need is sufficiently compelling, the two branches of government will be in accord that action must be taken. They may differ in important details—there is give and take—but between them national policy finally is hammered out or delicately wrought. A law is written, an organization created, and the course of governmental action set for years to come.

Legislative milestones in science and technology stand out more clearly after World War II. Two immediate postwar problems were: how to conserve the resources and sustain the momentum of war-induced scientific research; and, more pointedly, how to organize and control the future development of atomic energy. The problem of atomic energy was more urgent, because it carried not only the sinister potential of mass destruction but the bright promise of mass benefit through power development and other peaceful applications. The Atomic Energy Commission was established in the year after the war ended, but the debates which eventually established the National Science Foundation as the key support agency for basic research dragged on for 5 years.

In the 1950's there were other milestone enactments. These were, in a sense, the panic years, with missile and space technologies in the forefront. The National Aeronautics and Space Administration, created in 1958, was an immediate answer to Sputnik. The same year the Advanced Research Projects Agency and the Directorate of Re-

Copyright © 1968 by the American Association for the Advancement of Science. Reprinted from *Science*, Vol. 160, May 31, 1968, pp. 964-971, by permission.

search and Engineering were established in the Department of Defense. Spurring these new organizations was a quest for national security interwoven with national prestige and welfare. Although NASA's mission, for example, was stated in terms of peaceful space exploration, in the public mind Sputnik posed a military menace, because it denoted Soviet mastery of large booster technology and complex space systems which had military import.

Our crash programs for ICBMs achieved a formidable defense posture and the ambitious Apollo program demonstrated dramatically our entry into the space race. Thereafter, in the 1960's, the government became increasingly involved in the welfare field. The technologies, techniques, and resources applied to missile and space development were examined for their application to social problems. Aerospace contractors began to work on such prosaic problems as garbage disposal and traffic congestion or on such esoteric ones as an artificial heart or a teaching machine.

New government departments were established to deal with housing and transportation. Within existing departments new agencies and organizations sprang up to put science and technology to work for the Great Society. More and more research emphasis was given to traffic safety, urban transportation, air and water purification, public health, crime prevention, and scores of other problem areas which received legislative and executive attention. The larger action agencies created in response to acute national needs, develop vast technical infrastructures to support their missions. They build laboratories, let contracts, and acquire constituencies in business, professional, and academic circles. They attract community and regional support for the jobs and payrolls they provide and they gain advocates in Congress.

As technical empires expand, programs proliferate and agencies compete for technical talents and contractor resources, coordinating and control mechanisms are created to smooth working relationships, minimize overlap and duplication, concert policies and programs, develop joint projects, and identify gaps and omissions. The coordinating and control agencies, relatively small in size and consultative in nature, lack the money leverage and the operational impact of the action agencies. These coordinating devices vary in effectiveness, of course, depending on their locus and authority in governmental structure. The Office of Science and Technology, a coordinating agency at the apex of government, enjoys the prestige and authority of the President's Executive Office but undoubtedly suffers in its remoteness from the centers of technical power, where important

Congress and the Science Budget

decisions are made daily and agency heads report directly to the President.

Technical agencies of government, whether directly supporting or conducting research and development or more broadly concerned with coordination and review, participate in the allocation of technical resources. Congress does too, to the extent that it brings these agencies into being, defines their missions and authorities, provides money for their performance, and supervises administration. The definitive action in the first instance is the enabling legislation, which records congressional intent and carves out a technical area for performance. The legislative charter constitutes a mandate to do things, a commitment to create and use national resources for years, decades, even centuries.

Funding Process

Once an agency is created by the Congress, its career is shaped largely by the funding process. By law, the President is called upon to submit, during the first 15 days of each session of Congress, his budget plan for the government (1). From the executive side, budget making is practically continuous. The budget is the most worked over and worried over document in the government. By processes too intricate and probably too mysterious to detail here, the yearly budget estimates are developed, starting at lower levels in the agencies, working their way up through the administrative hierarchies to the agency heads and finally to the President. The higher the level of review, the more intense the competing policy and political considerations which require budgetary adjustment and constraint. The Budget Director said that the President approved for 1968 $27 billion less than the agencies wanted (2).

There is blood and sweat in this budget-making process, internal politics, bureaucratic in-fighting, arbitrariness, hunch, whim, and prayer, all of which finally comes out as informed judgment and rational decision. Whatever its faults as a means of allocating resources, the budget plan on the executive side does have a single source for final review and approval—the President, assisted by the Bureau of the Budget and, in technical fields, by the Office of Science and Technology.

In the Congress there is no comparable organization and single source of authority. All budgetary matters are handled by the appropriations committee in each house but these committees alone do not control the funding process; and even if they did, they cannot work on the budgets for all agencies at one time nor report out multiple budget requests in a single piece of legislation. The omnibus appropriation bill, not unknown, is impracticable. In any case, it would not yield a coherent statement or systematic analysis of scientific and technical activities any more than the huge budget document submitted by the President. For the government as a whole, R & D is not a budget category but a jigsaw puzzle of activities in many agencies. Each year some 13 or 14 appropriations subcommittees in each House work on the agency budgets assigned to them. Each subcommittee reports out its own bill; these, together with supplementals, add up to 15 or 16 appropriation statutes each year.

Appropriation bills are considered first in the House of Representatives. Agency heads, reinforced by their budget experts and other officials, make their way to the Hill with detailed "justification" documents, and are there challenged, questioned, criticized, and sometimes given short shrift by House subcommittee members sitting in closed session. The subcommittee, after the hearings are done, "marks up" the bill, assisted by its own staff, cutting here or there, writing statutory language to limit freedom of administrative action in certain expenditures. Less formal but no less constraining are the instructions, admonitions, cautions, and requests written into the subcommittee reports. Agency heads who ignore them will not be treated gently in future appropriations. The appropriation bills have privileged access to the floor and occasionally amendments on the floor will add to, or subtract from, the committee's recommendations.

The appropriations committee members tend to suspect that agency heads pad budgets in anticipation of cuts, and thus cuts are usually made. Budget-cutting signifies the committee is exercising judgment and control rather than rubber-stamping the Executive, and has the added value of demonstrating congressional interest in economy. If the cut is of the "meat-ax" variety, across-the-board, of course the aspect of economy becomes more important than judgments about the merit of specific programs. The House also knows by experience that the Senate usually will be more generous in appropriation matters, and so cuts may be made in the House for bargaining purposes or as an economy gesture.

The Senate in fact serves as a kind of appeals body in hearing reclamas (3) from agency heads. Even in the case of the National Institutes of Health, which rarely suffer a cut and usually get more than they request, the generosity of the House consistently is outmatched by that of the Senate. In the past 18 years, for which I reviewed the figures, the Senate unfailingly has increased the amount

Congress and the Science Budget

of NIH appropriations approved by the House. The final result, as worked out in committees of conference, is a split somewhere along the middle.

A similar pattern shows up in the appropriations for the National Science Foundation. The Senate generally does better by that agency than the House, and the differences are compromised. Occasionally the Senate is content to take the House figure without change. Only once in the Foundation's 18 years did the Senate go below the House recommendation, and the issue was minor—a proposed $1000 contribution by NSF to the President's Committee on Equal Employment Opportunity. The Senate objected, and that is how the NSF came to receive, instead of the House-approved $480 million, an appropriation of exactly $479,999,000 in 1966, repeated to the dollar in 1967.

Yearly Authorizations

The Constitution enjoins the spending of public funds without a legislative enactment. Legislative rules and procedures enjoin an appropriation without a prior authorization. Historically the rationale for the distinction between authorization and appropriation was to separate substantive policy and money matters and to expedite appropriation measures by freeing them of the inevitable clutter of legislative "riders" for pet projects. In practice, this dual legislative process of authorization and appropriation has introduced new complications and delays. The complicating feature is the statutory requirement for *annual* authorizations. Appropriations for NASA, the AEC, and the "research, development, test and evaluation" of the DOD, which together account for the bulk of the government's R & D outlays, are among those which must wait upon antecedent authorizing statutes. If the separate legislative processes for authorization and funding are not in proper sequence, the fiscal year may run out, and special resolutions are needed to permit agency spending at the previous year's rate until the new appropriation becomes law. The appropriations delays in the first session of the 90th Congress, which caused disruptions in agency planning and even held up paychecks of employees in some agencies, were attributed in large part to the lag between authorizing and funding measures (4).

This dual legislative process—authorization and appropriations—with the process being repeated in each house of Congress, makes for a great many hearings and heavy demands upon the time of busy executives and administrators. It may be inconvenient and even exasperating to high-level government witnesses to spend hours, days, or even weeks before congressional committees, substantially repeating their testimony from one to the other, sometimes suffering delays when other witnesses take up too much time, or merely sitting by when senators and representatives must interrupt the hearing sessions to answer roll calls on the floor. Typically the Executive Branch does not look with favor upon yearly authorization requirements, which now cover approximately one-third of the total federal budget.

Still, this is the way of the Congress, and not much can be done about it. Neither the Congress as a whole nor its legislative committees are content to let the appropriations committees monopolize funding—the most important part of the legislative process so far as the executive departments and agencies are concerned. There are, moreover, compensating advantages to these departments and agencies in the requirement for yearly authorizations. The legislative committees often serve as protectors or advocates of agency programs in opposition to seemingly unwise budget cuts by the appropriations committees. By their hearings and reports on yearly authorization bills, the legislative committees keep abreast of new programs, broaden congressional participation in policy formation, provide platforms for new ideas, and raise the statutory ceilings for future funding. They are more innovative, marking out new areas of technological and social advance. The appropriating committees tend to be more tightfisted, since they have to add up the total bill and pay the check.

Reviews and Investigations

The Congress not only enacts laws to establish agencies, authorize programs, and provide funds; it seeks to determine how well the laws are administered and how wisely the monies are spent. This is termed the congressional "oversight" or review function, and it ranges from full-dress investigations or hearings in public or private to informal and intermittent inquiries by congressional staffs.

All committees of Congress are enjoined by law and rule to exercise "continuous watchfulness" over the agencies within their jurisdictional orbits. Several committees have established special subcommittees for oversight or investigation, such as the House committees dealing with defense, space, and commerce activities. The Committees on Appropriations have the right to investigate any agency or matter involving public expenditures, and the House committee employs a special survey and investigations staff, with rotating personnel, for this purpose. The reports of the investigative unit are used by the chairman and members in questioning agency witnesses at closed hearings. Occasionally these reports are made public, after the various House subcommittees on appropriations have completed their examining work.

The Committees on Government Operations also have jurisdictional reach over all government agencies and activities from the standpoint of assessing economy and efficiency. The Committees on Government Operations historically have close affinity with the General Accounting Office, which serves as an investigative arm of the Congress and is on call to any committee for personnel detail or special investigations. The GAO also inquires on its own initiative into matters where the Congress shows, or is likely to show, particular interest.

It can be expected that auditors and investigators will be around from time to time to inspect books and ask questions in government laboratories and even in academic halls where federal monies are spent. Scientists and science administrators are not notably more efficient, and sometimes much less so, than workers or managers in other sectors. GAO investigations usually are not concerned with broad policy and program choice, but with more mundane housekeeping matters—effectiveness of procurement, utilization of equipment, inventory control and accounting, and the like.

Both the House and Senate Committees on Government Operations established special subcommittees in the research and development area. The House Committee on Science and Astronautics also has a standing subcommittee on research. During the 88th Congress, a House Select Committee on Government Research of limited tenure also was established. These several groups, organized around research and development policies and problems, have been active in gathering information, stimulating new studies and more systematic data collection, promoting improved management practices, and generating or sustaining a dialogue in science policy issues of government and academic concern.

The hearings and studies of these subcommittees, valuable and informative as they are, do not bear directly on the allocation of resources in the manner of a legislative committee which authorizes a specific program or the appropriations committee which provides the requisite funds. The authorization and funding processes provide the greatest leverage of congressional control. On the other hand, the several subcommittees on research and development, relieved of the yearly grind of legislation, are better suited to examining policies and programs which cut across agencies and disciplines. They bring broader perspective and greater depth of analysis.

The Committees on Government Operations have jurisdictional reach over all government agencies by their concern not only with economy and efficiency but with government reorganization. Sooner or later every agency of government is involved in reorganization of one kind or another. If these are to be accomplished by presidential reorganization plans, the plans are referred to the Committees on Government Operations for review. The Office of Science and Technology, for example, was created by a reorganization plan in 1962. The House committee held hearings on the plan and submitted a favorable report to the Congress. This is a kind of reverse legislative process in which the President drafts a law, so to speak, and the Congress has the veto power. The plan has the force and effect of law after a 60-day waiting period if the Congress does not pass a disapproving resolution.

The rationale for creating the OST was partly to moderate congressional concern about overlap and duplication of scientific agencies and activities and to counter proposals for a department of science as a possible solution; and partly to make the OST director more accessible to the Congress. By defining his powers and duties in statutory form, the director would be made something more than a presidential adviser and thus would be partly relieved of the obligation of confidence attaching to that role. In recent years the OST director has made frequent appearances before committees of the Congress and has provided helpful information and advice. It cannot be said, however, that OST is rid of the problem of executive privilege or that the Congress is making the most effective use of its services (5).

When to consult with the Congress is always a problem on the executive side. Apart from the narrow issue of executive privilege which the President asserts now and then, on a matter of great sensitivity, administrators are reluctant to publicize their mistakes or to seek prior congressional approval for every decision they make. It would be obviously impractical to run to the Hill every time a problem comes up, and premature publicity can be harmful to policies still in the making or issues still unresolved. Generally committee chairmen and agency heads consult frequently and cooperate well. Occasionally the communication breaks down, bad feelings are engendered, and caustic comments are traded.

The investigation of the disastrous Apollo fire by the House and Senate space committees was a case in point. When a House committee member obtained in some undisclosed way a copy of the so-called Phillips report criticizing the major Apollo contractor and released it to the press, there was great consternation. This was the first time the committees had heard of the Phillips report. The House Committee on Science and Astronautics recommended, and the House approved, a provision in the NASA authorization bill for FY 1968 requiring that agency to keep the space committees of each house "fully and currently informed" (6). This language was borrowed from the Atomic Energy Act, which places a similar obligation on the AEC relative to the Joint Committee on Atomic Energy. However, the Senate objected to the House language, and it was deleted from the authorization bill on the ground that NASA's obligation to keep the committees posted is implicit in its organic act (7). Later, the NASA administrator and the committee chairmen came to some less formal understandings that the flow of information in the future would be improved.

Politics and Personalities

Each year thousands of bills are introduced in the Congress, hundreds of them perhaps dealing with one aspect or another of science and technology. These bills are referred to committees having jurisdiction in the subject matter, and most of them are pigeonholed. Some may get a hearing in committee and a favorable report leading to action in one or both houses. If a measure is part of the President's legislative program for the year, it stands high on the priority list, particularly if the President and the majority in Congress are of the same party. If the Chief Executive is forceful in conveying his views, if the party majority in Congress is substantial enough to prevail against temporary coalitions, legislative proposals of the Administration will fare better. Bills concerned with yearly authorizations and appropriations demand timely attention lest the wheels of government grind to a halt. War and taxes and spending and approaching elections enhance the acerbity if not the profundity of the political debate.

Within a broad and sometimes vague context of party differences, each member of Congress records his personal and political preferences, his rough scale of priorities for legislative action. He may try, by action in committee or on the floor, by amendments and parliamentary maneuvers, to give some effect to his own readings of priorities. For the most part individual legislative forays, though they may gain publicity and please a given constituency, do not substantially change the outcome. In the Congress the legislative distribution of work is rather highly specialized and the members rely heavily upon the committees involved. They look to the chairmen of the committees and subcommittees for guidance in the more complex technical areas requiring legislative action.

Whether a committee or subcommittee chairman is strong or weak, liberal or conservative, interested or inactive, makes a big difference to the legislative result. The watchfulness and influence of a member of Congress strategically placed as chairman of a committee or subcommittee, or high in seniority, may well account for the sustained support in a given research and development sector. Certainly the National Institutes of Health owe much of their generous funding to the unflagging zeal and attention of the late John E. Fogarty in the House and Lister Hill in the Senate. Fogarty, a former bricklayer, became a member of the House Committee on Appropriations in 1947. By 1949 he became chairman of the subcommittee handling appropriations for the Department of Health, Education and Welfare, which includes the Public Health Service and NIH. Senator Hill, who recently announced his intention to retire, heads the counterpart appropriations subcommittee in the Senate and also is chairman of the Senate Committee on Labor and Public Welfare, which handles substantive legislation in the health field.

The influence of key chairmen, important as it is, should not be exaggerated. The prudent chairman does not step too far away from the dominant sentiment and prevailing mood in the Congress. He tempers his opinions and tailors his recommendations to gain support and acceptance for his legislative and appropriation bills. It follows that some programs fare better than others not only because the sponsoring committee or subcommittee chairman is respected or persuasive, but because the program itself has wide political appeal. For example, diseases that disable and kill are the concern of all; they are close-to-home problems, and so the Congress is unusually generous in support of health and medical research. During an 18-year span, yearly funding for NIH rose from $60 million to almost $1.2 billion, and in only 4 of those years did appropriations fall below the amounts requested. Even in the 1968 fiscal year, when the economy drive was strong, NIH received $1,178,924,000 or 99 percent of the budget request.

Cumulative appropriations to NIH, compared with budget requests, show an average excess of about 15 percent. The NSF ratio, by contrast, shows an average deficiency of about 15 percent. It is apparent that basic research specifically associated with health and medicine is more appealing than basic research in general. In no year since it

Congress and the Science Budget

was created in 1950 has NSF received from the Congress quite as much money as the Budget Bureau has requested.

Appropriations for NSF and NIH are handled by different subcommittees and chairmen, NIH being a component of a cabinet department, NSF an independent agency. In each case the House subcommittee chairmen virtually "grew up" with the agency. Fogarty, as we noted, was on the job for NIH since at least 1949. The late Albert Thomas of Texas, chairman of the Independent Offices Subcommittee, supervised the NSF appropriation for the first 15 years of that agency's existence (excepting 2 years when Congress had a Republican majority). These men developed an intimate knowledge of the agencies' operations and were instrumental in providing for stable and continuous growth. If, as usually happened, the NIH yearly budget request was added to, and the NSF's subtracted from, under their respective leadership, at least these responses were predictable, and adjustments could be made without too serious consequences. Both of these venerable chairmen now have passed from the scene, and undoubtedly there will be changes in approach and treatment of NIH and NSF from the congressional side.

The ill-fated Mohole project was an immediate casualty. Though NSF reported that some $57 million of work already had been done, and only another $21.5 million was needed to finish the drilling platform, the Mohole contract was terminated after Albert Thomas' death in 1966. It happened that the contractor originally chosen for the project was headquartered in Houston, the subcommittee chairman's home city. The choice of contractor by NSF never sat well with members from states with other prominent contenders for the drilling job who represented the oil, metal, and aerospace industries. Project costs also seemed to have a discouraging elasticity, so that what started as a $20- or $40-million job kept stretching with each new estimate. When the legislators were told that the total project costs for 3 years, including platform construction, drilling operations, and yearly maintenance, had risen from $70 million estimated in 1963 to $127 mil;ion estimated in 1966, they decided to call the whole thing off. Recorded in the subcommittee hearings was the concern of members about the substantial cost overruns, the uncertain scientific results, the seeming lack of practical benefits, and the need to put the money elsewhere. Unrecoverable costs of Project Mohole will approach $40 million (*8*).

The demise of Project Mohole points up an interesting issue of congressional funding procedure already mentioned. The NSF, sponsoring agency for the project, does not go through a yearly authorizing procedure. It looks to the Appropriations Committee alone for a yearly review of program content and project justification. Possibly Mohole would have survived and been completed on schedule if separately authorized by a legislative committee.

There is no guarantee of course that a program authorized will be funded. Occasionally an authorization remains a dead letter or funding activity is deferred for some years. What Project Mohole needed, however, was a broader base of congressional support—involvement by more committees, informed attention by more members. From time to time suggestions have been made for subjecting NSF to the yearly authorization routine. So far the Congress has not responded, possibly because of the differing jurisdictions of the space and science committees in the house and Senate, their preoccupation with NASA and the Apollo program, and recognition of NSF's need to have latitude and discretion in allocating its grant monies and other support to universities without undue congressional interference and control.

Expenditure Control

Holding the key to the national purse, the Congress always is sensitive to economy arguments. Research and development has become, in recent years, a subject of congressional interest and concern because of the increasingly large dollar outlays in this sector. When economy in general becomes a matter of overriding concern, as was manifest in the first session of the 90th Congress, then R & D programs are bound to suffer with the rest, depending on their relative vulnerabilities.

Congress, awed by a $135-billion expenditures budget and a looming deficit of $20 billion or more, was reluctant to raise taxes. It preferred to cut back expenditures. Retrenchment was the order of the day. Reflecting this posture, Chairman Wilbur D. Mills of the House Ways and Means Committee made it known that if a tax bill were to have any chance at all, spending must be curtailed. And not only spending for the current fiscal year, but spending for the future. When asked about unnecessary spending, Chairman Mills was reported as saying: "Any professor who wants a vacation in the woods can get a grant to make a study of the formation of leaves and then he may write a report or he may not" (9).

As the appropriation bill for each agency or combination of agencies made its tortuous way through the first session of the 90th Congress, cuts were made here and there, and research and development

Congress and the Science Budget

came in for their share. NASA, for example, was cut $500 million, spread over some 17 R & D categories and a number of construction projects. The only R & D category not cut was $21 million for "human factor systems." The Voyager program, budgeted for $71.5 million, went to zero. Apollo applications, which seeks to determine what usefully can be done with Apollo hardware beyond the moon landing, went down from $454.7 million to $315.5 million, a cut of $139.2 million.

Vulnerability of the civilian space program had been building up, of course, for some time. The large expenditure demands of the war in Vietnam, and the assorted ills of urban society underscored by the prevalence of rioting and crime, made it difficult for many members of Congress to justify to themselves or to their constituents a $5 billion yearly outlay for space exploration. Recognition that we were too deeply committed to a lunar landing program to turn back and that there were important values to be subserved by a vigorous program of space exploration only made the dilemma more painful and the rhetoric of criticism more eloquent.

The DOD was cut $1,647,380,000, about one-tenth of which was in the category of research, development, test, and evaluation. The 1968 budget estimate for this category was $7,273,000,000 from which $133,400,000 (or 1.8 percent) was subtracted by the Congress. Modest cuts showed up in studies and analysis, basic research (but not Project Themis, which is designed to spread funds to smaller universities and help build up more centers of excellence), and nonprofit organizations identified as Federal Contract Research Centers. The Advanced Manned Strategic Aircraft program, still in the study stage, got an increase from $26 million to $47 million. The Manned Orbital Laboratory, representing the Air Force's most ambitious space program, got no cut at all in contrast to NASA.

From the President's 1968 budget request for new obligational authority of $144 billion, the general goal of cuts in appropriation requests was $5 billion. Chairman George H. Mahon of the House Committee on Appropriations had committed himself to at least this figure, and the appropriation bills in the House easily made the mark. Many of the funds were restored in the Senate, but in conference, the House budget cutters largely prevailed, reflecting the pressure of the economy drive.

Cutting the appropriation requests, measure by measure, is of course the traditional way. But economizers in the Congress wanted more drastic action. The Congress was behind in its appropriation chores anyway. Continuing resolutions had to be enacted to permit spending at last year's rate, while the work on new appropriation bills was completed. No less than five such resolutions were enacted and these offered opportunities for amendments proposing across-the-board cuts.

The issue was joined by Representative Frank Bow, ranking Republican member of the House Appropriations Committee, who proposed a $5 billion cut in actual expenditures budget for the current fiscal year rather than in new obligational authority. A point made by Bow and his supporters was that since obligational authority is spread over a number of years, a $5 billion cut in obligational authority would mean only half that amount or less in reduced expenditures for 1968. The proposal for expenditures reduction came in the form of the Bow amendment to a continuing resolution.

Critics of the Bow amendment argued in the floor debates that if the President were compelled by law to cut $5 billion from the expenditure budget, this would be tantamount to an item veto by the President and an abdication of congressional control of the purse. The proper and constitutional way—these critics argued—was for the Congress itself to make the cuts.

Charges of political maneuvering were heard. There were those who believed the only reason why a conservative opposition wanted to invest the President with blanket budget-cutting responsibilities was to thrust on him any political liabilities that might ensue. Others were greatly put out when the President, through some of his department heads, began to explore the possibilities of a freeze in contract awards and new projects.

The Bow amendment in the first try fell on a point of order, but prevailed in the House when another resolution was passed after a complicated sequence of parliamentary maneuvering. The House also adopted the Whitten amendment, which would limit 1968 expenditures to 95 percent of the preceding fiscal year—with certain exceptions. The Senate Appropriations Committee tossed out the Bow and Whitten amendments, and the Senate sustained its committee. Efforts on the floor to reinstate these and other across-the-board budget-cutting proposals were defeated by close margins. The majority of the senators was persuaded by testimony of the Budget Director that a $5 billion cut in current expenditures would mean a $10 billion cut in programs, and the "controllable" programs represented about one-third of the budget. The Budget Director could identify only $38 billion of the budget which was not already "locked in" by military necessity, fixed obligations such as interest on the debt, payment of salaries, and the like.

Congress and the Science Budget

After repeated conferences, the House and Senate came to a compromise on a $9 billion reduction in obligational authority and a $4 billion reduction in actual expenditures for fiscal year 1968. Cuts were to be spread uniformly throughout the Executive Branch on a percentage basis. Each department and agency was required to reduce 1968 obligations incurred by 2 percent for personnel and 10 percent in other respects. These cuts would apply to the "controllable" portion of each agency's budget, taking account of reductions already made in appropriation acts. Obligations for defense spending were to be reduced by 10 percent except for special Vietnam costs. The Congress also recorded its preference that personnel reductions be accomplished as much as possible by not filling vacancies and that new construction projects be stretched out rather than eliminated.

How and where to cut agency budgets was left to agency heads except for decisions already recorded in specific appropriation acts. Thus the readjustment of priorities and the budget cutting resulting from the economy drive are jointly done by the legislative and the executive branches.

Outlook for Change

Congressional procedure is hallowed by tradition, and practices have evolved through the years to regulate and control the legislative process. The extent to which reforms are needed to oil the wheels of Congress and improve the legislative process is a subject of recurring debate. Certain it is that the Congress is slow to change its ways, and most proposals for reform fall by the wayside.

Perhaps the reformers want to make of the Congress an institution and assign it a role which it cannot and probably should not perform. The Congress participates in policy formation; it is a policy-making body in the broadest sense. But the Congress is not, in the nature of the case, an independent planning agency. It can challenge, question, criticize, modify or reject what the Executive proposes, but practically it cannot develop an alternative budget or plan all operations for the government as a whole. Some of the proposals for improving the capacity of Congress to be informed, to assess and choose, including the widespread use of computers and system analysis techniques, seem to assume that the Congress should duplicate the executive branch in planning and programming.

What the Congress needs to do, above all, is to preserve its pragmatic, common-sense approach to public affairs. The answer is not to develop a corps of technical experts on the Hill. Committees of

Herbert Roback

Congress—and individual members for that matter—can be as well informed about any subject as time and willingness permit. All the experts in government, in universities and elsewhere, even in the far corners of the earth, are available to them. For most matters demanding legislative attention, there is no dearth of information. In fact, there is usually too much information for the busy congressmen to assimilate and use.

It has been proposed that scientific advisers be attached to the Congress either in a special office for one or both House or as adjuncts to committee staffs. A realistic appraisal of the committee structure of the Congress discloses the difficulties which would confront any such proposal. There are a score of committees and 100 or more subcommittees in each House. While defense, atomic energy, space, and commerce committees (in addition to the appropriations committees) predominate in science legislation, many other committees have jurisdiction and interest in one phase or another of scientific affairs.

The jurisdictional areas cannot be neatly delineated. Assignment of legislative bills by direction of the Speaker of the House and the President of the Senate, or of committee tasks by direction of their respective chairmen, do not always follow the written rules. And even if they did, the rules cannot cover the variety of changing situations reflected in legislative processes.

To cite a few jurisdictional problems: A bill to create a Commission on Science and Technology is referred in the Senate to the Committee on Government Operations which has jurisdiction, among other things, over organizational matters in the government; the same bill in the House is referred to the Committee on Science and Astronautics which has jurisdiction in science matters. The Senate Committee on Aeronautical and Space Sciences does not have the same jurisdictional reach as the House committee. The bill to amend the National Science Foundation Act went to the House Committee on Science and Astronautics and was approved by the House, but in the Senate it went to the Committee on Labor and Public Welfare, where it still remains. So far, proposals to align the House and Senate committee jurisdictions in science matters have not prevailed (*10*).

With this complex committee structure and distribution of legislative work in the Congress, a small central group of scientific advisers could not hope to respond to the recurring or continuous legislative demands for information and advice on scientific affairs. And, on the other hand, if the scientific experts were attached to separate committees and subcommittees, it would not be practical, except in very limited and informal ways, for the staff expert on one committee to assist another committee.

Congress and the Science Budget

The consequences would be either that the scientific experts would have to build up a big bureaucracy of their own in the legislative branch, which the Congress would not sanction, or they would be bypassed in the hurry and scuffle of legislative work.

If the scientific advisers were full-time employees, they would be expected to do many chores beyond their special talents. If they were part-time consultants, then they would be too remote from the legislative process to respond to its exigent demands and hence no better placed than witnesses before the committees.

A modest alternative to these proposals was the creation in 1963 of the Science Policy Research Division in the Legislative Reference Service of the Library of Congress. This division, though small in size and subject to a variety of demands, has been very helpful to the Congress. The division serves committees and members by collating information, making special studies and analyses, and developing background material for legislation.

There are, of course, many other kinds of scientific advisory services available to the Congress. Expert witnesses can be called individually or in panels, in public or private session. Occasionally a committee contracts with a university group or nonprofit corporation for technical studies. The House Committee on Science and Astronautics has a standing advisory panel established in 1960 of 16 members drawn from major scientific areas. Reports of the yearly panel proceedings are presented by the committee to the House of Representatives. The National Academy of Sciences and the National Science Foundation have undertaken special studies for the Congress. The Office of Science and Technology, as noted above, also has been represented before many congressional committees. Finally, particular committee staffs will include a few persons with good technical background when the legislative work requires them.

Since the Executive Branch, through the Office of Science and Technology and other instrumentalities, has introduced coordinating mechanisms for research and development, it is common to inquire whether similar steps can be taken in the Congress. Changes in committee structure and organization are extremely difficult to make, and even if they were easy, we still have a government of many agencies and many functions, for which legislative divisions of labor have to be devised.

Typically and inevitably, the legislative and the appropriations committees are agency- rather than function-oriented. After all, agencies are going institutions of government and must be provided for. Members become familiar with their key personnel and styles of work and their administrative problems. As a function or category of

Herbert Roback

concern, research and development is much discussed but little affected, except as it is involved in the agency's budget.

From time to time, the special committees and subcommittees concerned particularly with research and development will examine such activities as they cut across agencies and as they enter into general policy considerations. However, the committees have their own jurisdictional outlook, work schedules, staffing arrangements, and methods of carrying on the legislative business. For these and other reasons, joint committees or joint hearings of separate committees are not ordinarily a convenient way to do business in a bicameral Congress. Perhaps this is why the idea of a Joint Committee on Research Policy, proposed by the Elliott Committee a few years ago as the legislative counterpart to OST in the Executive Branch, has not taken root.

Department of Science and Technology

A re-sorting of responsibilities and better integration of committee activities on the congressional side possibly could be achieved if a Department of Science and Technology were created. The case for a new department rests not on bringing together a multitude of governmental research and development functions torn from their agency settings, but on the availability of large relatively self-contained technical agencies to serve as basic components of the new organization.

There is good logic in establishing a Department of Science and Technology not only to house older more mature technical agencies but new ones, such as oceanography, which has not yet found a permanent home, as well as other technical agencies and bureaus which may now be in less congenial surroundings. It would not be wise to break out research and development functions from old established departments and agencies except if attachment to the parent organization is tenuous in terms of mission or readily adaptable to serving multiple agency missions.

In the event such a department came to pass, committee changes likely would be in order (*11*).

Notes

1. Title 31, *U.S. Code*, sec. 11.
2. Bureau of the Budget press release, Oct. 6, 1967.
3. "Reclama" is a government term which refers to the budgetary appeal procedure.

4. *Congressional Rec.* Dec. 7, 1967, p. H16458.
5. "The Office of Science and Technology," *Science Policy Research Div., Legislative Ref. Serv., Library of Congress Rep.* March 1967.
6. House Rep. No. 338, 90th Congress, 1st session, June 6, 1967, p. 148.
7. House Rep. No. 535, 90th Congress, 1st session, Aug. 3, 1967, p. 9. Section 303 of the National Aeronautics and Space Act contains a proviso "that nothing in this Act shall authorize withholding of information by the Administrator from the duly authorized committees of the Congress."
8. Hearings before Independent Offices Subcommittee of the House Committee on Appropriations, 89th Congress, 2nd session, Feb. 1, 1966, pt. 2, p. 115; also hearings before Independent Offices Subcommittee of the Senate Committee on Appropriations, 89th Congress, 2nd session, June 13, 1966, p. 1726.
9. L. J. Carter, *Science,* 158, (Oct. 13, 1967), 233.
10. Jurisdictional changes were recommended by the Joint Committee on the Organization of the Congress, but no legislative action has been taken on these particular recommendations.
11. In this article the author is expressing his personal views and they are not necessarily those of any member of the Congress.

15. Scientists on the Hill

BARRY M. CASPER

> D: Yes, Sir. I spent some years on the Hill myself and one of the things I always noticed was the inability of the Congress to deal effectively with the Executive Branch because they never have provided themselves with adequate staffs, had adequate information available...
> P: Well now they have huge staffs compared to what we had.
> D: Well, they have huge staffs, true, as compared to what they had years ago. But they are still inadequate to deal effectively...
> P: (expletive deleted) Don't try to help them out!
>
> President Nixon and John Dean,
> Oval Office, Feb. 28, 1973[1]

As these words were spoken, two efforts were already underway to help Congress in an area where it most needed help—science and technology. Several professional societies were organizing fellowship programs to send scientists and engineers to work in congressional

Reprinted by permission of the *Bulletin of the Atomic Scientists,* November 1977, pp. 8-15. Copyright © 1977 by the Educational Foundation for Nuclear Science.

offices, and Congress was preparing to open its own technological research arm—the Office of Technology Assessment (OTA).[2]

In terms of *numbers,* these efforts have had a remarkable effect. When the 93rd Congress opened in 1973, there were only two PhD scientists on the permanent congressional staff and only a handful of staffers with any significant scientific background. Now there are more than 50 PhD scientists and engineers on the Hill.

With five years of experience, one can begin to assess the *impact* of these programs on Congress and on the scientific community. With this as the goal, over 40 interviews with congressional scientists and other staff members were conducted during the summer of 1976. This report will focus on the new staff scientists.

The fellowship programs brought scientists directly to the places where decisions are made in the Congress—the offices of its Members and Committees. The programs began at a critical time. Soon after they started in 1973, the Arab oil embargo focused the nation's attention on the energy crisis. As the Congress began to grapple with the diverse and complex aspects of energy policy, the scientist-fellows were much in demand. Given the almost total lack of in-house expertise four years ago, it is striking today to find scientists nearly everywhere energy R & D and related issues are dealt with in the Congress.

By the most primitive measure, the programs have been judged a success by the Congress: with very few exceptions, the Fellows have been asked to join the permanent staffs of their offices after their fellowship year. It is an ideal hiring arrangement for the congressional offices. They have a year to look the Fellows over and observe their abilities and their politics at close range without spending a penny. The fact that so many decide to offer their own payroll money to keep the Fellows indicates that they have generally been pleased with what they have seen. However, this probably says at least as much for a Fellow's personal and forensic skills as for his or her scientific competence.

In describing one electrical engineer's success on the Hill, another staffer explained: "His value is less his ability in technical matters and more his substantial ability as a political operator." Whatever their salient characteristics, it is clear the fellowship programs have brought in first-rate staff, useful to the Congress.

One intriguing effect that has emerged is the strong coupling between the Congress and one particular kind of scientist—the physicist. Of the 36 scientists and engineers who held fellowships during the first four years, 13 stayed on in the Congress—and 10 of these are physicists. The overall retention rate was 36 percent; but for physicists, it was a remarkable 83 percent, as 10 of the 12 physicists elected to remain.

Several published accounts have described at length the activities of the scientist-fellows.[3] It seems clear that they have given Congress an enhanced capacity for in-house technical analysis.

Reflecting on their impact, physicist Allan Hoffman of the Senate Commerce Committee staff has noted that "having a scientifically trained staff member gives a Committee greater confidence in dealing with issues with technical components. It could also mean picking up on issues that would have been avoided in the past due to fear of technological complexity." For example, another Fellow, Haven Whiteside, credits Hoffman with just such a role in the auto fuel economy standards provisions enacted by the Congress in 1975. "Separating the emission and economy standards probably would not have happened without Al there," Whiteside said.

Staff scientists enhance the ability of Congress to filter and evaluate the enormous flow of technical information they receive from agencies, industry and other interest groups. One Fellow with the Senate Energy and Natural Resources Committee remarked: "I hate dealing with the nuclear industry; every issue you raise, they bury in paper."

Another Fellow with the House Armed Services Committee described his dealings with the Pentagon: "If you ask them a question in the morning, you get reams of response in the afternoon."

Scientists can also at least help to keep the information sources honest. AAAS Fellow Bill Moomaw, a physical chemist on the personal staff of Senator Dale Bumpers, related that "when a lobbyist would try to slip some technical mumbo-jumbo past, I would respond with a technically-worded question. Most were startled to discover that I was knowledgeable in a substantive way, suggesting a rather abysmal ignorance of technical matters in other offices."

In-house scientists can also provide Congress with discriminating access to the scientific community. Moomaw, for example, worked on the ozone depletion provisions of the 1976 Clean Air Act. He was able to follow the preprints from the very active research in progress. He recalls: "Because of my technical background as a spectroscopist and photochemist, . . . I was able to play an extremely useful role by maintaining direct contact with industry, university and government scientists who were working on the problem."

One must be careful, however, not to overstate the impact of scientists on the capacity of Congress to deal with technical information.

Barry M. Casper

Before the scientist-fellows arrived the specialized committee staffs were by no means devoid of competence to cope with technical issues. Commenting on his experience with a Senate subcommittee in dealing with such matters as auto emissions and pollution control devices, physicist Haven Whiteside emphasized:

> We have a political scientist, an historian, and an economist on the subcommittee staff who can talk to lobbyists just as well as a scientist. They have developed technical expertise over the years.... It is a myth that there was no scientific expertise here before. If you pick up the phone here, anyone in the country will talk to you. Access to the best technical support is readily available.... People with a scientific background will be comfortable with numbers, quantitative data; but you cannot say scientists are unique in this regard.

Thus, having scientists on the committees is useful; but, generally speaking, it has increased only incrementally the capability of congressional committees to analyze and evaluate technical aspects of legislation.

One thing is certain. Having the scientists on the scene has not much affected the level of congressional dialogue on technical subjects: superficial appeals to scientific authority are still prevalent.

For example, two of the 1976 Fellows, Ernest Johnson and Arthur Silverstein, were instrumental in developing the Hart-Kennedy bill to impose a tax on cigarettes based on their tar and nicotine content. Silverstein, explaining his approach to these issues in the Senate, stated that during the hearings he arranged a "Nobel Prize Winners Day in the Senate." With television cameras present, several Nobel laureates in medicine appeared before Senator Kennedy and testified to their support for his bill. That evening, excerpts appeared on Walter Cronkite and other television news programs. "That's the level to work on with the public," Silverstein asserted. The basic strategy, he explained, would be to have Senator Kennedy say "I'm not a scientist, but I have consulted experts. Your experts are the Tobacco Institute people; mine are Nobel Prize winners."

Such uncritical appeals to scientific authority evidently extend to the private dialogue within congressional committees and among the staff as well as to the public dialogue in hearings and on the floor. Physicist Kevin Cornell explained the problem faced by the new staff scientists: "You have to find a way to convince people, but you can't argue on technical grounds. You have to use authoritative sources to convince Members and staff." He did add this caveat: The continued reliance on authoritative sources may be because there has not been sufficient time for trust and confidence in the new staff scientists to build up. With such trust and confidence, Cornell suggested, Congress may eventually come to rely on the technical judgments of its own staff scientists.

Of course there is always the reverse danger: that is, scientists in the Congress will be believed too much, that they will misuse the authority of science. Physicist John Andelin, one of the two scientist-staffers before these fellowship programs began, asserts that abuse of this sort is already becoming a problem with some of the new staff scientists. According to Andelin, some of the Fellows "have taken advantage of their positions to influence policy. They have taken positions that are philosophical rather than technical, but have couched them in technical language."

Rules of the Game

In assessing their impact, it is important to appreciate the highly political milieu in which the scientist-fellows have been placed. Congress is the focus of the intense attention of many parties—in the Executive Branch, the private sector and those claiming to represent "the public interest." Describing the function of Congress in technology policy, physicist Ben Cooper noted, "long term planning is done in the Executive Branch; Congress is where you run it by the interests." In this milieu technical considerations generally take a back seat to political compromise and accommodation to particular interests. The staff scientists are by no means free agents; they are significantly constrained in the activities they may undertake. Diane Wyble, an electrical engineer working with Congressman Mike McCormack, summed it up nicely:

> The longer you are here the more you realize there are bounds over which you may not step. You do background work, etc., but it is the Members who make the decisions. The overriding constraint is there is politics involved in everything you do.

Arthur Silverstein, an immunologist who served with Senator Kennedy's Health Subcommittee, described how politics can intervene. One health issue that came before the Subcommittee in 1976 was President Ford's proposal for a nationwide program of inoculation against swine flu. With his background in epidemiology, Silverstein gave Congress the capacity to subject this program to independent critical examination. He and other subcommittee staffers had grave reservations about the program on medical grounds. But there was never any question of opposing it. According to Silverstein, "from the very first, everyone in Congress stood in awe of what was generally thought to be an impressive political coup by the President in this election year."

One of the first things the new staff scientist learns is that Congress has its own special "rules of the game." One physicist described his coming to work there as "like leaving the twentieth century and being thrust into a bizarre feudal system." To function effectively in this system requires that scientists be attuned to certain sensitivities and adopt certain modes of behavior. They soon learn that the staff qualities most valued on the Hill are loyalty to the Member, sensitivity to actions that might put Him in a politically vulnerable position, compatibility with His interests and politics and a willingness to maintain a low profile, meaning the staff does the work and the Member gets the credit.

Just how this socialization takes place is a subtle process, but it is evidently quite speedy and effective. The results are often striking. As Kevin Cornell has described it:

> I see a real difference in the Congressional Fellows in just one year, especially in terms of communication. They (especially those who stay on) adopt the Capitol Hill style, become less open and more closed-mouthed, and more "political" in their approach to issues.

What was particularly striking in the interviews conducted for this study is the difference in openness between those scientists who have decided to stay on and those who plan to return to their former careers. The former are much more circumspect and guarded in commenting about their activities. They are particularly sensitive about being quoted on the record as personally taking credit for influencing a Member's decision or even for work they did on his behalf.

Perhaps the most important question about the influx of scientists to the Congress is whether it has changed the politics of technology policy. A principal rationale advanced by proponents of both the fellowship programs and OTA was that in-house technical expertise would enable Congress to make independent assessments and, possibly, to challenge Executive Branch programs. However, there is a crucial flaw in this argument.

Cozy Triangles

If there were truly an adversary relationship between Congress and the Executive Branch in technology policy, then the addition of in-house congressional scientists might redress the balance of power. But this is frequently not the case. Instead there tends to be a sympathetic relationship among congressional committees, the Executive agencies and corporate interests. And these *cozy triangles* have traditionally been the principal locus of technology policy decision-making.

Within the Congress, technical competence and congressional receptors of technical information have traditionally been concentrated in the staffs of the specialized committees, who serve the chairmen and a few other senior members. This concentration of expertise and information contributes to the power of both the committee chairmen and these cozy triangles.

Thus the most relevant question about the political impact of the new scientist-fellows is not whether they augment Congress vis-à-vis the Executive Branch, but whether they support the cozy triangles or the competing interests in Congress. In fact, most of the Fellows have associated with the congressional committees concerned with technology policy or with senior members of those committees. This is only natural since that is where policies are decided in the Congress. As one physicist working with a House committee put it, "that's where what you do makes a difference." It does mean, however, that to a first approximation the Fellows have augmented, not challenged, the traditional centers of power.

If one looks more closely, however, there are some intriguing second order effects. For example, in a few instances, Fellows have associated with junior members and their presence has produced some limited challenges to the traditional order. One notable example in 1976 was in the Senate Interior (now Energy and Natural Resources) Committee, where Senator Dale Bumpers with physical chemist Bill Moomaw and Senator John Glenn with physicist David Hafemeister and electrical engineer Len Weiss had far more influence on the ERDA authorization bill than might have been expected of freshman Senators. One member of the Energy Committee's permanent staff described their effect:

> The presence of the Fellows throws sand in the system so far as the cushy relationship between agencies, industry, and the Congress is concerned. They slow the system down, prevent things from happening. For example, ERDA has more trouble with Bumpers because Bill Moomaw is here with many questions. The prime example is John Glenn, who feels free to tinker with details of the ERDA program. Glenn amends the shit out of things. Glenn and his staff have a mind of their own.

This suggests that increased efforts to fill the technical vacuum, which still exists in the Congress outside the committee staffs, could result in competing centers of technical competence, and that, in turn, could have an effect which could be of political significance in the long run.

The Scientific Community

When asked to evaluate the impact of the new staff scientists, one Fellow, Michael Telson, asserted: "The only certain thing is that it has opened up Congress to the scientific community; it has placed people sympathetic to the concerns of the scientific community within the walls of Congress." Of course, there is no such thing as a monolithic "scientific community" with a single agenda of public policy concerns. But the thrust of the assertion is valid. Various parts of the Congress have been opened up to many scientists as a result of the placement of a few scientists in offices on the Hill.

While a principal motivation for the fellowship programs is public service, it is clear that the scientific and engineering professions and their professional societies are benefiting as well. Links between the professional societies and the Congress have been forged and strengthened. And, in the case of some Fellows, individuals sensitive to the concerns of scientists and engineers have ended up in positions where they have influence on congressional activities directly related to those concerns.

For example, AIAA Fellow William Widnall went to work in 1975 for the House subcommittee that had authority over the NASA budget. Widnall, an aerospace engineer, was disturbed by what he felt was a lack of imagination, excitement and long-term vision in NASA planning for future space activities. He convinced the subcommittee chairman to allow him to organize a series of hearings in the summer of 1975 on the long-term promise of the space program. The resulting three volumes, *Future Space Programs 1975,* effectively lay out the case for a continued U.S. space program at a high level of funding. Widnall pointed out in his final report to the AIAA how this effort might benefit the profession:

> The impact of these hearings, in my opinion, was to expose some members of the Congress to some very exciting future program possibilities, backed up with the practical view that these would help the economy significantly.
>
> NASA planning by comparison appeared to be conservative and shorter range. The hearings strengthened the case for a significant Space budget increase.... If an outcome of the Future Space Program Hearings is a more understandable and ambitious set of space objectives, with an increased budget commitment, then aerospace engineers will directly benefit from increased employment opportunities.

Scientists on the Hill

Paul Horwitz, an APS Fellow from the Avco-Everett Research Laboratory who spent 1976 on the personal staff of Senator Kennedy, brought an industrial scientist's perspective to the Congress. For example, he feels the tightening of the academic job market means that, increasingly, bright young scientists are going to work in industry, but that federal funding of research has not reflected this change. Horwitz maintains that a disproportionate share of federal research dollars goes to academic institutions and national laboratories while private laboratories, increasingly the repository of the creative minds, are being neglected.

Senator Kennedy serves as chairman of the subcommittee that oversees the National Science Foundation (NSF). Not surprisingly, having Horwitz on the staff led to particular attention being focused on the question of NSF funding of research in private industry.

The committee report on the 1977 NSF authorization bill points out: "At no point in the Act is the Foundation prohibited from funding profit-making research institutions, such as industrial laboratories or individual inventors and entrepreneurs. However, in practice it has largely refrained from doing so." The report directs the National Science Foundation to undertake a study that would include the anticipated problems and benefits of a policy that would broaden the funding patterns of the Foundation to include more support of nonacademic institutions. The authorization bill directs the National Science Foundation to establish an Office of Small Business R & D, to act as an ombudsman for small businessmen within the National Science Foundation.

These illustrations show how scientists and engineers on the Hill naturally enhance the sensitivity of Congress to the problems of their professions. Of course, as is certainly the case in these examples, there may not be general agreement, even within the "scientific community," with the particular policies they promote.

Potentials for Abuse

In the course of the interviews for this study, it became apparent that certain practices associated with some of the congressional fellowship programs create at least the possibility for conflict of interest. The potential for abuse seems sufficiently serious that the sponsoring professional societies should institute procedures to protect against it. Each of the practices is, by itself, quite defensible. It is only when taken together that they create disturbing patterns of potential abuse. Consider first the *practices,* then the potential *problems* and, finally, some possible protective *safeguard* procedures.

Barry M. Casper

Three Practices

—*Supplemental salaries.* In several instances, Fellows are being paid by their employer while they are working in the Congress. For each of the engineering society programs, the usual practice is to have the normal employer pay approximately half the Fellow's salary. Other professional societies have also permitted high-salaried Fellows from industry or universities to receive supplemental funding from their employer during the fellowship year.

—*Overlap of Fellow assignments and employer interests.* In some instances, the congressional office in which the Fellow chooses to work deals with legislation affecting his or her employer and/or oversees federal agencies whose policies affect the employer. For example, last year one Fellow employed by a defense contractor spent the year on the staff of the House Armed Services Committee, which oversees the defense budget. Another Fellow, employed by a major manufacturer of light water nuclear reactors and developer of breeder reactors, served on the personal staff of an influential member of the Joint Committee on Atomic Energy and chairman of its breeder subcommittee.

—*New jobs for industry Fellows.* Several Fellows from industry indicated in interviews that they had no intention of resuming their prior positions when they returned to their companies. They expected to be promoted to jobs that would enable them and their companies to make use of their experience and contacts in Washington. A year on the Hill evidently can substantially increase a Fellow's value to his company.

Possible Patterns of Abuse

Individually, none of these practices necessarily poses a problem. But they suggest some potential patterns that should be of concern. For example, two obvious ones are:

—*"Our Man in Washington."* It would be possible under the present programs to have an individual working in the Congress while partially on the payroll of a private employer, writing legislation that affects his or her employer or influencing a federal agency to adopt policies that would benefit his or her employer. The combination of having Fellows paid by an outside source while in a congressional office that deals with matters of interest to that source *could* lead to abuse.

—*Lobbyist Traineeships.* It would be possible under these programs to have individuals sent to Washington explicitly to serve in particular congressional offices in order to make contacts that would be used to benefit the employer by having the Fellow serve as a lobbyist when he returns. The combination of having the Fellow working in a congressional office that deals with the interests of his employer and the possibility of having him use his Washington contacts when he returns *could* lead to abuse.

Safeguards

Serious thought must be given to safeguards if the sponsoring professional societies, who are lending their good names to the fellowship programs, are to insure that they will not be perverted to purposes other than those intended. Even the appearance of conflict of interest might seriously compromise the reputation of a sponsoring society.

One possibility is to restrict the assignments of the Fellows. This seems a practical response to the "Our Man in Washington" problem. In cases where Fellows are being paid by private sources, they should not be permitted to work in congressional offices that deal with matters of interest to those sources.

On the other hand, in the majority of cases, where Fellows are being paid exclusively by the professional societies, this would seem an unduly restrictive requirement. In such cases, the best interests of Congress might well be served if the Fellows were permitted to use their background and experience to the fullest in choosing a congressional office, even if there were some overlap with employer interests. Certain safeguards should be required, however:

—To avoid the "Lobbyist Traineeship" problem, the employer should be required to stipulate that the Fellow will not take part in any transactions on behalf of the employer with the Congress or agencies he or she dealt with while a Fellow, for a specified period of time—at least two years.

—There is still the possibility of a variant of the "Our Man in Washington" problem if Fellows coming from an institution and planning to return spend a year in a congressional office dealing directly with matters that affect the institution, even if they receive no stipend from the institution during the fellowship year. To avoid this problem, the sponsoring professional society should negotiate beforehand a detailed agreement between the Fellow and the congressional office, explicitly setting out those areas of the office's business that will be out of bounds to the Fellow because of potential conflict of interest.

Thus, what is needed now is a formal code of ethics, applicable to the Fellows, their permanent employers, and the Congress, explicitly designed to protect against conflict of interest situations.

Barry M. Casper
The Place to Be

To conclude this discussion of the impact of scientists in the Congress, an illuminating perspective is provided by the comments of two Congressmen who are themselves scientists, Rep. Mike McCormack and Rep. George Brown.

The question, "Is Congress the place to be if a scientist wants to influence public policy?" elicited strikingly different replies.

> Rep. McCormack: Yes. Key members of committees are close to the action, second only to the White House offices as the place to influence policy.
> Rep. Brown: No. If a scientist wants to make an impact on public policy, he should work with public interest groups. The key to major changes in policy is public opinion. When public opinion changes, the Congress and the Executive Branch follow along.

The reason for their disagreement, I would suggest, is basically political. McCormack and Brown differ over what are desirable ends in technology policy and this colors their evaluation of congressional means of influencing those ends. If a scientist is generally comfortable with the policies espoused by the "cozy triangles," as McCormack is, then Congress is indeed the place to be. On the other hand, if he or she feels less comfortable with these policies and feels "major changes" are in order, as does Brown, then the possibilities of accomplishing this while working within the confines of the congressional committee structure are indeed limited.

But there is more that must be said. On hearing of Brown's remark, McCormack immediately responded: "George is only partially right. Popular opinion can do a lot of things, but it can't handle amendments."

What McCormack meant, presumably, is that Congress often deals with issues in such an esoteric fashion that the public cannot follow what is going on. In this way Congress is effectively insulated from public opinion on many issues concerning technology.

The long term hope for democratic control of technology rests on reducing that insulation. This will require better public understanding of the substance of technology policy issues and timely public awareness of what Congress is doing about them. In the meantime, however, Congress continues to make decisions that critically influence the course of technology policy. Under these circumstances, physicist Allan Hoffman reflects the sentiments of many of the staff scientists in Congress when he says: "I can have more impact on issues I care about here than any place I know of."

Notes

1. New York Times, *The White House Transcripts*, (New York: Bantam Books, 1974), p. 80.
2. The programs began in 1973 with 7 Fellows from 4 professional societies: the American Association for the Advancement of Science (AAAS), the American Physical Society (APS), the American Society of Mechanical Engineers (ASME), and the Institute of Electrical and Electronic Engineers (IEEE). In 1977, there are 17 Fellows from 11 professional societies: AAAS (3), APS (2), ASME (2), IEEE (2), the American Institute of Aeronautics and Astronautics (AIAA) (2), the Optical Society of America (1), the Acoustical Society of America (1), the American Psychological Association (1), the Federation of Societies for Experimental Biology (1), the American Society for Microbiology (1), and the American Geophysical Union (1).
3. See, for example, B. S. Cooper and N. R. Werthamer, "Two Physicists on Capital Hill," *Physics Today*, January 1975, p. 63; P. Horwitz, "Congressional Interactions at Very Small Impact Parameter," *Physics Today*, December 1976, pp. 28-32; A. Hoffman, T. Moss, and H. Whiteside, "Helping Shape Legislative Policy," *Physics Today*, August 1977, pp. 42-48.

16. The Rhetoric and Reality of Congressional Technology Assessment

BARRY M. CASPER

> Inspector Gregory: Is there any point to which you would wish to draw my attention?
> Sherlock Holmes: To the curious incident of the dog in the night-time.
> Inspector Gregory: The dog did nothing in the night-time.
> Sherlock Holmes: That was the curious incident.

Congress created the Office of Technology Assessment (OTA) in 1972 with much fanfare and optimistic predictions. Now, five years later, one can begin to compare the reality with the rhetoric and to assess the impact of OTA on the politics of technology. OTA's record to date, particularly in the area of military technology, will be examined here. In that area especially it is largely, but not totally, a record

Reprinted by permission of the *Bulletin of the Atomic Scientists*, February 1978, pp. 20-31. Copyright © 1978 by the Educational Foundation for Nuclear Science.

of omission. That record and the story behind it illustrate both the limitations and the potential of this new office.

Like the dog in Sir Arthur Conan Doyle's story, OTA has been strangely silent. In reviewing the list of OTA reports to date, it does not take a Sherlock Holmes to recognize gaping holes in the coverage of technology policy topics. The record contrasts with the conception of OTA enunciated by OTA Board Chairman Edward Kennedy:

> The impetus for OTA was congressional inability to assess major federal programs involving complex technology. The executive branch, with its vast resources, would develop complex programs which Congress had no capability to properly evaluate. So when the time came to vote on issues such as the ABM, the SST, or the Space Shuttle, it was extremely difficult for Congress to marshall the facts and arguments effectively.[1]

The most striking feature of OTA's record so far is the systematic exclusion from its agenda of controversial new programs of the sort mentioned by Senator Kennedy. In 1976 and 1977, for example, several such issues came to the floor, including the B-1 Bomber, the Liquid Metal Fast Breeder Reactor, and a number of provisions of the Clean Air Act. OTA studies could have helped members of Congress to marshall the facts and arguments effectively. But none were prepared. There is little indication in OTA's record that Congress will be any better off the next time it has to vote on a controversial major program involving complex technology.

A prime example of an area where OTA *might* contribute is new military technologies. None of the next generation of multi-billion dollar replacements for the U.S. strategic arsenal—the B-1 bomber, the Trident submarine, the MX mobile missile, the controversial cruise missile program—have received any attention from OTA. No "early warning" assessments have been performed on precision-guided munitions, which may revolutionize conventional warfare, or on anti-submarine warfare, which may erode the invulnerability of the strategic submarine—the keystone of the U.S. and Soviet nuclear deterrents.

What Sherlock Holmes deduced in *Silver Blaze* is that the dog did not bark because the culprit was his master. Reviewing the background and development of OTA, we shall discover that its reticence has an analogous explanation.

The first director of OTA, Emilio Q. Daddario, resigned last July and his successor Russell W. Peterson has just assumed office. To be sure the new director will be severely constrained by forces within the

Congressional Technology Assessment

Congress. But even with those constraints, he could move forcefully to rectify many missed opportunities of OTA's early years: assembling a first-rate staff, tackling the really controversial technology policy issues facing the Congress and providing "early warning" of the implications of new technologies and technology policies.

Two principal rationales were advanced for the creation of OTA. The first was the need to redress an imbalance of technical expertise between the Congress and the executive branch. Rep. Charles Mosher made this argument persuasively when the OTA bill was being debated on the House floor in 1972:

> Let us face it, Mr. Chairman, we in the Congress are constantly outmanned and outgunned by the expertise of the executive agencies. We desperately need a stronger source of professional advice and information, more immediately and entirely responsible to us and responsive to the demands of our own committees, in order to more nearly match those resources in the executive agencies.
>
> Many, perhaps most of the proposals for new or expanding technologies come to us from the executive branch; or at least it is the representatives of those agencies who present expert testimony to us concerning such proposals. We need to be much more sure of ourselves, from our own sources, to properly challenge the agency people, to probe deeply their advice, to more effectively force them to justify their testimony—to ask sharper questions, demand more precise answers, to pose better alternatives.[2]

The notion was that Congress was dependent on executive agencies for technical information and expertise but, as Sen. Kennedy has noted, those are "the very agencies having the most to gain or lose by the decisions made by Congress."[3] To challenge executive branch programs, Congress needed its own independent technical capability.

A second rationale for OTA was to introduce a unique new element into technology policy development. One of its principal functions would be an aspect of "technology assessment" termed "early warning." OTA would attempt to ascertain early in the development of a new technology what its full range of impacts was likely to be and, in particular, provide an early warning of potentially adverse consequences.

More generally, OTA was to attempt to anticipate future national problems such as the one we face today in energy and direct long-term planning of technology so as to avoid or at least mitigate those problems. In the language of the Technology Assessment Act, signed into law on October 13, 1972, "It is essential that to the fullest extent possible, the consequences of technological applications be anticipated, understood, and considered in the determination of public policy on existing or emerging national problems."

Among those who promoted the creation of OTA, two groups stand out in light of subsequent events. Within Congress, the initiative came from a subcommittee of what was then the House Science and Astronautics (now Science and Technology) Committee. The chairman of this subcommittee, Rep. Emilio Q. Daddario, and his staff developed the legislation over several years before Daddario left Congress in 1970 in an unsuccessful bid to become the governor of Connecticut.

Outside Congress, the intellectual underpinnings of OTA came from the academic community and, in particular, an eminent group of former White House science advisers, including Harvard's Harvey Brooks and MIT's Jerome Wiesner. From 1957 to 1972 the President's Science Advisory Committee (PSAC) and its expert panels brought in many of America's most eminent scientists to serve as part-time confidential advisers to the highest levels of government. PSAC was a relatively in-grown group that initially drew heavily upon scientists who had worked on the World War II atomic bomb and radar projects, and who subsequently maintained expertise in strategic weapons technologies, and acted as advisers to the Defense Department.

By all accounts the period of PSAC's greatest influence was 1957–63, the Eisenhower and Kennedy years, when it served as a counterveiling force within the government on strategic weapons policy. Later the concerns of PSAC broadened to encompass emerging global problems but its influence declined markedly under Lyndon B. Johnson. When some former members publicly opposed the Nixon administration on the ABM and the SST, PSAC found itself on the "enemies list" and it was banished from the White House at the beginning of Nixon's second term.

As their influence in the White House diminished, some of the PSAC scientists turned their attention elsewhere in government, most notably the Congress. A 1969 National Academy of Sciences study led by Harvey Brooks (*Technology: Process of Assessment and Choice*) laid out the case for technology assessment and urged both an expansion of the White House science advisory office to include this function (hardly a likely prospect under Richard Nixon) and the creation of a parallel apparatus in the Congress. A congressional mechanism was essential, the study argued, to insulate the assessment process from the influence of interested parties in the executive branch:

Congressional Technology Assessment

We do not believe that a viable technology assessment mechanism can be built if it is restricted to the executive branch. There are many arguments for this point of view, but most of them ultimately center on the undue influence of the President and his executive agencies on the course of the assessment process. Complementary organizations more directly accessible to the Congress are needed. We need to provide within Congress or close to Congress an effective public forum for responsible assessment activities of individuals or groups operating outside present governmental and industrial institutions.[4]

Thus, from the beginning OTA had a variety of constituencies, including leaders in the Congress and leaders in the scientific community. These constituents had a variety of visions, including technical expertise in exclusive service to the Congress; a new forum for the policy concerns of the ex-science advisers; and an early warning of unintended consequences of new technologies. Among these different constituencies and diverse visions are serious tensions. To appreciate those tensions, one need only consider how Congress really deals with technology policy.

Reality: Cozy Triangles

In order to cope with the wide diversity of policy issues that come before it, Congress delegates responsibility to its various specialized committees. And it is at the committee level where technology policy is usually decided upon in the Congress. A striking feature of the operation of many of those committees is the phenomenon of "cozy triangles." The congressional committees often develop close working relationships with the federal agencies and sub-agency units they oversee and with related industry. An obvious example is the strategic weapons triumvirate, consisting of the Armed Services and Defense Appropriations committees in the Congress, Pentagon planners, and defense contractors. The chief promoters of the ABM were not just the Army and Western Electric; Congress' defense committees were close, powerful allies.

Thus, one often finds not an adversary relationship at all between Congress and the executive agencies, but rather close cooperation in pursuit of mutually agreed-upon goals. A principal source of a committee's power, vis-à-vis the rest of Congress, is its access to technical expertise and information about new technologies through its agency and industry contacts. For example, former congressional staffer Alton Frye has described how Congress dealt with the development of multiple, independently targetable warheads (MIRVs) for intercontinental missiles:

> Throughout this period [1962-68], the fact of an American MIRV program had registered but slightly in Congress, the significance of the effort not at all. MIRV had reached the flight stage with an invisibility scarcely imaginable for a program expected to cost upwards of $10 billion. The members cognizant of the Minuteman III and Poseidon programs were those on the Armed Services Committees and the Joint Committee on Atomic Energy—men who, it must be said, inclined to skepticism about the possibilities for international arms control and who evaluated MIRV less as a threat to the stability of mutual deterrence than as a hopeful means of protecting U.S. strategic superiority. Outside the still closed circles of the national security cognoscenti, members of Congress were generally ignorant of MIRV, although sizeable sums had already been appropriated for the program in previous defense bills.[5]

Similar remarks would apply to almost any new technology. The vast majority of members of Congress are simply unaware of the implications of most new technologies as they pass through the research and development stage. Awareness and sometimes controversy comes (if at all) only years after the critical decisions have been made to proceed to the advanced development stage of a new technology such as MIRV, ABM, cruise missiles, the B-1, the liquid metal fast breeder reactor, the space shuttle, the SST, and so forth. At that point, of course, reconsideration is much more difficult. Muscle and money have been invested, jobs are at stake; a politically potent constituency for the technology has developed to resist the economic and bureaucratic dislocation that would then accompany significant modification of the program.

On the other hand, some members of Congress—those serving on the specialized committees—are aware of new technologies and are party to the decisions as to whether to proceed to advanced development. As the MIRV example illustrates, however, they tend to be close to the agency promoting the development and sympathetic to its objectives and outlook. This is the reality in Congress: the institutional setting in which OTA has been placed.

This reality calls into question some of the tacit assumptions underlying the creation of OTA, including the notions that Congress would *want* to

—Reduce its dependence on the executive agencies in many areas of technology

—Create a new center of authoritative information and expertise in these areas

Congressional Technology Assessment

—Probe deeply the secondary consequences of new technologies before they are deployed

—Provide a public forum for individuals and groups, such as the science advisers, operating outside present governmental and industrial institutions

The tensions between rhetoric and reality began to be reconciled when the way OTA was to be organized was debated in the Congress. The part of the original bill dealing with how OTA would be controlled had been written by Harvey Brooks, working with the staff of the Science and Astronautics Committee. It called for an OTA governing board consisting of 13 members: two Senators, two Representatives, the directors of the General Accounting Office and the Congressional Research Service, and seven public members appointed by the President. That is, the "public members" (presumably including eminent scientists) would be a majority of the board.

But Congress, acutely sensitive to the allocation of political power, would have none of that. When the OTA bill was debated on the floor of the House in February 1972, Rep. Jack Brooks asserted, "I am convinced that in our representative system the experts should be on tap, not on top."[6] His colleagues ratified his proposal to substitute a board composed entirely of members of Congress. The final version of the bill established a bipartisan 12-member Technology Assessment Board (hereafter the Board) composed of six Senators and six Representatives, with an equal number from each political party. The bill contained only vestigal remains of the public board; it established a 12-member Technology Assessment Advisory Council (hereafter the Advisory Council), including 10 representatives of the public. The responsibilities of the Advisory Council, clearly subsidiary to the Board, were not spelled out.

During the debate, a concerted attempt was made to allay the concerns of those who feared that OTA might be a Frankenstein monster that would rise up and challenge the authority of the committees. It was emphasized that OTA (1) would remain small (less than about 100 persons), (2) would be a servant of the committees, and (3) would not become an independent center of authoritative expertise. Rep. Charles Mosher, a senior member of the House Science and Astronautics Committee and later the first vice-chairman of the OTA Board, described the relation of OTA to the Congress:

> It is absolutely fundamental to our entire concept, and it is the very essence of this bill, that the OTA shall not in any way usurp any of the intrinsic powers or functions of the Congress itself, nor of any of the

Congressional committees; it will only be supplemented. The OTA shall be solely a servant of the Congress.[7]

In particular, OTA would serve the committees. In Mosher's words:

> Now it will be asked, who shall have the privilege of asking or demanding assessment reports from the new Office of Technology Assessment? As we conceive it, in the bill before us, the OTA shall be responsive only to official requests from the committees of the Congress, acting through their respective chairmen; or, of course, the OTA could also respond to requests and instructions from the Congressional leadership.[8]

Fears that OTA might itself become an independent center of authority were assuaged by assurances that the OTA staff would not itself perform assessments, but rather would merely administer contracted studies that congressional committees requested. Witness this exchange between Rep. James Rhodes of Arizona and Science and Astronautics Committee member Rep. Marvin Esch:

> *Mr. Rhodes.* Mr. Chairman, as I read the committee report on this bill, there is no intent to allow this office to develop an in-house scientific capability. In other words it is not intended to be a scientific resource. It is to be a clearing-house and a purveyor of knowledge which has been gathered by other governmental bodies and which is made available for this office for that particular purpose.
> *Mr. Esch.* The gentleman from Arizona is correct. It is intended that it would be a filtering or channeling of information down to specific committees rather than conducting intensive research 'in-house.'[9]

Rep. Mosher spelled out the limited function envisioned for the staff:

> It is very important to recognize that the staff of the new OTA will not themselves do the actual assessment studies and reports that Congress will ask of them. The essential function of the OTA staff will be administrative. They will be expected to identify, recruit, and employ the best available expert talent, wherever it may be found, to do the actual assessment studies and reports.[10]

It took nearly a year and a half following passage of the Act before OTA actually went into operation in early 1974. During that time, the 12 members of the Board were named, with Sen. Kennedy chosen as the chairman for the 93rd Congress. The Board then selected former Rep. Daddario to serve as director. Several individuals formerly associated with the White House Science Advisory office in the Kennedy era were named to the Advisory Council and became its leaders,

Congressional Technology Assessment

including the first chairman Harold Brown (now Secretary of Defense), his vice-chairman Edward Wenk, and the present chairman Jerome Wiesner.

So far, the forces most responsible for shaping the nature and direction of OTA have been the personal interests of the Board members, the constraining influence of the congressional committees, and the extreme caution of its first director.

The Board has played a major role in both personnel and program choices. In reading the transcript of the first substantive meeting of the Board, it is disconcerting to find that the initial concern of several of the members was acquiring additions to their own staffs. The following exchange is illustrative of the discussion:[11]

> *Sen. Humphrey:* I serve on five subcommittees of the Congress and don't have a single staff man assigned to one of those subcommittees. Now, by God, I've worked long enough and hard enough so that I ought to get one out of this Congress. And I am not about ready to permit that to happen any longer.
>
> *Sen. Schweiker:* Here, hear! I feel the same way exactly.
>
> *Sen. Humphrey:* There was a time when I had them running out of my ears, but I lost all my seniority. I know how the ballgame is played around here.

Sen. Kennedy, who was chairing the meeting, felt compelled to remind his colleagues that they might be subject to criticism "if the first thing all of us did was just go out and get someone." Nevertheless, that is just what they did and loyalty to a Board member or to Director Daddario was a principal criterion for the selection of OTA group leaders and other key staffers.

This has led to an OTA staff of mixed quality—a handful of outstanding individuals, with a core of hard-working and dedicated scientists, who are, however, decidedly less impressive. Given the support for OTA in the scientific community when it first started and the pool of literally thousands of scientists who applied, the staff quality is most regrettable. OTA's staff could have been highly-qualified, first-rate people.

What is most striking is the extent to which OTA's program has been tied to particular interests of certain Board members. The whole organization reflects this influence. Initially it was decided to organize around working groups in six program areas: Oceans, Food, Health, Transportation, Materials, and Energy.

The Oceans Group is closely identified with Board member Sen. Ernest Hollings, to the point where the group functions like an extension of his staff for the National Ocean Policy Study. Sen. Humphrey did get a staffer out of OTA, the leader of the Food Group, and to a large degree this group's activities have been oriented to serve Humphrey's interests. "Mr. Health" in the Senate is Edward Kennedy; he named the leader of the Health Group and, so far, the group has worked mostly on his requests.

In part these close associations represent a necessary transition phenomenon, with Board members contributing requests to get the OTA groups started, with the expectation of a much wider constituency in the Congress eventually. In part, however, OTA has clearly been used for empire-building. This has started the organization off in the service of much narrower and more limited ends than the language of the OTA Act would suggest.

Furthermore, the program areas chosen encompass only a subset of technology policy issues. For example, the two largest categories in the federal R & D budget, military and space, are not included. Many environmental issues, such as air pollution and ozone depletion, do not fit either. Thus, at the same time Sen. Kennedy was stating publicly that the impetus for creating OTA was "to assess major federal programs involving complex technology... such as the ABM, the SST, or the Space Shuttle," the OTA Board was creating an organizational structure that in effect defined as out of bounds for OTA many of the most controversial and costly programs.

From the beginning, a powerful constraining influence on OTA's activities has been congressional committee interests. My strong impression from a series of interviews with committee staffers is that outside the few committees that have worked very closely with OTA groups, most technologically-oriented committees are suspicious of OTA and a little bit nervous. They are mostly concerned that OTA stay off their "turf," or at least that it tread lightly and in a closely-supervised manner if it should enter their territory.

On one extreme is the hostile reaction of a House Armed Services Committee staffer, who said: "Who in the defense establishment would touch OTA run by Kennedy and Humphrey. I can't imagine OTA putting out material that Kennedy disagrees with." More typical are skeptical comments of staffers on energy and environment committees about the utility and quality of OTA studies: "If OTA had a fine track record, we might turn to them." The exception is the reaction to those studies in which the OTA staff maintained close consultation with the committee staff: "We are exceedingly pleased with the quality of their work," said the staff director of Sen. Hollings' National Ocean Policy Study. He added: "We have a strong professional rapport with the OTA staff."

Congressional Technology Assessment

Within OTA, several key personnel have served as protectors of the primacy of the committees, most notably Kennedy's successor as chairman of the OTA Board in the 94th Congress, Rep. Olin Teague, and former Director Daddario. Teague, Chairman of the House Science and Technology Committee, has stated his view of OTA's relation to the committees in no uncertain terms: "The Office of Technology Assessment has only one function. That is to aid the committees of the Congress as they go about their legislative and investigative duties."[12] This principle has been institutionalized by the Board. Although the OTA Act permits OTA studies to be initiated in three possible ways—by the Board, by the director, or by the request of a congressional committee—the Board has adopted a policy of awaiting committee requests.

Naturally, many of the committees have not been eager to invite OTA into their territory. For example, none of the four committees primarily concerned with defense—the House and Senate Armed Services Committees and the House and Senate Defense Appropriations Subcommittees—have submitted a request. They are not likely to in the future, either. One staffer of the House Armed Services Committee stated flatly: "The Office of Technology Assessment is invisible for defense issues."

Daddario's Style

The most powerful protector of committee primacy was Director Daddario. Perhaps the best word to describe his leadership is "cautious." He ran OTA like a congressional office, personally keeping close tabs on even the most pedestrian details. This imbued everything OTA did with a large measure of Daddario's brand of political sensitivity and caution, often resulting in significant delays for even minor steps, while the political waters were tested and the appropriate political litmus tests were applied. In particular, Daddario took great care to avoid offending the committees. His response to committee apprehensions about OTA was to avoid "sensitive" topics. OTA's own in-house technology assessment theorist, Joseph Coates, has articulated the conceptual basis for the Daddario style: "Not all important problems can be addressed at any particular time in a given institution or institutional environment because the political costs of even addressing the question, much less the political risks of possible outcomes, may be too great."[13]

Thus, for example, the volatile issue of nuclear reactor safety has been the subject of two requests to OTA. In 1974, Sen. Henry Jackson requested that OTA assist in funding the American Physical Society's reactor safety study. OTA declined. In 1975, Senators Ribicoff and Glenn requested that the OTA cooperate with the General Accounting Office on a comprehensive, independent assessment of ERDA's reactor safety program. OTA adroitly stepped aside and permitted the GAO to do the entire study itself.

A counterveiling, but less powerful, influence on OTA has been the Advisory Council, an articulate, activist group which clearly has a different, broader vision of what OTA might do. Institutionally, it is an anomaly. Its function is not clearly spelled out in the OTA Act; that was left to the Board to define. This has led to considerable tension.

Among the Council there is a strong feeling that the Board's policy of awaiting requests from congressional committees will lead to neglect of "early warning" technology assessments and too much emphasis on short-term policy studies. It has urged alternative avenues for initiating OTA studies. In 1975, the Council recommended to the Board that OTA institute procedures to uncover and bring to the attention of Congress important emerging issues and technologies that might not surface in the congressional process. It was suggested, for example, that OTA "have funds available for use by individuals and groups that come to OTA with ideas for assessments on early-warning type issues."[14] This obviously runs counter to the Board policy of committee primacy. In 1976 a tentative step in this direction was taken; under the aegis of Sen. Clifford Case a small group on New and Emerging Technologies was created.

While the science advisers' original vision of OTA as PSAC reincarnated has dimmed, it has not died. The Council has urged tapping the resources of the scientific community in the same ways PSAC did: through greater use of panels of experts on specialized topics and more studies performed in-house by an augmented OTA staff, with less reliance on large studies contracted out to private think tanks. In his valedictory letter of December 1975, out-going Council chairman Harold Brown argued forcefully for this approach. For some Board members, like Chairman Olin Teague, this raised the old specter of a prestigious OTA staff with independence and authority to challenge the committees. In a heated reply to Brown, Teague recalled:

> OTA was sold to the Congress from start to finish, House and Senate, as a *contract* operation. It was also sold on the basis of a small, but highly capable in-house staff. I can say in all candor, as one who must justify OTA's budget to the Appropriations Committee each year, that OTA would be unfunded today without those assurances.[15]

Congressional Technology Assessment

Nevertheless, it is now clear that OTA will *have* to move, however reluctantly, in the direction urged by the Council. The initial operating mode on conducting assessments through large contracts to private contractors has been, by all accounts, a disaster. Time after time, OTA has received attractive, promising proposals from those organizations, then, hundreds of thousands of dollars later, uncritical, superficial reports.

One early study of solar electric power generation, for example, cost OTA over $300,000 and, according to one staffer, "even as an interim report, the study would have been an embarrassment." That particular study was essentially restarted from scratch as an in-house project by a small but highly competent group of OTA staffers led by physicist Henry Kelly. Released last June, it promises to be an important contribution to the energy policy literature, effectively challenging the present emphasis of ERDA's solar electric program on large central generating facilities.

Referring to large contracted studies, one OTA staffer remarked: "We don't do that anymore: we will contract, but only for small and well-defined tasks." This trend away from contracted work is clearly reflected in OTA's budget figures:[16]

	Fiscal Years (percent)		
	1974	1975	1976
Contract obligations	72	59	42
Personnel compensation	20	31	45

This change in philosophy is now a part of OTA's public posture. Near the end of his tenure at OTA, director Daddario explained to a reporter, "It was never envisioned that we could contract everything out."[17]

As for the role of the Advisory Council, it has in effect been given its own OTA group, R & D Priorities. During the latter days of the Ford administration, one observer likened this group to a Democratic White House Science Advisory Office in exile. Indeed, while a Republican task force in the White House was mapping potential issues for the new Office of Science and Technology Policy, three panels of the R & D Priorities Group (under Harvey Brooks, Lewis Branscomb and Edward Wenk) were conducting what looked very much like a parallel exercise at OTA.

Barry M. Casper

Organized by Sen. Kennedy's appointee Ellis Mottur, a highly respected science policy specialist formerly at George Washington University, the R & D Priorities program has a style and tone quite different from the rest of OTA. Its three panels are composed primarily of prestigious scientists and science policy analysts. It also has a task force on Appropriate Technologies, composed of citizen activists from outside the science establishment and led by consumer advocate Lola Redford. The panels and task forces have been given considerable latitude to critique the current R & D enterprise and propose new science and technology policy ideas and initiatives for Sen. Kennedy and other interested members of Congress.

One intriguing aspect of the R & D Priorities program is its potential as an important technology policy link between Congress and the Carter White House. Under the Executive Office reorganization plan, presidential science adviser Frank Press will have a minuscule staff of perhaps a dozen people. It would be natural for Press to consult closely with his former PSAC colleagues on the OTA panels for ideas and studies in support of his office.

OTA's efforts to steer clear of military issues have largely been successful. It has avoided "early warning" assessments of new military technologies. However, one policy study completed in 1975, the effects of a limited nuclear war, demonstrates well the potential of this new office.

In September 1974, Defense Secretary James Schlesinger testified before a subcommittee of the Senate Foreign Relations Committee on behalf of a new strategic doctrine. Under this doctrine the United States would target its nuclear weapons on Soviet military installations, not cities, and hopefully the Soviets would find it in their interest to do likewise. In support of this doctrine, Schlesinger claimed that Pentagon computer studies showed that fewer than a million Americans would die in a Soviet attack on the United States under these conditions.

The committee requested that OTA examine the Pentagon analyses, including the nature and validity of their underlying assumptions, and carry out independent analyses for the Congress.

When the matter came before the OTA Board, however, Board member Rep. Charles Gubser reported the strong reservations of the House Armed Services Committee about OTA's undertaking this study. Ordinarily that would have been sufficient to kill the proposal. But in this case OTA Board member Sen. Clifford Case, who had been responsible for the committee's original request, with the support of Chairman Kennedy, insisted that OTA take it on. It was a ticklish situation, but some backroom maneuvering by Director Daddario avoided a showdown. OTA would not formally accept the assessment, but instead, in an unprecedented action, would refer the matter to the Advisory Council.

The Council then assembled a panel of experts, chaired by former Presidential Science Adviser Jerome Wiesner, to review the findings of the Defense Department. The panel concluded that the Defense Department's casualty estimates were substantially low because of unrealistic assumptions in the analysis; the Panel suggested that the committee require Defense to redo the calculations, substituting alternative assumptions, which it enumerated. When this was done, the estimates were dramatically higher, with more than 20 million U.S. fatalities in a "limited nuclear war." The result was the 1975 report by the Senate Foreign Relations Committee, *Analyses of Effects of Limited Nuclear Warfare,* which included Secretary Schlesinger's original testimony, the OTA panel's critique, and the revised Defense Department calculations. These findings provided the basis for the Senate committee hearings, and later received wider public attention when two OTA consultants who had worked on the project, physicists Sidney Drell and Frank von Hippel, published them in the November 1976 issue of *Scientific American.*

As for "early warning" assessments of military technologies, OTA's record is a veritable vacuum. Even when asked by a congressional committee, it failed to deliver. Among the topics the Senate Government Operations Committee requested OTA to consider in an assessment of nuclear proliferation and safeguards was a technology currently being developed, laser enrichment of uranium, potentially a much less expensive means of obtaining fuel for nuclear power plants, but also possibly a significant new path to the proliferation of nuclear weapons. The study was released in April 1977, but the few superficial pages devoted to laser enrichment could hardly be termed a technology assessment. Although it had the opportunity, OTA certainly did not advance significantly the information available to decision-makers and the public about the costs and benefits associated with this emerging technology.

Reflecting on the OTA experience suggests a potential largely untapped. In particular, the record suggests that a few topics are really out of bounds for OTA, that a few scientists can make a difference in Congress' ability to challenge executive agency programs, and that a reassessment of OTA assessment procedures is in order.

Out of Bounds? To be sure, many committees would prefer that

OTA stay off their turf. But this may not be nearly the constraint that OTA's record of omission suggests. In most areas of technology policy, Congress is not totally monolithic; jurisdiction is frequently shared among a number of congressional committees. Thus, in the case of strategic weapons doctrine, not only the Armed Services committees, but also the Senate Foreign Relations Committee had a legitimate interest. Similarly, in the case of nuclear proliferation, the Government Operations Committee could claim some jurisdiction. This suggests that even if OTA were to maintain a strict policy of awaiting a committee request before taking on an assessment, few topics would really be out of bounds. As one senior OTA official put it: "I don't see that as an insuperable constraint. All you need is one committee to get it through."

In the area of military technologies, there is a natural way OTA *could* come into play. The executive branch is supposed to submit arms control impact statements to the Congress, describing the arms control implications of new technologies. In 1976, the first year the statements were received, they were clearly inadequate, leading to vociferous complaints from many members of Congress.[18] But Congressmen could do more than just complain. They (for example, the Senate Foreign Relations Committee) could ask that OTA assemble expert panels to review and comment on the statements, requesting additional information and analyses from the agencies where needed. Adding routine review under OTA auspices of arms control impact statements could convert pro forma responses into meaningful assessments in the future.

Reviewing arms control impact statements with the purpose of insuring their adequacy would permit OTA to perform its basic function as mandated by law—"to provide early indications of the probable beneficial and adverse impacts of the application of technology"—a function it has thus far neglected in the vital area of military technologies. As a matter of course it would bring those technologies that constitute the largest fraction of the federal R & D dollar under OTA scrutiny at just the point in their development that a technology assessment would be of most use to the Congress. To signal its receptivity to this assignment, OTA could begin by filling a most conspicuous gap in its organization by creating a Defense Technologies Group.

Challenging Agency Programs. The limited nuclear war study illustrates the *potential* OTA has brought to the Congress for challenging the technical assertions of executive agencies. It demonstrates the fallacy in the statement that a few scientists in the Congress cannot hope to match the technical expertise of the executive branch and are therefore bound to be overwhelmed. Of course, OTA cannot hope to match the executive agencies in numbers of specialists or in facilities for carrying out detailed analyses, but as the limited nuclear war study clearly shows, that is not necessary. By providing the capability for independent critical scrutiny of technical information, OTA can assist the Congress to comprehend the underlying assumptions and, if it chooses, to challenge those assumptions and require the agencies to come back with revised analyses.

Assessment Procedures. So far, the most successful mode of operation for OTA has proved to be the ad hoc expert panel. For OTA to take on many important topics with optimal links to the technical community, the large ratio of outside experts to in-house staff and the flexibility to choose appropriate personnel for each topic makes ad hoc panels attractive. They are particularly well suited to short-term policy issues, as exemplified by the Wiesner panel critique of the Defense Department calculations. Such panels can also aid in "early warning" assessments. The panels can propose questions congressional committees might put to agencies and industry, challenging assumptions and structuring the inquiry, but leaving detailed analyses to the respondents.

Another way of using ad hoc panels would seem particularly apt for OTA. It resembles the "mediation" mechanism proposed in the April 1977 *Bulletin* by ex-OTA staffer Nancy Abrams and OTA consultant Stephen Berry. One way to avoid charges of bias is to be superficial and eschew controversial issues; that has been the OTA approach. Another is to employ panels as Abrams and Berry suggest. Members of a panel would be chosen purposefully to represent the full spectrum of views on policy alternatives. The charge to the panelists would not be to reach a consensus, but rather to come to understand each other's positions and then prepare a document clarifying what is agreed and what is really at issue. Of course, considerable time would be required for such an exercise; OTA might develop a regular program of summer studies in this mode.

Such a panel could also provide a heretofore neglected, but very important function in the case of classified technologies. A regular mechanism for resisting excessive classification of new technologies and bringing now-secret information into the public domain is much needed. A principal purpose of the panel would be to produce an unclassified document that transmitted to the entire Congress and the American people the maximum amount of information consistent with legitimate "national security" requirements.

The emphasis in this discussion on ad hoc expert panels is not meant to imply that there is no useful place at OTA for in-house or contracted studies. Ad hoc panelists tend to have little time to spend on detailed analyses. They are good for posing perceptive questions that can help Congress keep executive agencies honest. But it is unreasonable to expect the agencies to respond with "early warning" assessments highly critical of their own new technologies.

Independent sources of assessments are also required. Some could be performed by the OTA staff itself. The on-site solar study shows that OTA can hire first-rate staff and turn out first-rate assessments from within the Office. But with a small staff, there is no way in-house studies could handle the already heavy load of assessment requests, much less extend OTA coverage to include those vast territories now off-limits.

As noted, almost without exception, OTA's large contracted studies with private "think-tanks" have been mediocre or worse. In part, this results from the contractor's assumption, validated by long experience with executive agencies, that the documents are not being produced to be acted upon, but rather to be placed on the shelf and pointed to as justification of the contracting agency's policies. The new elements needed for "early warning" and critical policy studies are "clean money," available from sources other than the interested agencies and contractors dedicated to critical analyses. OTA could be a source of clean money; contractors who will do the job can be found. In particular, OTA could provide the kind of support "public interest" groups need to hire adequate personnel for such studies and make it possible for university-based assessment groups to return the universities to their traditional and vital role as independent critics of government policies, from which they have strayed so far in an age of dependence on federal agency support.

Reflections

There is an immense gap between the rhetoric that surrounded the creation of OTA and its record of accomplishment. Some of the most significant areas of technology have been effectively out of bounds; the choice of assessment topics has been strongly coupled to the interests of a few Board members; the emphasis has been heavily weighted toward short-term policy studies, while the long-term "early warning" function has largely been ignored; and, with a few notable exceptions, OTA reports have been bland and superficial.

To be sure there are reasons for this disappointing start. It is a non-trivial problem to reconcile the academic concept of uncon-

Congressional Technology Assessment

strained "let the chips fall where they may" technology assessment with the practical constraints of calculating, high-stakes congressional politics. OTA is still groping to find its way. But, as our discussion suggests, it has the potential to do much more. This will require a major shakeup in the present organization, including putting together the outstanding staff OTA could have had in the first place and taking on the technology policy issues that really matter in ways that cannot be ignored.

Sherlock Holmes' evaluation of his bewildered colleague Inspector Gregory in *Silver Blaze* surely applies to OTA as well: "Were he but gifted with imagination, he might rise to great heights." What has been lacking so far is leadership with vision. Perhaps, OTA's new director, Russell Peterson, can provide this imagination and leadership.

Notes

1. Edward M. Kennedy. "Toward the Year 2000," address before the World Future Society, June 3, 1975, reprinted in the *Congressional Record,* June 17, 1975, p. S10787.
2. *Cong. Rec.,* Feb. 8, 1972, p. H867.
3. U.S. Senate Appropriations Committee. Hearings on Fiscal Year 1976 Legislative Branch Appropriations, 94th Congress, 1st session (Washington, D.C.: Government Printing Office, April 1975), part 1, p. 666.
4. H. Brooks and R. Bowers, "Technology: Process of Assessment and Choice," summary of National Academy of Sciences report in A. Teich, ed., *Technology and Man's Future,* (New York: St. Martin's Press, 1972), p. 231.
5. A. Frye. *A Responsible Congress: The Politics of National Security* (New York: McGraw-Hill, 1975). p. 51.
6. *Cong. Rec.,* Feb. 8, 1972, p. H873.
7. *Cong. Rec.,* p. H868.
8. *Cong. Rec.,* p. H867.
9. *Cong. Rec.,* p. H882.
10. *Cong. Rec.,* p. H867.
11. Technology Assessment Board Meeting, Feb. 20, 1974.
12. Senate Appropriations Committee, 1976 Hearings, p. 664.
13. *National Journal,* Sept. 21, 1974, p. 1425.
14. Technology Assessment Board and the Technology Assessment Advisory Council, minutes of joint meeting on Feb. 25, 1975.
15. Rep. Olin Teague to Harold Brown, Feb. 10, 1976: reprinted in OTA Annual Report (Washington, D.C.: GPO, March 15, 1976), p. 105.
16. *The Office of Technology Assessment: A Study of its Organizational Effectiveness,* communication from the chairman, House Commission on Information and Facilities (Washington, D.C.: GPO, June 18, 1976), p. 44.
17. *Physics Today,* 29 (November 1976), 102.
18. B. R. Schneider, "Stonewalling on the Arms Control Impact Statements." *Bulletin of the Atomic Scientists,* 33 (January 1977), 5.

G. The Courts

Having looked at the executive and legislative branches, we now turn to the judicial. How do the courts enter into the management of technology? Two important roles are identified by the authors of Chapters 17 and 18—the role of common law in deciding disputes between private parties and the role of the courts in monitoring the actions of executive agencies.

In Chapter 17, Laurence H. Tribe focuses on the common law as a means to control the consequences of technological development. He perceives a growing trend in contract, tort, and nuisance law to hold the technologist liable for damages sustained—even if these were unintentional and unforeseen. The effects of lawsuits go beyond direct damages to exerting influence for more responsible technology by dramatizing injustices and influencing public values.

Yet common law is not a solution in itself. There are inherent difficulties in trying to make technologists pay for the negative impacts of their technology. It is hard to fully internalize such costs, and partial internalizing may do no good. In addition, complex causal relationships and widely dispersed and diffuse impacts of technological change are poorly suited to legal judgments.

In Chapter 18, David L. Bazelon addresses a specific issue associated with advancing technology—risk regulation. He finds "no bright line between questions of value and of fact" and so determines it inappropriate to entrust risk assessment to "disinterested" scientists. While it is tough for judges to cope with complex science and technology issues, the courts can properly hold executive agencies (and their experts) accountable to explain the bases for their determination.

17. Towards a New Technological Ethic: The Role of Legal Liability

LAURENCE H. TRIBE

A recurring theme in contemporary thought and policy has been a growing unease with the human and ecological costs of advancing technology. Particularly in the most industrialized nations, it has become a commonplace that the by-products of technological progress have been none other than the ravaged countryside, the poisoned biosphere, the threatened loss of man's psychic and biological integrity. All are said to derive in some measure from the headlong rush of the engineer and the applied scientist; heedless of the price their progress exacts, the technologists plunge ahead.

This is, of course, a vastly oversimplified view. It makes science and technology villains when they are, in truth, victims—victims of a social and economic structure that encourages their thoughtless exploitation rather than fostering their more responsible development. But, wherever the blame may lie, and whether or not the search for "blame" is even a meaningful one, the fact remains that technological development has brought, and will continue to bring, significant and occasionally unacceptable injuries in its wake. Especially as the momentum of technology increases, and as its interpenetrating consequences become increasingly complex and even irreversible, the fear of its untoward effects become ever more justifiable.

Yet this contemporary concern with the undesirable "side-effects" of technology reflects much more than the unprecedented magnitude and seriousness of those effects. It reflects as well the fact that individuals and organizations have perceived their uses of science and technology to have wider consequences than they had previously imagined, have begun to learn how science and technology might themselves be turned to the exploration and alteration of those con-

Copyright © 1971 by UNESCO. Reprinted from *Impact of Science on Society*, vol. 21, no. 3, with permission of UNESCO.

Laurence H. Tribe

sequences, and have started to wonder whether they should assume greater responsibility for their control.[1]

I wish to address myself here to the social and ethical significance of this potentially expanding horizon of responsibility, with particular emphasis on the role of law and of legal institutions in stretching private decision-making to encompass a growing concern for remote technological effects.

Civil-Liability Trends in American Law

In the development of legal doctrine throughout the United States over the past half-century or more, there has been a pervasive (if occasionally faltering) trend towards the imposition of civil liability upon economic enterprises for the reasonably foreseeable adverse consequences of their technological choices.

Contract Law

Some of the earliest cases signalling this development arose as suits for breach of contract, in which the injured buyer successfully sought damages from the seller of a defectively manufactured product on a "breach of warranty" theory. Gradually, innovative courts relaxed the requirement of an explicit contract or warranty and permitted injured consumers to recover damage from the original manufacturer, a party with whom the consumers themselves of course had no direct contractual relationship.

As the citadel of contract technicalities crumbled, the circle of legal responsibility widened to include first the remote buyer, then the members of his personal or economic "family," and, most recently, those "innocent bystanders" who happen to be injured when a defective product malfunctions.

Tort Law

Tort law is that branch of law which governs wrongdoing (apart from breach of contract) which is not necessarily legally punishable, but for which civil action can be brought.

The movement of tort law has closely paralleled, and has not infrequently overtaken, the contractual trend described above. Increasingly, manufacturers and other technology-using individuals and institutions have been held liable in tort, quite apart from any contractual undertaking, for the injuries they have been found to cause—even in the absence of any showing of *intentional* wrongdoing or negligence. In this respect at least, tort doctrine has little more than returned to its pre-industrial shape, in which such defences as innocent intent and reasonable care were essentially unknown.

So long as the products or processes of an enterprise have in some sense been "defective" for their intended use, the enterprise has been required to compensate all individuals who could show they had been injured by the defect—under gradually loosening standards of causation, and with slowly liberalizing criteria of injury. Particularly when the doctrinal state of contract law precluded the imposition of liability on a warranty theory, tort law often furnished another route for the plaintiff.

The development of tort law to accommodate the plaintiff's interest in requiring the technological enterprise to compensate him, and hopefully to search out technological alternatives more compatible with his well-being, may be expected to continue.[2]

Property Law

A similar direction of development has been discernible as well in doctrines of American law governing the uses of property.

Among the most enduring (if malleable) principles of Anglo-American law has been the maxim that every person should so use his property as not to cause unreasonable injury to the interests of another. One who would use his property in violation of this standard has long been subject to an injunction at the behest of any threatened neighbour who asks a court for preventive judicial relief. When such an injunction has seemed to the court unduly harsh in light of the competing equities, the violator has traditionally been required at least to compensate his neighbour financially for the "nuisance" that his conduct has caused.

The law of nuisance, as even this necessarily simplified description should make clear, is capable of taking many turns, depending chiefly upon how broadly or narrowly courts define such concepts as unreasonable injury, how ready they are to balance the equities for or against preventive relief in the form of a creatively framed injunction or judicial decree, and how constricted or expansive a view they take of the sorts of interests the nuisance claimant should be allowed to vindicate through a private lawsuit.

The period of most energetic economic growth in the United States was understandably accompanied, and was to some degree fostered, by a set of nuisance doctrines quite hospitable to the relatively untrammelled exploitation of property. More recently, as the emphasis upon quantitative economic development has begun to give way to a new stress upon qualitative improvement in various aspects of national life, the law of nuisance has begun a corresponding movement (in some respects a return to much earlier doctrines) towards limiting the economic exploitation of private property to protect in advance, or at least to compensate after the fact, an ever-broadening set of "neighbouring" interests. Increasingly, for example, courts—occasionally spurred on by legislation—are holding that individuals demonstrably injured in some tangible way by environmental pollution are entitled to collect money damages on nuisance grounds from the industrial enterprise causing the pollution, and are in some instances entitled to a judicial order requiring that the enterprise take certain technological steps to abate the pollution in question.

Policies Underlying These Trends

Throughout American law, then, we have witnessed a broad movement towards the fastening of liability for technological harm upon the party causing it. The reasons for this movement have been many and diverse. Among them have been an emerging suspicion that the effective immunity from legal liability granted in the infancy of the industrial era is no longer necessary; a growing desire to provide restitution for individuals injured through no fault of their own in an increasingly technological society; an evolving belief that efforts should be made to spread the costs of such injury more widely and equitably; and a mounting conviction that it would be both just and useful to bring such costs home—to "internalize" them—to the enterprise best able to avoid or at least minimize those costs by changes in technological design or deployment.

It is this last reason that is of greatest interest here, for it is the reason that most responds to the emerging sense that the technologist should be "socially responsible"—that the designer and maker of tools to manipulate the natural order should reflect in his choices a deeper and more dominant concern for the human and ecological effects of all he does.

Criticisms of "Cost-Internalizing"

It has been fashionable of late to criticize this "cost-internalizing" function of liability—particularly on economic grounds, but on legal grounds as well.

The Economic Critique

The economic criticisms of cost-internalizing as an objective of rules imposing legal liability assume that such cost-internalizing is designed to obtain the greatest satisfaction for the most people—to maximize aggregate human satisfaction or to achieve the most efficient allocation of resources (which are often different ways of saying the same thing). The thrust of the economic critique has been to deny that rules of liability in fact further any such goal.

Economists have pointed out that, if bargaining were free (that is, if arranging and enforcing transactions among all affected parties required no expenditure of capital or energy), then resources would be allocated just as efficiently if liability rules gave injured parties absolutely no redress against the technological source of their misfortunes as they would be if such rules uniformly favoured the injured. A series of self-interested bargains or "bribes"—that is, settlements—among injurers and injured would yield the same minimum-cost solution in either case, with no demonstrable increase in aggregate human satisfaction attendant upon the judicial imposition of liability.[3]

Conceding that bargaining is *not* in fact free but actually entails substantial costs (such as, for example, the cost of excluding "freeloaders" who seek to benefit from social arrangements without sharing their burdens), economists have gone on to observe that, once such costs are adequately taken into account, it is far from obvious which rules of liability or non-liability would maximize the total bundle of human satisfactions.[4]

This difficulty is aggravated by the perverse economic law of the "second-best," according to which resources may not in fact be allocated more efficiently, or aggregate satisfaction increased, by a system that satisfies *more* of the conditions for optimal resource allocation than by a system that satisfies *less,* given that neither system satisfies *all*[5] Internalizing some costs, therefore, while not internalizing others, and while leaving unredressed some other forms of market imperfection, such as incomplete information (as in advertising or on product labels) and oligopolistic power, may be a "second-best" solution that is not better than no solution at all, and could in fact be worse.

The Legal Critique

Lawyers might be expected to join the critique in their own rather different terms, arguing that the judicial imposition of liability at the behest of private litigants possesses numerous procedural and institutional defects as a device for channelling economic and technological development in socially and environmentally responsible directions.

Like any other method of control that demands the measurement of unwanted effects and their evidentiary linking with specific technology-related causes, judicially assessed liability confronts intractable problems of proof, even when various aspects of the burden of such proof can fairly be shifted to the technological enterprise. This is so especially when the effects in question are intangible and thus difficult to measure, or when they are cumulative or interacting and thus difficult to trace to their origins; the problem of proof, however, remains a troublesome one even in the most straightforward situations.

The preceding problem, of course, is not peculiar to the private lawsuit; it plagues public modes of control as well. Several difficulties, however, inhere particularly in private litigation as a regulatory technique. Specifically, voluntary private enforcement of damage claims, whether brought by individuals or filed by groups in the form of class actions, tends to be "ill-suited to technological effects too weakly associated with presently existing and identifiable individuals, or too thinly spread among such individuals, to arouse their organized opposition in a timely way."[6] Broad-ranging ecological disturbance, damage to future generations, or the gradual erosion of what it means to be human—such consequences as these cannot readily be associated with any existing plaintiff or group of plaintiffs, and could not in any event be translated meaningfully into dollars.

Furthermore, the burdens imposed upon selected defendants by private lawsuits do not ordinarily fall into patterns that can readily be justified either as completely fair or as congruent with critical social priorities. It is often a matter of chance, or a corollary of opportunism, that one enterprise is sued and another spared. Private lawsuits, moreover, are typically too episodic and unpredictable in distribution and in impact to exert a timely and continuous influence on technological decision-making; they may be too prone, in addition, to trigger efforts to delay or circumvent the deterrent and compensatory effects of judicial cost-internalization.[7]

The threat of private litigation may thus represent little more than a thorn in technology's side, a nuisance to be avoided whenever and however possible, rather than a source of continuing economic pressure to change the direction of research and development.

To these specific indictments there might be added the observation that damage claims (or, indeed, cost-internalization generally, even if implemented through taxing schemes) are inappropriate when one's objective is not simply to make an activity bear all of its costs in the interest of improved resource allocation but is to require the prior consent of all affected individuals, or perhaps to forbid the activity altogether on moral or ethical grounds.

Towards a New Technological Ethic

Finally on a more institutional level, one might stress the inherent limitations of *all* judicial lawmaking. For courts are lacking in technical expertise and they must remain lacking in either the analytic tools or the political mandate to engage in such comprehensive planning as may be required or to fashion compromise solutions in areas necessarily charged with profound conflicts of value and clashes of preference.

The Social Value of Private Suits for Liability

On one level, both the economic and the legal critiques may well be right. The case for private damage actions, armed with even the broadest theories of liability, as primary tools for the direct social control of applied science and technology seems a tenuous one at best. But to end the inquiry there would be to ignore the complexity of legal process and subtlety of its social effects. Law operates on expressive and symbolic levels no less important than the direct instrumental level at which the analysis has thus far proceeded, and although the role of law at these more complex levels is less well understood and perhaps less amenable to empirical investigation, no exploration of legal liability as a tool for channelling technology would be complete without some analysis of these other facets of that tool.

Participation

I have in mind first the matter of participation: that is, the role of private damage actions as expressions of the litigant's involvement in those processes of technological decision that most deeply affect his life.

It may well be, as the now conventional economic wisdom would have it, that the internalization of costs within the technological enterprise will not invariably enhance aggregate welfare, and that it cannot yield economic goals unconnected with the maximization of aggregate utility. It may also be, as the earlier discussion indicates, that such cost-internalization as is none the less desired can rarely if ever be achieved satisfactorily through the device of private litigation.

But neither of these conclusions, however valid, denies the intangible benefits to the litigant, and to those in the society whose interests he represents, of being given a formal role in the adjudicatory process that determines whether a technological development affecting him may proceed and, if it may, on what economic terms. To have been accorded an opportunity to participate in that determination, even if only by presenting one's case to a judicial arbiter, has value in itself—as an affirmation of the litigant's right not to be treated as merely a passive object of another's technological strategy.

To some extent, indeed, the demand for improved social control of technology is little more than a demand by individuals that they not be reduced to means toward others' technological ends, that their claim to be respected *as ends in themselves* be recognized by whatever processes shape the technological choices that will alter their lives. Privately enforced legal liability, whatever its many shortcomings, at least has the merit of giving explicit recognition to this basic moral claim.

Publicity

A second quite easily overlooked facet of the private lawsuit that seeks to impose technological liability is its potential role as a catalyst for needed changes elsewhere in the system. Even if one were to deny completely the immediate utility of private lawsuits in this area, one could not disregard the capacity of such lawsuits—particularly in a litigious society that treats many of its judicial proceedings as occasions of great public interest—to dramatize injustices and to channel executive and legislative attention toward areas in need of more systematic reform or more comprehensive regulation. Just as much of the most creative civil rights legislation in the United States was encouraged by dramatic, if themselves ineffectual, judicial pronouncements, so too much of the legislation that is required to bring technological development under more sensitive human control might well be sparked—at least in the United States—by private lawsuits imposing (or, indeed, failing to impose) liability on technological enterprises.

Attitude

Third and finally, I think it important to consider the possible consequences of broadened technological liability as an ethical "lightning rod": as a focal point for the gathering, evolution and dissemination of new professional attitudes and new entrepreneurial assumptions with respect to the obligations that accompany the use of science and the development and application of technology.

The language of the law is often as revealing as its content—sometimes more so. Thus, it is not without significance that, even as the substantive *bases* for liability have broadened far beyond the intentional wrongdoer, the *rhetoric* of liability has retained a flavour of moral opprobrium. Courts still speak in such terms as "nuisance" and "defective product," even when they assess damages without requiring any showing of malevolence or neglect.

The wisdom of such a course, it seems to me, is that it can enable legal doctrine to induce cultural and moral change, to serve as an educational channel for the transmission of advancing ethical demands. To require an industry to "internalize the costs" of its technology says little or nothing to the community; but to hold the same industry liable for the "nuisance" it has caused, or to hold it legally responsible for the "defect" in its products, says a great deal and may even contribute, gradually and over time, to a new notion of technological responsibility. This new notion may alter, however slowly and subtly, the constraints by which technological decision-makers, concerned with what is legally permissible, feel bound, and the weight they attach to the many and varied indices by which the success or failure of their ventures might be judged.

The common law has often developed in just this way—through a complex interaction of changing judicial doctrines of liability with maturing concepts of obligation and ethical conduct, expressed in evolving practices and professionally articulated standards from which courts can extrapolate still further levels of common-law development. It is precisely such an interaction, difficult as its contours may be to define with precision and elusive as its measurement may be for the tools of behavioural science in their present imperfect state, that I think possible with respect to contemporary technology and the expansion of liability for its deleterious consequences.

Commentators who describe the legal trend identified in this essay as a trend toward "liability without fault" would do well to ponder the merits of a quite different description, one that views the cases as representing not a movement to liability *without* fault but a movement to a *wider concept of fault itself*. Such a description might capture, as the conventional description cannot, the profound ethical significance of the continuing expansion of technological liability that has come to characterize so many areas of the development of American law.

Notes

1. I developed this theme at greater length in a report written for the National Academy of Sciences: *Technology: Processes of Assessment and Choice* (Washington, D.C.: United States Government Printing Office, 1969), p. 15.
2. M. Katz, "The Function of Tort Liability in Technology Assessment," *The University of Cincinnati Law Review,* 38 (Fall 1969), 587–662.
3. R. Coase, "The Problem of Social Cost," *Journal of Law & Economics,* 3 (October 1960), 1–44.

4. 'Symposium—Products Liability: Economic Analysis and the Law', *The University of Chicago Law Review*, 38, (Fall 1970), 1-141.
5. R. Lipsey and R. Lancaster, "The General Theory of Second Best," *Review of Economic Studies*, 24, (1956), 11-32.
6. Laurence Tribe, "Legal Frameworks for the Assessment and Control of Technology," *Minerva*, April 1971.
7. Ibid.

18. Risk and Responsibility
DAVID L. BAZELON

Risk Regulation: A Problem for Democracy in the Technological Age

In 1906, Congress enacted the Pure Food and Drug Act, the first general food and drug safety law for the United States. Commenting on the provisions of the act, the House committee observed: "The question whether certain substances are poisonous or deleterious to health the bill does not undertake to determine, but leaves that to the determination of the Secretary... under the guidance of proper disinterested scientific authorities, after most careful study, examination, experiment and thorough research."

This statement reflected a deep faith in the ability of "disinterested" scientists to determine for society what substances posed an unacceptable risk. More than 70 years of regulation have called into question that naïve faith. We are no longer content to delegate the assessment of and response to risk to so-called disinterested scientists. Indeed, the very concept of objectivity embodied in the word disinterested is now discredited. The astounding explosion of scientific knowledge and the increasing sophistication of the public have radically transformed our attitude toward risk regulation. As governmental health and safety regulation has become pervasive, there is a pressing need to redefine the relation between science and law. This is one of the greatest challenges now facing government and, indeed, society as a whole.

Risk regulation poses a peculiar problem for government. Few

Copyright © 1979 by the American Association for the Advancement of Science. Reprinted from *Science*, Vol. 205, July 20, 1979, pp. 277-280, by permission.

favor risk for its own sake. But new risks are the inevitable price of the benefits of progress in an advanced industrial society. In order to have the energy necessary to run our homes and our factories, we incur risks of energy production, whether they be the risks of coal mining, nuclear reactor accidents, or the chance that a tree will fall on a man felling it to produce firewood. In order to have mobility, we risk auto accidents and illness from air pollution. In order to have variety and convenience in our food supply, we risk cancer or other toxic reactions to additives.

Ironically, scientific progress not only creates new risks but also uncovers previously unknown risks. As our understanding of the world grows exponentially, we are constantly learning that old activities, once thought safe, in fact pose substantial risks. The question then is not whether we will have risk at all, but how much risk, and from what source. Perhaps even more important, the question is who shall decide.

In our daily lives we do not confront the trade-off between dollars and lives very directly or self-consciously. But when we make societal policy decisions, such as how much to spend to eliminate disease-producing pollutants, we are painfully aware that we must make what Guido Calabresi has called "tragic choices."

In primitive societies these choices were often made by the tribal witch doctor. When the need to choose between cherished but conflicting values threatened to disrupt the society, the simplest path was decision by a shaman, or wizard, who claimed special and miraculous insight. In our time shamans carry the title doctor instead of wizard, and wear lab coats and black robes instead of religious garb.

But ours is an age of doubt and skepticism. The realist movement in law effectively stripped the judiciary of its Solomonic cloak. So, too, the public has come to realize the inherent limitations of scientific wisdom and knowledge. We have been cast from Eden, and must find ways to cope with our intellectual nakedness. To the basic question of how much risk is acceptable—a choice of values—we have learned that there is no one answer. To the problem of how much risk a given activity poses, we have learned that even our experts often lack the certain knowledge that would ease our decision-making tasks. Often the best we can say is that a product or an activity poses a "risk of risk."

Who Decides? Scientists v. the Public

Under these circumstances, the questions of who decides and how that decision is made become all the more critical. Since we have no

David L. Bazelon

shaman we must have confidence in the decision-making process so that we may better tolerate the uncertainties of our decisions.

Courts are often thrust into the role of authoritative decision-makers. But in recent years there has been growing concern about the ability of the judiciary to cope with the complex scientific and technical issues that come before our courts. Critics note, quite correctly, that judges have little or no training to understand and resolve problems on the frontiers of nuclear physics, toxicology, hydrology, and a myriad of other specialties. And the problem is growing. Hardly a sitting in our court goes by without a case from the Environmental Protection Agency, the Food and Drug Administration, the Occupational Safety and Health Administration (OSHA), or the Nuclear Regulatory Commission (NRC). These cases often present questions that experts have grappled with for years, without coming to any consensus.

But the problem, of course, is not confined to the judicial branch. Legislators are daily faced with the same perplexing questions. They, too, lack the expertise to penetrate the deepest scientific mysteries at the core of important issues of public concern. This problem ultimately strikes at the very heart of democracy. The most important element of our government, the voter, simply cannot be expected to understand the scientific predicate of many issues he must face at the polls.

Some well-meaning scientists question the wisdom of leaving risk regulation to the scientifically untutored. They wonder, to themselves if not aloud, whether the public should be permitted to make decisions for society when it cannot understand the complex scientific questions that underlie the decisions. Some scientists point with relish to the contradictory and seemingly irrational response of the public to risk. They observe the public's alarm at the prospect of nuclear power and note that the same public tolerates 50,000 automobile deaths a year. They decry the Delaney clause, which singles out cancer among all serious risks and imposes a rigid ban, regardless of countervailing benefits.

Scientists are also concerned by the growing public involvement in decisions that, in the past, were left entirely to the scientific community. Many scientists believe that regulation has intruded too deeply into the sanctum sanctorum. The controversy ranges from the periphery of scientific pursuits, such as OSHA regulation of laboratory work conditions, to the heart of the scientific enterprise, such as the conflict over recombinant DNA research. Regulators are accused of stifling creativity and innovation in the name of the false god of

Risk and Responsibility

safety. Science, once invoked as an ally to progressive government, more and more views the political process with hostility and disdain.

In reaction to the public's often emotional response to risk, scientists are tempted to disguise controversial value decisions in the cloak of scientific objectivity, obscuring those decisions from political accountability.

At its most extreme, I have heard scientists say that they would consider not disclosing risks which in their view are insignificant, but which might alarm the public if taken out of context. This problem is not mere speculation. Consider the recently released tapes of the NRC's deliberation over the accident at Three Mile Island. They illustrate dramatically how concern for minimizing public reaction can overwhelm scientific candor.

This attitude is doubly dangerous. First, it arrogates to the scientists the final say over which risks are important enough to merit public discussion. More important, it leads to the suppression of information that may be critical to developing new knowledge about risks or even to developing ways of avoiding those risks.

It is certainly true that the public's reaction to risk is not always in proportion to the seriousness of the threatened harm discounted by its probability. But the public's fears are real.

Scientists must resist the temptation to belittle these concerns, however irrational they may seem. The scientific community must not turn its back on the political processes to which we commit societal decisions. Scientists, like all citizens, must play an active role in the discussion of competing values. Their special expertise will inevitably and rightly give them a persuasive voice when issues are discussed in our assemblies and on our streets. But the choice must ultimately be made in a politically responsible fashion. To those who feel the public is incapable of comprehending the issues, and so unable to make informed value choices, I respond with the words of Thomas Jefferson:

> I know no safe depository of the ultimate powers of the society but the people themselves; and if we think them not enlightened enough to exercise their control with a wholesome discretion, the remedy is not to take it from them, but to inform their discretion.

Scientist, regulator, lawyer, and layman must work together to reconcile the sometimes conflicting values that underlie their respective interest, perspectives, and goals. This cooperation can be achieved only through a greater understanding of the proper roles of the scientific, political, and legal communities in addressing the public regulation of risk. Only then can we achieve a program of risk regulation that accommodates the best of scientific learning with the demands of democracy.

Sorting Out Scientific Facts, Inferences, and Values in Risk Regulation

The starting point is to identify the fact and value questions involved in a risk regulation decision. In determining questions of fact, such as the magnitude of risk from an activity, we as a society must rely on those with the appropriate expertise. Judges and politicians have no special insights in this area. Where questions of risk regulation involve value choices such as how much risk is acceptable, we must turn to the political process.

But even this formulation leaves many problems unanswered. There is no bright line between questions of value and of fact. Even where a problem is appropriately characterized as one of scientific fact, consensus and certainty may very often be impossible even in the scientific community. Many problems of scientific inference lie in the realm of "trans-science" and cannot be resolved by scientific method and experimentation.

The recent National Academy of Sciences (NAS) report on saccharin vividly illustrates the problem of separating fact from value in risk regulation. Although there is a reasonable scientific consensus on the effects of saccharin in rats, the important question of human risks and the appropriate response to those risks remain controversial. On the basis of uncontroverted animal experimental data, the NAS panel could not conclude whether saccharin should be considered a substance posing a "high" risk of cancer, or only a "moderate" risk. Yet this lack of consensus should not surprise us. As Philip Handler, president of the NAS, observed in his preface to the report, "the difference of opinion which led to this ambivalent statement is not a differing interpretation of scientific fact or observation; it reflects, rather, seriously differing value systems."

Handler's statement reveals a critical issue in risk regulation. When the debate over saccharin is couched in terms of the degree of risk, it sounds as though there is a scientific issue, appropriate for resolution by trained scientists. In fact, however, the terms moderate and high do not conform to any differences in experimental data, but rather correspond to the scientists' view of the appropriate regulatory response.

The growing use of analytic tools such as cost-benefit analysis magnifies the chance that unrecognized value judgments will creep into

Risk and Responsibility

apparently objective assessments. Even the most conscientious effort by experts not to exceed their sphere of competence may be inadequate to safeguard the validity of the decision-making process. Outside scrutiny may be imperative.

The Role of Courts

It is at this point that courts can make their contribution to sound decision-making. Courts cannot second-guess the decisions made by those who, by virtue of their expertise or their political accountability, have been entrusted with ultimate decisions. But courts can and have played a critical role in fostering the kind of dialogue and reflection that can improve the quality of those decisions.

Courts, standing outside both scientific and political debate, can help to make sure that decision-makers articulate the basis for their decisions. In the scientists' realm—the sphere of fact—courts can ask that the data be described, hypotheses articulated, and above all, in those areas where we lack knowledge, that ignorance be confessed. In the political realm—the sphere of values—courts can ask that decision-makers explain why they believe that a risk is too great to run, or why a particular trade-off is acceptable. Perhaps most important, at the interface of fact and value, courts can help ensure that the value component of decisions is explicitly acknowledged, not hidden in quasi-scientific jargon.

This role does not require, as some have suggested, that courts intrude excessively into an agency's processes. The demands of adequate process are not burdensome. Surely it is not unreasonable to suggest that agencies articulate the basis of their decisions or that they open their proceedings and deliberations to all interested participants and all relevant information.

These requirements are in everyone's best interest, including decision-makers themselves. If the decision-making process is open and candid it will inspire more confidence in those who are affected. Further, by opening the process to public scrutiny and criticism, we reduce the risk that important information will be overlooked or ignored. Finally, openness will promote peer review of both factual determinations and value judgments.

Coping with Uncertainty

Risk regulation in itself carries risks. No problem of any significance is so well understood that we can predict with confidence what the outcome of any decision will be. But there are two different kinds of uncertainty that plague risk regulation. Some uncertainty is inherent in regulating activities on the frontiers of scientific progress. For example, we simply do not know enough about the containment potential of salt domes to know with confidence whether they are adequate for storing nuclear wastes for thousands of years. In the face of such uncertainty society must decide whether or not to take a chance—to wait for more information before going ahead with nuclear production, or to go forward and gamble that solutions will be found in the future.

The other kind of uncertainty that infects risk regulation comes from a refusal to face the hard questions created by lack of knowledge. It is uncertainty produced by scientists and regulators who assure the public that there are no risks, but know that the answers are not at hand. Perhaps more important, it is a false sense of security because the hard questions have never been asked in the first place.

In the early days of nuclear plant licensing, for example, the problem of long-term waste disposal was never even an issue. Only after extensive prodding by environmental and citizens' groups did the industry and regulators show any awareness of waste disposal as a problem at all. Judges like myself became troubled when those charged with ensuring nuclear safety refused even to recognize the seriousness of the waste disposal issue, much less to propose a solution.

I expressed these concerns in *Natural Resources Defense Council v. Nuclear Regulatory Commission (1)*. In that case our court was asked to review the NRC's quantification of the environmental effects of the uranium fuel cycle, including the "back end" of the cycle, waste disposal and reprocessing.

The NRC concluded that those effects are "relatively insignificant." Yet the only evidence adduced in support of its assessment was the testimony of a single NRC expert. Most of the testimony was conclusory and the expert gave little or no explanation of the underlying basis for his optimism.

To my mind, that testimony, without more, provided an inadequate basis for making critical nuclear plant licensing decisions. My objection was not founded on any disagreement with the expert's conclusions. For all I knew then or know now, he may have been accurate in minimizing the risks from nuclear waste disposal. Nor do I criticize the NRC for failing to develop foolproof solutions to the problem of waste disposal. What I found unacceptable was the almost cavalier manner with which the NRC accepted the sanguine predictions and refused to come to grips with the limits of the agency's knowledge. I stated (2):

Risk and Responsibility

To the extent that uncertainties necessarily underlie predictions of this importance on the frontiers of science and technology, there is a concomitant necessity to confront and explore fully the depth and consequences of such uncertainties. Not only were the generalities relied on in this case not subject to rigorous probing—in any form—but when apparently substantial criticisms were brought to the Commission's attention, it simply ignored them, or brushed them aside. Without a thorough exploration of the problems involved in waste disposal, including past mistakes, and a forthright assessment of the uncertainties and differences in expert opinion, this type of agency action cannot pass muster as reasoned decisionmaking.

The "thorough exploration" that I found lacking is particularly important in technically complex matters such as nuclear waste disposal. Since courts lack the expertise to assess the merits of the scientific controversy, "society must depend largely on oversight by the technically trained members of the agency and the scientific community at large to monitor technical decisions." There were a number of avenues open to the NRC for the kind of exploration that permits meaningful oversight—but the agency adopted none of them.

The Supreme Court unanimously reversed our decision (3). They felt that we had imposed extra procedures on the NRC beyond those required by law for so-called informal rule-making under the 1946 Administrative Procedure Act. They returned the case to our court, however, to determine whether the record supported the substantive conclusions of the NRC.

Whether the Supreme Court's decision represents a fair reading of what our opinion in fact required the agency to do, I leave to the legal scholars. My own view is that the Supreme Court's decision will have little impact because many of the new laws governing risk regulation explicitly direct agencies to use decision-making procedures that supplement the minimal requirements of informal rule-making under the Administrative Procedure Act. Statutes such as the Clean Air Act Amendments of 1977, the Clean Water Act of 1977, and the Toxic Substances Control Act of 1976 include procedural and record-enhancing features that will contribute substantially to the quality and accountability of agency decisions.

A Structured Approach to Decision-Making under Uncertainty.

I have never believed that procedures per se are a cure-all for solving regulatory problems. Rather, procedural safeguards serve an instrumental role, and it is the fullness of the inquiry that is paramount. If the inquiry is comprehensive and conscientious without additional procedural safeguards, it provides the best record we

David L. Bazelon

can hope for in making the difficult choices we now face. Conversely, even when all the procedural niceties are observed, if there is no commitment to a candid exploration of the issues, the predicate for good decision-making will be lacking.

Agencies are now revising their procedures to increase the availability of expert advice without abdicating agency responsibility for value decisions. Agencies have begun to encourage and fund public intervenors. These steps have increased the range of the administrative process, and have forced the agencies to wrestle with the difficult questions which might otherwise escape public scrutiny. Restrictions on ex parte contacts have increased our confidence in agencies' impartiality and fairness. The visibility of decision-making processes and decisions themselves has been enhanced by Congress' and the courts' commitment to openness, through the Freedom of Information Act, the Advisory Committee Act, and the Sunshine Act. I am confident that the courts will continue vigorously to carry out Congress' mandate that decision-making be honest, open, thorough, rational, and fair.

The Problem of Delay

Considering all relevant data and viewpoints is essential to good decisions. This is why I am concerned by recent proposals to shorten the decision-making process for licensing nuclear reactors. I have no doubt that some of the current delay is unnecessary, and it may be that current proposals do not affect critical deliberative processes. I do not express any views on specific proposals. I only want to caution that in speeding up the process, we must take care not to sacrifice the valuable and productive safeguards that have come to be built into the decision-making process.

I do not favor delay caused by an unthinking rejection of progress. Delay from unjustified fear of the future can in the long run cause more harm than the risks it prevents. But delay that is necessary for calm reflection, full debate, and mature decision more than compensates for the additional costs it imposes. The Alaska Pipeline was embroiled in extensive controversy in our courts, primarily by environmental groups who questioned whether sufficient attention was given to safety issues. The litigation imposed substantial costs, both the rising expenses for building the pipeline and the cost of postponing a major source of domestic energy. But in the subsequent attorneys' fees proceedings the companies themselves conceded that the litigation produced substantial safety improvements in the pipeline

Risk and Responsibility

that Congress ultimately approved. Sometimes the benefits of delay can be dramatic. The American experience in avoiding the tragedy of thalidomide is a poignant but not unique example.

By strengthening the administrative process we provide a constructive and creative response to the inherent uncertainties of risk regulation. Approaching the decision to take or to step back from risks such as nuclear power is like coming to a busy intersection with our view partially obscured. Our instincts tell us to proceed with caution, because intersections are dangerous. Ultimately, the importance of our journey and the desirability of our goal may lead us to brave the traffic and pull out into the highway. But even when we decide to proceed, we should not omit the moment of reflection to observe the passing cars, and look both ways.

Notes

1. *Natural Resources Defense Council v. Nuclear Regulatory Commission*, Fed. Rep. 2nd Ser., vol. 547, p. 633 (D.C. Circuit 1976), reversed *sub nom. Vermont Yankee Nuclear Power Corp. v. Natural Resources Defense Council (3)*. Nothing in these remarks should be taken to intimate any views of the merits of this case in its present posture, on remand from the Supreme Court.

2. *Ibid.*, p. 653.

3. *Vermont Yankee Nuclear Power Corp. v. Natural Resources Defense Council,* U.S. Rep., vol. 435, p. 519 (1978).

H. State and Local Government

In our federal system of government considerable authority is retained by the states. Many of the issues confronting state executive and legislative branches in particular are of a highly technical nature—for example, highway development, environmental protection, and public health and safety. Yet the consideration of scientific and technological matters and the role of scientists and technologists in influencing state policy remains minuscule in comparison with that at the federal level.

In Chapter 19, Harvey M. Sapolsky tracks the growth of state interests in science and technology to the corresponding expansion of the federal role after World War II. The example of the federal government stimulated interest of both scientists and politicians at the state level. And the lure of federal R & D money prompted states to try to get their share. But state interests are very different. National defense and space exploration, or support of basic R & D, are not major state functions. So while state scientific advisory bodies proliferate, their advice on social issues lacks the aura of expertise of national advisers on high technology and defense matters. After careful examination of case experiences in nine states Sapolsky concludes, "No state has found a useful and recognized role for science advisers in the formation of public policy."

In Chapter 20, J. David Roessner considers various federal mechanisms to promote public technology—that is, the development and use of new technology by state, regional, and local governments to meet rising demands to solve increasingly complex public problems. He notes both demand weaknesses (public organizations tend to be less innovative than private organizations) and supply weaknesses (local governments are not usually promising markets for ventures based on technological innovation). Building the analytical and evaluative capabilities of state and local governments appears more helpful than supplying them with specific, federally supported technologies.

19. Science Policy in American State Government

HARVEY M. SAPOLSKY

Major legacies of the mobilisation of scientists in the United States during the Second World War were the post-war recruitment of scientists as policy advisers to the federal government and the federal government's massive support of research and development activities. Scientists had been called upon to serve in prior national emergencies, but each time the passing of crisis left no marked change in relationships between government and science.[1] Until the 1940s government was a reluctant patron of science, confining its support largely to its own laboratories and standing essentially indifferent to the needs of science in general. Before the Second World War scientists had little and sought little influence in the formation of public policies in the United States.[2]

The war significantly altered these relationships because science-based technological and operational advances achieved during the conflict dramatically affected its outcome and the ensuing peace.[3] With their direction of weapons development projects, particularly that which produced the atomic bomb, and with their calculation of optimal strategies for aerial bombardment and anti-submarine warfare, scientists had demonstrated the practical utility of science to the country's political and military leaders. The unprecedented level of research support available during the war, permitting as it did the opening up of new areas of investigation and the undertaking of large-scale research efforts, also brought a full awareness to the scientific community of the potential opportunities which governmental financing of science could provide. With the advent of an uncertain peace, both federal officials and scientists sought to strengthen the ties between government and science.[4]

Reprinted with permission from *Minerva*, vol. IX, no. 3, July 1971, pp. 322–348.

Their efforts were obviously quite successful. Arguments that progress in science and technology contributed to national power were the main considerations which caused federal research and development expenditures to rise from 1.5 percent of total federal expenditures in 1946 to about 12 percent in 1970 and to total nearly $200,000 million for this 25-year period.[5] Scientists have continued to work on new weapons systems and other military problems. Research in universities and industrial firms now receives direct support from the federal government although the network of government laboratories has also been expanded. At every level of the federal bureaucracy, science advisory committees exist dealing with questions ranging from the allocation of resources for science to those which C. P. Snow has called the cardinal issues of government.

This post-war revolution in government-science relations was not without important political effects. Almost unconsciously the American political system has been changed by the federal government's involvement in the support of science and the scientists' involvement in the formation of national policies. Most of the federal science funds have been allocated through project contracts and grants to universities and industrial firms, and, since the initial appropriations were based on beliefs about the needs of national security, these allocations were not subject to the equity and geographical considerations which usually constrain federal programmes. As Dean Don K. Price points out, because of the growth of the research contracts and grants system, a new form of federalism has developed, a form in which state governments have been by-passed in the distribution of federal funds.[6] The extensive use of scientists as policy advisers has also had political consequences. No longer can it be assumed that governmental decisions represent simply the judgments of politically responsible elected and appointed officials. Important governmental actions are now often subject to prior analysis by independent research groups and can carry the authoritative endorsements of self-selected scientific panels. There has arisen, in the terms used by Dr. Robert C. Wood, an apolitical elite of scientists who expect to be, and frequently are, asked to participate in the making of public policy.[7]

The "New Federalism"

The significance of the "new federalism" can be seen in the analysis of federal expenditures. For example, in the fiscal year 1966 federal funds expended in the support of extramural research and development activities (i.e., work conducted in non-federal research institu-

Science Policy in American State Government

tions) and federal grants-in-aid to state and local governments (e.g., programmes in road construction, health, welfare and education) were approximately equal at about $12,000 million each,[8] but were allocated by quite different methods. Extramural research and development funds were allocated by federal government agencies to universities and industrial firms through project contracts and grants awarded on the basis of the anticipated excellence of their scientific contribution to agencies' missions. Only a relatively small amount of extramural research and development support, probably less than $100 million and mainly in the field of agricultural research, was allocated by the geographical and political criteria which provide for fixed awards for each state irrespective of the merits of the projects to be supported and which often require matching contributions. In contrast, the vast bulk of the grants-in-aid to state and local governments, close to 80 percent of the sum involved, was allocated through the standard formulae.[9] Only recently have grants awarded on the basis of the merits of the proposal in competition with other proposals begun to be used on a larger scale in federal programmes for the states.[10]

The allocation of federal research and development and grants-in-aid funds have had contradictory effects on the states (Table 1). Equity, defined politically, is usually a prime consideration in establishing the distributive formulae for these funds. Except as modified by the matching of public assistance expenditures of the industrialised states and certain requirements of minimum allocation, the grants-in-aid for state and local governments generally vary inversely with per capita income.[11] The research and development allocations, however, guided as they are by expectations of performance, tend to vary directly with per capita income. The states with the greatest scientific and technical resources, usually the states with the highest per capita incomes, have received the greatest research and development support from the federal government. Moreover, an analysis of variance within the allocations shows that the grants-in-aid are more equally distributed among the states than are the research and development funds. The research and development allocations tend to negate the equalising tendencies of the grants-in-aid.

The political significance of the scientists' new role as the apolitical policy adviser is less quantifiable, but potentially more profound.[12] The pragmatic arrangements designed to facilitate the military contribution of scientists in an uncertain peace appear to have long-term consequences. Analytical decision-making techniques and policy-study approaches developed for the Department of Defense by scien-

Harvey M. Sapolsky

Table 1. Per capita income, grants-in-aid and extramural support by state, fiscal year 1964

State	Per capita income calendar year 1964*	Rank	Federal extramural R & D 1964 per capita*	Rank	Federal grants per capita fiscal year 1964*	Rank
Alabama	1,781	48	57.14	14	63.15	21
Alaska	3,159	4	26.43	25	324.92	1
Arizona	2,287	31	33.61	23	60.36	24
Arkansas	1,712	49	2.09	49	78.60	12
California	3,112	5	250.48	3	51.84	30
Colorado	2,583	19	126.29	5	65.98	17
Connecticut	3,232	3	40.56	20	48.99	34
Delaware	3,091	7	11.55	36	47.50	38
District of Columbia	3,485	1	203.02	4	77.26	13
Florida	2,294	30	68.06	12	43.67	42
Georgia	2,003	41	8.50	37	57.77	26
Hawaii	2,787	12	5.32	44	49.02	32
Idaho	2,114	39	42.14	19	71.98	14
Illinois	3,042	9	19.00	28	43.48	43
Indiana	2,588	18	19.20	27	34.07	50
Iowa	2,356	28	7.40	40	44.96	40
Kansas	2,513	21	5.65	43	48.50	35
Kentucky	1,893	44	2.01	50	69.33	16
Louisiana	1,940	42	82.49	10	85.64	7
Maine	2,093	40	3.49	46	61.24	23
Maryland	2,829	11	105.50	6	37.86	47
Massachusetts	2,874	10	95.13	7	48.42	36
Michigan	2,764	13	17.79	30	45.51	39
Minnesota	2,432	26	18.60	29	50.97	31
Mississippi	1,493	51	1.98	51	63.37	20
Missouri	2,446	24	86.59	8	57.29	27
Montana	2,295	29	7.51	38	104.14	4
Nebraska	2,361	27	2.96	48	52.53	29
Nevada	3,261	2	429.26	1	118.53	3
New Hampshire	2,447	23	29.10	24	55.46	28
New Jersey	3,084	8	50.25	16	30.65	51
New Mexico	2.121	37	262.66	2	83.01	9
New York	3,108	6	62.18	13	37.59	48
North Carolina	1,923	43	4.66	45	40.14	45
North Dakota	2,122	36	49.55	17	84.50	8
Ohio	2,641	15	16.89	31	44.51	41

tific consultants are said to be spreading throughout the federal bureaucracy.[13] New domestic programmes now seem to be accompanied by contractual support from independent policy groups, summer studies, programme evaluation teams, and research grants. And the distinction between public officials and the selected outside

Table 1.—*continued*

State	Per capita income calendar year 1964*	Rank	Federal extramural R & D 1964 per capita*	Rank	Federal grants per capita fiscal year 1964*	Rank
Oklahoma	2,116	38	6.45	41	88.29	5
Oregon	2,613	17	7.50	39	70.36	15
Pennsylvania	2,571	20	35.83	22	42.26	44
Rhode Island	2,641	15	12.04	34	64.18	18
South Carolina	1,690	50	6.13	42	40.04	46
South Dakota	1,881	46	14.88	33	79.42	11
Tennessee	1,876	47	37.79	21	58.40	25
Texas	2,222	34	53.02	15	47.82	37
Utah	2,273	32	45.99	18	80.48	10
Vermont	2,135	35	11.74	35	86.27	6
Virginia	2,270	33	24.70	26	49.00	33
Washington	2,707	14	84.97	9	61.54	22
West Virginia	1,885	45	3.15	47	64.04	19
Wisconsin	2,507	22	16.03	32	35.72	49
Wyoming	2,444	25	81.14	11	137.96	2

SOURCES: U.S. Department of Health, Education, and Welfare, *Health, Education and Welfare Trends*, 1965 Edition, Part 2, State Data and State Rankings. Charts S-46, S-72. Report of the House Select Committee on Government Research, *Study 9: Statistical Review of Research and Development*. U.S. Department of Commerce, Bureau of the Census, *Current Population Reports*, Series P-25 No. 30 (24 November, 1967), Population Estimate.

Research and Development per capita
  Correlation with per capita Income * = +.37
  Standard Deviation * = 78.1
  Mean * = 50.5
Grants-in-aid per capita
  Correlation with per capita Income * = −.23
  Standard Deviation * = 21.8
  Mean * = 60.9
* in dollars

NOTE: These statistics are derived from per capita income, per capita research and development and per capita grants-in-aid by state with the omission of Alaska and Washington D.C. The latter cases are considered "federal wards" and, therefore, the purposes of allocation within the above programmes are not consistent with those of the country at large. It should be noted that inclusion of these "wards" alters the correlation for grants-in-aid per capita quite substantially, while not affecting the grants-in-aid statistics to any great extent.

policy advisers has become more difficult to draw as agency missions are increasingly defined through their collaboration.

The diminution of the military threats which initially involved scientists in governmental policy-making has not apparently affected the scientists' interest in public affairs. The programmes of the meetings

Harvey M. Sapolsky

of the major scientific societies are usually filled by public policy issues as more technical information is communicated through other methods. It is now frequently suggested that the mobilisation of scientists to work on domestic problems could be even more valuable to the society than was the wartime mobilisation.[14] And the prevailing view among scientists is that they have a social obligation to serve whether called upon or not.

The Response of State Governments

Science advisory mechanisms which have been established recently in state governments are the product of these political changes. The new federalism challenges state governments to think of ways by which they can expand or at least protect their share of federal research and development expenditures and the scientists as an apolitical elite seek new challenges as those of war become less salient and less palatable to them.

State government interest in federal research and development activities began in the late 1950s and early 1960s when federal appropriations for science and technology increased sharply in response to Sputnik. (Federal research and development expenditures grew from $4,500 million in the fiscal year 1957 to $12,000 million in the fiscal year 1963.) In these years economic growth had been restricted by federal fiscal policies and the only industries which appeared to be expanding rapidly were those such as the aerospace, electronics and computer industries where the federal research and development investments were large. Although data on the geographical distribution of federal research and development support was for a long period strangely missing in the many statistical surveys of the support of research which the federal government prepared, it was widely recognised that these expenditures were highly concentrated, benefiting directly only a few states. (When this data finally became available at the urging of Congress it revealed that California alone received nearly 32 percent of total federal budgetary obligations,[15] for research and development while the leading five states—California, New York, Maryland, Massachusetts and Texas—accounted for 57 percent).[16] The local economic effects of federal expenditures were visible along Route 128 near Boston and in Palo Alto, California, where rows of modern research and manufacturing facilities employing thousands sprang up, it was said, as "spin offs" from government-sponsored research at neighbouring universities.[17]

Science Policy in American State Government

Governors envisioning 128s and Palo Altos in their own states received much encouragement during the early and mid-1960s.[18] Federal research and development expenditures continued to climb during this period. The newly enlarged space programme and the potentially large national oceanographic programme were frequently described then as having significant benefits in terms of economic growth for the country as a whole and for particular regions where facilities were to be located. Congress passed the State Technical Services Act of 1965 which was designed to promote the transfer of new technology throughout the economy and which had as its advertised model the highly successful Agricultural Extension Service. The President issued a directive calling for federal agencies concerned with the allocation of funds for science to promote the development of new centres of academic excellence which meant that universities located in places other than Cambridge and the San Francisco Bay area could expect to receive a larger share of federal research and development grants and contracts. All that seemed to be needed for sustained science-based economic growth was some initiative on the part of the states.

By the mid-1960s, scientists' concerns had broadened to include the social problems such as welfare, education, crime and environmental pollution, which were becoming the major domestic political issues of the nation. To some extent their interest in these problems can be attributed to the need to find new applications and justifications for research activities as the old applications and justifications lost political favour.[19] But it should also be pointed out that it was the reports of federal science advisory groups which first called official attention to such now popularly discussed problems as population control and the degradation of the environment[20] and that many scientists see themselves as leading a drive to change what they consider to be a badly outdated set of national priorities.

Since the responsibility to act on social problems, unlike the case of military problems, is shared among levels of government in the American federal system, scientists have come to recognise that they must deal with state governments if they are to contribute to the solution of these problems. The resource constraints and political complexities of state governments, however, are new to most scientists and thus far it has been only a pioneering few who have ventured to the state capitals to offer their advice and assistance. Before examining their efforts we should take note of the extent to which state governments are involved in science.

Harvey M. Sapolsky
Science and State Governments

In the nineteenth century when science was in the process of becoming a professional pursuit and when the sources of its support were limited, state governments' involvement in science was proportionately greater than it is today and was probably as extensive as that of the federal government.[21] The same pressures for natural resource exploration, agricultural development and the promotion of commerce which caused the federal government to establish its first scientific agencies and to begin to support research occurred in the states but their governments were then more able to respond than was the limited federal government. During this period, state governments often initiated programmes which were later accepted and advanced by the federal government as it grew in scope and power. Programmes in geological mapping, agricultural experimentation, aid to higher education and public health are examples.

But as the federal government's role in the support of research and development activities has expanded in the post-Second World War period, state government involvement in science has lost most of its significance and originality.

Expenditures by state agencies on research and development present not much more than half of one percent of total national research and development expenditures and less than a quarter of one percent of total state expenditures for all purposes[22] (Tables 2 and 3). In contrast, federal support of research and development activities is approximately 66 percent of total national research and development expenditures and 12 percent of the total federal budget.

Moreover, an increasing share of what state agencies spend on research and development activities is provided by the federal government through contracts and grants from federal agencies. The figures for total research and development expenditures, for example, rose from an estimated $51 million in the fiscal year 1954 to $159 million in the fiscal year 1968, but since the federal share of these expenditures increased from an estimated 21 percent in the fiscal year 1954 to about 50 percent in the fiscal year 1968, the actual increase in state appropriated research and development expenditures during this period was only $39 million, not $108 million. Between 1964 and 1968, years for which complete state surveys are available, state appropriated research and development expenditures rose at an annual rate of 15 per cent or somewhat above the 12 per cent annual rate of increase the states experienced in their total expenditures for all purposes. Stimulated, however, by numerous social wel-

Table 2. Research and development expenditures of state agencies, fiscal years 1954, 1964, 1967 and 1968

State	1954*	1964	1965	1967	1968
Alabama		717	681	700	372
Alaska		2,361	2,319	1,397	2,623
Arizona		246	262	325	425
Arkansas		341	388	804	835
California	9,874	11,163	12,820	25,039	28,926
Colorado		1,027	1,108	931	1,085
Connecticut	488	844	1,220	1,294	2,257
Delaware		9	33	80	84
Florida		1,696	2,293	3,130	3,221
Georgia		1,169	1,435	1,290	1,771
Hawaii		767	807	2,092	1,667
Idaho		403	461	1,053	814
Illinois		3,957	4,454	8,902	9,418
Indiana		751	1,319	1,188	1,368
Iowa		416	423	1,402	1,752
Kansas		222	399	1,154	1,288
Kentucky		884	1,134	2,502	2,536
Louisiana		959	885	1,573	1,325
Maine		764	843	747	708
Maryland		475	613	735	1,187
Massachusetts		693	711	1,722	1,665
Michigan		1,943	2,913	3,519	3,457
Minnesota		1,117	1,626	2,019	2,368
Mississippi		1,232	1,530	630	858
Missouri		2,393	2,886	1,412	1,801
Montana		257	396	910	1,106
Nebraska		578	875	354	867
Nevada		58	45	78	82
New Hampshire		58	64	221	269
New Jersey		3,344	6,036	2,322	3,148
New Mexico	690	128	193	938	832
New York	8,368	17,832	18,806	33,287	36,836
North Carolina	1,191	476	2,233	4,314	7,222
North Dakota		452	529	311	301
Ohio		1,471	1,744	2,245	2,846
Oklahoma		938	1,084	1,388	1,905
Oregon		735	949	1,066	1,371
Pennsylvania		3,879	3,771	4,203	6,820
Rhode Island		130	191	510	1,070
South Carolina		259	312	355	570
South Dakota		587	658	379	511
Tennessee		264	356	277	371
Texas		2,545	3,299	5,767	7,008
Utah		94	169	1,799	2,006
Vermont		51	93	509	608
Virginia		2,397	2,811	2,690	2,875
Washington		1,527	1,832	2,899	3,281
West Virginia		844	738	1,155	1,133
Wisconsin	419	1,653	2,233	2,249	2,428
Wyoming		246	276	462	440
Total	48,622*	77,325	93,256	136,299	159,214

*Estimate

Table 3. Comparative analysis of federal and state agency research and development expenditure, fiscal years 1954, 1964, 1967 and 1968

	1954	1964	1965	1967	1968
State agency R & D expenditures	48,622*	77,352	93,256	136,299	159,214
Federal share of state R & D expenditures	10,697*	31,056	38,618	68,149*	79,607*
Federal share of state R & D expenditures—percentage	22*	40	41	50*	50*
Total state expenditures for all purposes	18,686,380	42,583,494	45,639,000	58,760,000	66,254,000
State agency R & D expenditures as a percentage of total state expenditures	0.26	0.18	0.20	0.23	0.24
Total national R & D expenditures all sources	5,738,000	19,215,000	20,449,000	23,680,000	25,330,000
State agency R & D expenditures as percentage of total national R & D expenditures	0.84	0.40	0.45	0.57	0.62
Federal R & D expenditures	3,148,000	14,694,000	14,875,000	16,842,000	16,865,000
Total federal expenditures	67,537,000	97,684,000	96,507,000	125,718,000	135,033,000*
Federal R & D expenditures as a percentage of total federal expenditures	4.7	15.0	15.4	13.4	12.1

*Estimate
SOURCES: See Notes 22 and 23.

fare programmes enacted by the Johnson Administration, the federal share of state agency research and development expenditures was rising at an annual rate of 26 per cent during the same five years.[23]

National Science Foundation surveys have classified the research and development expenditures of state agencies by functional area, field of science, character of work and by performing organisation. In both 1967 and 1968 over 40 per cent of the research and development expenditures of state agencies was concentrated in the area of health care; other functional areas with large research activities were natural resources with 25 per cent of the expenditures and highways with 15 per cent. By field of science nearly 60 per cent of the state agencies' expenditures were classified as being in the life sciences. Engineering accounted for an additional 15 per cent and the social sciences were next at 13 per cent. State agencies emphasised applied research which represented about 50 per cent of the expenditures; basic research and development shared equally in accounting for the remaining 50 per cent. In contrast with the pattern in the federal government, the state agencies performed over 80 per cent of the research and development work themselves, contracting less than 20 per cent to universities, non-profit organisations, and industrial firms.[24]

One might expect that the states would have a much more extensive involvement in science through their state universities and colleges than through state agencies, but this does not generally seem to be the case. To be sure, the states spend billions of dollars annually financing public institutions of higher education and many of these institutions conduct significant research, but with the exception of agricultural research activities which are directly oriented toward the solution of local problems, there is little indication that the states deliberately support much research at their universities and colleges. In the fiscal year 1964, the year for which the most usable data are available, the expenditures for research and research facilities at state universities and colleges were $792 million. Of this amount over $450 million was accounted for by federal research contracts and grants. The state government contribution from state tax funds was only about $200 million, over two-thirds of which were allocated to agricultural experimental stations and agricultural colleges.[25] Since agricultural research accounts for less than 30 per cent of the total research expenditures of state universities and colleges, the bulk of the research activities at these institutions is obviously not dependent upon state appropriations.

The general reluctance of the states to support much research and their tendency to emphasise work on applied problems in fields such as health care and agriculture when they do, is not surprising. Research, particularly basic research, is, after all, an uncertain undertaking the outcome of which in terms of benefits, costs, applicability and timing cannot be accurately predicted. Except in the case of the most applied activity, conducted on a modest scale, there can be no assurance that the state which pays for a successful research project will be able to claim sufficient direct benefits within a reasonable period of time to justify its investment. Since discoveries of general relevance financed by one state are likely to be available to all states relatively quickly and relatively inexpensively, no state has the incentive to undertake with its own funds research which is not uniquely related to the interests of its citizens. For example, if research supported by the state of Connecticut were to provide a solution to the state's air pollution problems, it would not take long or cost much for Massachusetts to adapt and adopt the solution for its own air pollution problem. The burden then of supporting research for even the most obvious state functions largely falls upon the federal government which encompasses within its jurisdiction all who, at least in the United States, might potentially benefit from new discoveries.[26]

Perhaps too well, the states have recognised the primacy of the federal government in science and technology. Their boldest efforts relating to research activities in the post–Second World War period have been simply intended to capture a larger share of the federal research and development allocations or to adopt without modification programmes and approaches which were initiated by the federal government.

During the business recessions of the late 1950s and early 1960s, for example, several states sought to aid their economies by investing directly in research facilities. The assumption which is still considered as unquestionably valid by some educators and scientists was that state assistance for the improvement of a university physics department, the establishment of a not-for-profit research organisation, or the development of a research-oriented industrial park would guarantee the attraction to the area of substantial research and development contracts and grants from the federal government which would, in turn, assure through "spin" off" and the multiplier-effect, an accelerating rate of local economic growth.[27] Ignored frequently in the state's efforts were the shortages in entrepreneurial talent, risk capital, and marketable ideas which necessarily limit the impact which scientific activities can have on economic development. No special successes were recorded.[28]

Or consider the widely publicised contracts which the State of California made with major aerospace firms in 1965 for the application of systems-engineering and systems-analysis techniques to four public policy problems. Accompanied by the unconcealed hopes of new employment opportunities for the aerospace industry, the expectation at their initiation was that these contracts would demonstrate that the methods and procedures used in approaching national defence problems could be adopted by state government.[29] The completed studies, although acclaimed as valuable educational experiences for those who participated in them, produced neither many politically valuable recommendations for policy nor many subsequent awards. The promise of a ready transference of defence-generated skills and experience to domestic problems seemed illusory.[30]

State Science Advisory Bodies

The same seductive visions of federal research and development dollars flowing into their states and the easy acquisition of the federal government's administrative sophistication in science apparently encouraged governors to establish general science advisory bodies. General science advisory bodies include committees, councils, boards, etc., formed to obtain advise from scientists outside the state civil service on topics of broad concern to a state, in contrast to those formed to obtain the advice of scientists on topics related to a single technical function of state government, such as public health or forest management. By any standard, the growth of advisory bodies of this type in state government has been most rapid. New York State created what was apparently the first general science advisory committee in 1959; 10 years later 47 states had some formally designated person or group responsible for general science advice. This growth has been aided by a new National Science Foundation programme designed to promote the use of science and technology by state and local governments, the State Technical Services Act, and the explicit and implicit suggestions of scientists and policy analysts that the federal science advisory system should be emulated at other levels of government.[31]

Initially few governors sought to bring scientists into state government as advisers and those who did viewed them mainly as a sort of scientific chamber of commerce which might influence the particular location decisions for major new federal and industrial research institutions which were being created with increasing frequency in the early 1960s. But several more state science advisory committees were formed after Dr. James R. Killian, Jr., the first Special Assistant to the President for Science and Technology, spoke at the Governor's Conference of 1963 urging their establishment and warning the governors of the danger of the obsolescence of federalism in an era of rapid technological change and significant national research and development support.[32] Others were formed in 1965 when States Technical Services Act was passed with provisions requiring, for participation, the establishment of state plans and annual programme reviews by advisory councils.[33]

In 1967 when the governors of all the states were asked in a survey if they had a science advisory group similar to those in the federal government, 22 said they did and another five asked, in turn, how they might establish one.[34] Two years later the National Science Foundation in announcing a programme of state grants for science policy studies could report that nearly all the governors, having named a state science adviser equivalent, were ready to participate in it.[35]

The Nine State Studies

The survey of 1967 described the formal structures of the state science advisory units then in existence. It was based on a self-administered questionnaire and did not reveal the origins and operations of these units. Many governors had designated certain scientists to act as advisers to state government on matters which apparently went beyond their own technical expertise, but it was not known under what conditions the scientists had been asked to give advice, what advice they had given, and what had been the effect of their advice on state policies. With this limited information little could be said about the function or value of general science advisory groups in state government. Consequently it was thought useful to examine the origins and operations of general science advisory groups within the states, in a more intensive manner. Case studies were required.

There were several criteria for selecting the states for the case studies. First, since the survey of 1967 had found that in obtaining general science advice governors had used a variety of organisational forms ranging from personal committees to independent boards to groups attached to cabinet departments, states had to be chosen to present each major organisational type. In a few states the local governments and even the state legislature had set up general science advisory bodies similar to ones serving the governors; such states too were regarded as worthy of study. Third, because the establishment of the general science advisory groups appeared to be linked to concern over a state's standing in federal research and development allocations, states which would represent the extremes in the allocations had to be studied. Fourth, both to check a natural tendency to attribute all recent state policy failures or successes to the exis-

Science Policy in American State Government

tence of formal science advisory bodies and to discover how science advice flowed in states without general science advisory groups it was decided to include among the states studied a few which had not established a formal advisory system or at least had not reported that they had. Fifth, geographical balance was also a criterion of choice. Finally, financial limitations required that the number of states studied should not exceed nine.

The states selected were California, Kansas, Kentucky, Massachusetts, Michigan, Mississippi, New Mexico, New York and North Carolina (Table 4). Among them were states which had advisory units

Table 4. Case study states

State	Science advisory organisation reported in 1967 survey	Type*	Year founded
California	none	—	—
Kansas	Research Foundation of Kansas	mixed	1963
Kentucky	Kentucky Science and Technology Council	PSAC	1965
Massachusetts	Governor's Advisory Committee on Science and Technology	PSAC	1966
Michigan	none	—	—
Mississippi	none	—	—
New Mexico	Governor's Science Adviser	special assistant	1963
New York	New York State Advisory Council for the Advancement of Industrial R & D	Dept. of Commerce	1959
	New York State Science and Technology Foundation	NSF	1965
North Carolina	North Carolina Board of Science and Technology	NSF	1963

State	Legislative or local committees	Per capita rank 1964 federal R & D distribution
California	yes	4
Kansas	no	46
Kentucky	no	49
Massachusetts	no	7
Michigan	no	30
Mississippi	no	44
New Mexico	no	3
New York	yes	17
North Carolina	no	47

*NSF—National Science Foundation.
PSAC—President's Science Advisory Committee.

representative of each major type recorded in the 1967 survey, together with three—California, Michigan and Mississippi—which apparently lacked a formal science advisory mechanism at the state level. California and New York were of special interest because they had legislative or local government science advisory units and New Mexico was unique because it was the only state at the time which had a science adviser to the governor rather than a multi-member committee or council system. Three states—New Mexico, California and Massachusetts—ranked among the leading 10 states in per capita federal research and development awards, while four—Mississippi, Kansas, North Carolina and Kentucky—were among the lowest 10. States representative of each major geographic region and economic base were included.[36]

Failure and Success in Science Advice

A brochure prepared by a state science advisory committee to describe its functions and activities quotes Victor Hugo—"Mightier than the tread of mighty armies is the power of an idea whose hour has come"—and implies that the hour for science advice in state government has arrived.[37] In view of the proliferation of science advisory bodies, the establishment by the states of centrally organised bodies through which general science advice might be obtained, this would certainly appear to be the case. In California and Michigan, states which were included in the study as controls because incumbent governors had reported that no general science advisory committee had yet been established, it was discovered that such committees had been formed under previous administrations. In Mississippi where the Research and Development Centre clearly had more limited functions, there was some feeling within the state that the centre could broaden its mission to include science advisory activities. And in the other states two, three or four groups were discovered where only one or two had been expected when the field-work began.

But in terms of organisational survival and achievement it would seem that the hour for state science advisory bodies has yet to be sounded. Among the many advisory bodies observed in the case studies, few survived a change in administration and only two (those in Kansas and New Mexico and both precariously) a change in political party control of the state government. It might be argued, of course, that the demise of advisory groups results from the close identification of a particular advisory group with a prominent political leader or a particular political party and thus could be considered as an indication of their strength rather than weakness. When a new leader or the opposing party assumes office in a state there is a tendency to curtail programmes or organisations closely identified with their predecessors. And it is clear that Governor Brown of California, Governor Campbell of New Mexico and Governor Sanford of North Carolina were closely identified with science advisory groups which their successors ignored or eliminated. Yet it is also clear from the experience of states such as Massachusetts where the same political party has ruled during two consecutive gubernatorial administrations and New York where the same governor has held office for a decade that science advisory bodies have regularly been ignored.

Those science advisory committees reporting to the governor which were examined originated almost without exception in a promise of economic gain for the state. The precipitating action was a public speech or a letter to the governor by a prominent scientist or public administrator or a report by an outside economic policy task force or committee whice raised the possibility that the state could benefit from some science-related development activity. In New Mexico it was an indication that NASA might locate within the state a "spaceport" or "hard" landing-area for its spacecraft. In Kentucky it was the hope that the state could utilise nuclear energy to develop its coal resources. In Massachusetts and New York it was the desire to continue the expansion of science-based industries, while in Kansas it was the hope to begin such an expansion. The governor's response, often proposed to him, was to establish a committee of scientists and industrialists to explore the matter. The committee was composed so as to provide a balanced representation of the state's scientific institutions and in several instances included among its members ranking public officials. To husband his own time and to ensure that staff resources were made available, the governor often placed the committee under the jurisdiction of the state department of commerce or development.

Quick, substantial achievement has eluded all the committees. NASA decided against building a "hard" landing-area in New Mexico or elsewhere; applying nuclear energy to harness coal resources was a long-term if not illusory task; and there were no magic formulae to guarantee science-based economic growth. No state has found a useful and recognised role for science advisers in the formation of public policy.

Attempts by the committees to promote economic development were apparently neither novel nor particularly productive. The development conferences and symposia which they organised and sponsored could not be distinguished from the efforts of numerous other groups both public and private pursuing the same topic. In no state could a link be established between any recent economic progress and science advisory activities.

The committees were tempted to demonstrate their usefullness in government by means of analysis of social problems but seldom followed through with actual studies. The jurisdictional complications these problems raise and the degree of political involvement their solutions require were apparently sufficiently inhibiting to prevent much initiative. Only in Massachusetts was the role of the advisers in social problems clearly tested and there the committee was rebuffed by state officials. A plausible though unproven hypothesis is that unsolicited policy advice which is unpredictable and uncontrollable by the recipient is not welcome within government.

The governors apparently felt little obligation to follow closely the efforts of their science advisers to define a mission for general science advice in state government. Among them only Governor Campbell of New Mexico was described as having attended regularly the meetings of his science advisory committee and only Governor Volpe of Massachusetts was reported as having requested the committee's assistance on a matter beyond its initial charge.

If the governors ignored the science advisory committees, so did the state administrative agencies and legislatures. The Kansas committee was supposedly to coordinate the research activities within the state government but this task was never carried out because of the hostile indifference of the state agencies and most particularly of the state educational institutions. The existence of Kentucky's committee was largely unknown within the state government and few of the legislators and state officials who were aware of its existence turned to it for advice.

With the exception of those in New York State, state advisory committees received meagre financial and staff support. Visions of major undertakings faded as governors and legislatures discouraged or reduced requests for significant appropriations. Although the advisers themselves gladly served on a part-time basis without compensation they generally felt that their work was constrained without the assistance of a full-time staff and contract services.

The most common recommendation of these committees and often the only one implemented was a proposal for the establishment of a state science and technology foundation. Limited in scope and overwhelmed by the complexity of the economic development task, the committees were attracted to the concept of a grant-dispensing foundation which would largely be controlled by scientists and which would promote local economic and social progress through "seed money" awards to research groups within the state. Such foundations have in fact been formed in Massachusetts, New York, North Carolina and several other states.[38]

The effectiveness of these foundations is difficult to assess. Most information is available about that established in North Carolina. In the period 1963-69, the first seven years of its existence, the North Carolina Science and Technology Foundation awarded $2,253,000 in grants, over 90 per cent of which it allocated to four institutions— Duke University, the University of North Carolina, North Carolina State University and the Research Triangle Institute. The foundation presents no direct measure of the economic impact of these grants, but it does claim that their "multiplier" or "leverage" effect—the ability of the grants to attract supplementary or "follow on" awards—was large. At first glance this would seem to be the case since the foundation reports that on the average over four dollars of outside awards (mainly from the federal government) were generated by each dollar it allocated to a project. But even without assuming that the outside awards would have been forthcoming in the absence of the initial "seed" grant, there are serious doubts about the usefulness of this uncertain indicator of economic effectiveness. (Most of the outside grants were made for a few "research facilities" during the first two years of the foundation's existence: these first grants were not followed by significant further grants.) It is also doubtful whether the foundation could have had a major direct role in providing science advice in the state since only 4 percent of its dollar awards have been made to projects concerned in any way with state problems. Thus no matter how useful the North Carolina Science and Technology Foundation's more than $2 million might have been in furthering research in science and technology, it is not clear that it contributed at all to the economic and social progress of North Carolina.

Other than noting their unusual location within state government, there is little that can be said about the two legislative science advisory committees which were studied. The New York State science advisory council which reported to the leadership of the two houses rather than to the entire legislature was dissolved less than two years after it was established on the suggestion of a staff aide when neither the speaker of the assembly nor the president *pro tem* of the senate, its key legislative patrons, were returned by their local constituencies. The equivalent California state legislative advisory council which has just been established also on the suggestion of a staff aide is officially linked only to the assembly and reports to a committee of the whole

Harvey M. Sapolsky

known as the General Research Committee under the chairmanship of the speaker. Both the New York and California legislative science advisory groups were given broad charges to consider opportunities and problems raised for state government by advances in science and technology. Since the New York council had an early demise and the California council has only recently been created, it is not possible to describe the consequences of an interaction of Nobel laureates and state legislators. The incongruity of the image such an interaction evokes, however, is only exceeded by a report that the California State Assembly has just received a research grant from the federal Department of Transportation to demonstrate the feasibility of a steam-propelled bus.[39]

Taken together it appears that state governors, agency officials and legislators do not lack what they consider to be science advice, technical information and support. Their preference, however, in obtaining this advice is to use channels other than a general science advisory committee. In Massachusetts, for example, when the governor wanted a report on fluoridation he turned to the State Department of Health and staff members of local hospitals and medical schools. In Kentucky it was clear that the state legislators rely upon constituents, colleagues and staff aides whose biases are known. And state agencies in North Carolina, Kentucky and Michigan were found to be connected with functionally specialised networks of advisory groups which extended from federal agencies to the cities and towns. When asked if they could use more scientific advice most officials the investigators surveyed said they could but few said they wanted a single, centralised, science advisory committee to provide this advice.

Why then does the idea of a general committee persist? When one is disbanded it almost seems that there is another in the process of being established within the same state. Current influences include what has been called in another context the "Flying Feds," federal officials with money or the promise of it, who promote social innovations, in this instance, science advisory mechanisms, within the states, in an effort to promote their own programmes.[40] But the pressures for these committees seems to be more basic because interest in them preceded the establishment of federal programmes in economic development and science policy which now endorse the idea and because their members seem to be able to endure continued frustration and failure.

The committees continue to be formed, largely because political leaders find that they can be somewhat useful politically. Politics involves style as well as substance, the appearance of action as well as action itself. From the perspective of those they are to advise, science advisory committees can offer protection and support as well as aid in substantive policy. The establishment of a science advisory committee can be used to demonstrate a political leader's great concern for science and technology and the committee's reports can be used to endorse, i.e., legitimate, programmes and plans he has already formulated. Thus in the crisis of national self-confidence which followed the Soviet Union's launching of Sputnik, President Eisenhower found it useful to appoint a special assistant for science and technology and, at presidential level, a science advisory committee and to announce their support for the administration's decisions in space exploration.[41]

Although the stakes are certainly smaller, similar political opportunities exist in state government. By establishing the advisory committees, the sponsoring state official looks progressive, modern, concerned in the eyes of a small but articulate segment of his constituency, largely confined to university campuses, which believes that science and scientists have an important role to play in the formation of public policy. Since the establishment of a science advisory committee in itself can arouse no opposition and since there is no compulsion either to solicit or to heed its advice, the venture is nearly without risk. In terms then of the net political benefits they produce for their sponsors, the committees can be considered a minor success and can be thought worth initiating despite their failure to contribute substantially to public policy.

For some advisers the reward of being involved in government, if only peripherally, of serving an important political official, if only marginally, may be sufficient to continue their participation.[42] But for most it is likely the hope that they might at some time aid in the formation of more rational public policy which sustains their participation despite frustration and failure.

The question then to ask is, why have state science advisory committees been so ineffective? The answer would seem to lie in the limitations of the scientist as adviser in public policy problems relevant to the states.

The Limits of Advice in Science Policy

To describe the scientist-advisers as an apolitical elite is not to say that scientists are disinterested participants in the political process of determining public policy; they hate or love war, privilege, blacks, etc., as passionately as—though probably in different proportions from—the rest of the population. Just as those who are advised can be said to seek the maximum political advantage from an advisory relationship, those who advise can be said to seek the maximum influence for science and themselves in the formation of public policy. It is, of course, in the interest of both sides of the advisory relationship to claim that the scientist is devoid of any political interest. For the politician this claim clothes the scientists' advice with the legitimacy of objective science and thus enhances the political protection and support advisory groups can provide him. For the scientists this claim dissociates their advice from that of the special interests and thus enhances the value and effectiveness of their participation in governmental decision-making.

The science advisory relationship can be conceived of as a relationship in which scientists offer officials assistance politically and substantively in return for the opportunity of having their views on public policy issues heard. The weight which will be assigned to the political assistance or substantive assistance offered depends upon such factors as the needs of the politician and the persuasiveness of the scientist's policy positions, but in any case the price extracted for the assistance is present and future access to policy-making councils where arguments for the support and utilisation of science and the scientists' general preferences in matters of policy could be made. The degree to which the politician will grant scientists access to these councils and the degree to which he will be favourably disposed toward their arguments, in turn, probably depends at least partially upon the amount of the protection and support offered him by the scientists in their capacity as advisers. However, since such a relationship if openly admitted is likely to be viewed simply as being self-serving even though it can also have wide benefits for society, both politicians and scientists are inclined only to stress the technical-scientific justification for the advisory relationship and any policy statement which may emerge from them.

Scientists gained access and influence within the post–Second World War federal government through their wartime contributions. The atomic bomb and similar technical accomplishments established the authority of science and scientists in national security affairs. At the end of the war the country committed itself to the strongest possible defence force and scientists were perceived within and outside of government as being experts in the means of maintaining and increasing the strength of this force. The assent of scientists, particularly of those who were project leaders or held other important positions in the wartime mobilisation, quickly became recognised as a necessary if not sufficient condition for the approval of new weapon projects.[43] Through competitive example, defense agencies learned the advantages, both of bringing scientists into their policy-planning councils and of increasing their support and use of science.[44] Since decisions affecting weapon technologies determined the future growth opportunities of defence agencies, scientists obtained considerable influence within these agencies and thus came to affect national security policies. The scientists' substantive policy contributions in this area included the advancement of particular political values such as disarmament, but their entry into security policy councils was dependent upon their publicly recognised possession of technical expertise which provided them with politically exchangeable resources.

In domestic affairs, however, the situation is quite different. Race relations, poverty, crime, transportation, education, pollution and the other "cardinal" issues of our time are perceived by most politicians to be almost exclusively questions of conflicting political values or ends rather than questions of alternative technological means, and in questions of values, politicians and the public consider themselves to be at least as expert as the scientists.[45] Moreover, even when debate turns to questions of means, no special deference is shown the scientists since there has been no domestic equivalent of the atomic bomb to establish their authority and that of science itself in the domestic issues.[46]

Few stand in awe of the nuclear physicist, the biologist, the city planner, the sociologist or the political scientist when they discuss "bussing", the negative income tax, or open housing, while many stand in awe of the nuclear physicist discussing a chain reaction, warhead configurations or the blast effects of a nuclear weapon. And few seriously expect, despite hopes to the contrary, that "technological fixes" will close the gaps among persons within the society as simply as they closed the bomber, missile and space gaps.[47]

Scientists without the aura of expertise, without a believable promise of providing solutions to domestic problems have few political assets to exchange with public officials; as a result they gain little access to and influence in the formation of public policy in domestic affairs. State science advisory committees are relegated to the extreme periphery of policy-making because politicians can see no advantage in a closer relationship. The establishment of these committees is meant to symbolise the politicians' belief that science and scientists have an important role to play in the formation of public policy. But the consistency with which the committees are ignored once they are formed is probably a better indicator of the politicians' real assessment of their potential value in government.

Opportunities and Prospects

It is obvious that state governments need new scientific and technical knowledge in order to carry out their responsibilities. State agencies are called upon to measure the quality of the water, to protect against communicable diseases, to build more durable roads and to perform numerous other governmental tasks which have a large scientific or technological component. Hence, these agencies employ thousands of scientists, engineers and technicians and maintain professional contacts with federal, local and private scientific and technical organisations.[48]

But it is less than obvious that state governments and state officials need in the same sense general science advisory arrangements. Although the proposals which suggest the establishment of state science advisory committees and similar groups are unusually vague as to their specific purposes, one can identify three substantive functions such central science advisory mechanisms are intended to perform: (1) aid the state's economic development, (2) help anticipate and solve the state's social problems and (3) coordinate the state's research and development activities. None seems compelling.

Economic development surely has been a paramount goal of the states, and scientific and technological progress has been a prominent factor in our national economic development. There is no reason to believe however that science and technology provide a unique impetus to local economic development. Economists point out that there is no theoretical support for the favoured assumption among scientists that the demand generated by research and development activities has a greater power to stimulate local development than has the equivalent demand generated by any other extra-regional investments.[49] No special economic benefits have been demonstrated to flow from large-scale federal technology programmes.[50] Appropriately, the interest of the states in federal research and development activities and research and development in general seems to rise and fall directly with the rise and fall of these allocations. To encourage them to do otherwise is not necessarily to contribute to their economic prosperity and might only stimulate inefficient federal allocations.

One man's utopia is another man's hell. Scientists enter the lists of social policy debates with no fewer though perhaps different biases than the rest of the population.[51] To claim for them privileged access on the grounds of objectivity or special insight is not to enhance the application of reason to governmental policy-making but to make such an application more difficult since the scientists' human limitations will become quickly apparent in policy discussions. Futurology is as yet entertainment, not science. The application of science to social problems is as yet only a promise, the failure of which could be as harmful to the support and practice of science as its successful fulfilment is said to be beneficial to society.

Science Policy in American State Government

The coordination of state research and development activities would seem to be both a superfluous and impossible task. As state research budgets are likely to remain quite small relative to total state budgets and highly dependent upon federal allocations, there is really not much to coordinate at the state level. Moreover, since these budgets are distributed among dozens of independent mission-oriented agencies in each state their coordination would require the coordination of a major share of a state government's programmes. Federal experience in science and other fields reveals there are neither the criteria nor the political will to accomplish such a task.

Perhaps, if the public comes increasingly to believe that the problems of environmental pollution have reached the point of crisis and if it believes as well that scientists are experts in such environmental matters, state science advisory groups in the United States will find their position in government strengthened. State governors may then discover great advantages in having their policies guided and endorsed by scientists. But if the public believes, as it does with so many other problems, that environmental issues involve choices among conflicting values in which the public itself is as well qualified as the scientists, the role of the state science advisory bodies will continue to be limited.

Notes

1. During the Civil War, for example, the Navy established a "permanent commission" of scientists to aid its war effort and the National Academy of Sciences received a charter from Congress to advise all government agencies, but by the end of the hostilities the commission was no longer permanent and the Academy was in a near permanent torpor. See A. Hunter Dupree, *Science and the Federal Government* (Cambridge: Harvard University Press, 1957), for a history of government science relations up to 1940.

2. *Ibid.* President Roosevelt's Science Advisory Board (1933-35) was an unproductive though perhaps far-sighted attempt to involve prominent scientists in the analysis of major public problems. Its origins and activities are discussed in Don K. Price, "Federal Money and University Research." *Science*, 151, no. 3708 (Jan. 21, 1966), 285-290, and Lewis E. Auerbach, "Scientists in the New Deal," *Minerva*, 3, no. 4 (Summer, 1965), 457-482.

3. The scientists' role in the Second World War is extensively examined in James Phinney Baxter III, *Scientists against Time* (Boston: Little Brown and Co., 1946). See also Julius A. Furer, *Administration of the Navy Department in World War II* (Washington: Department of the Navy, 1959).

4. In late 1944, President Roosevelt requested Vannevar Bush, the Director of the Office of Scientific Research and Development, to prepare a report on how the lessons of the wartime science effort could be applied in years of peace. Bush's report, *Science: The Endless Frontier,* though never formally adopted as government policy, provided the basic principles for the post-war

support and utilisation of science by the federal government: Vannevar Bush, *Science: The Endless Frontier,* reprinted by the National Science Foundation with an introduction by Alan T. Waterman, (Washington: National Science Foundation, 1960).

5. The Organisation for Economic Cooperation and Development, *Review of National Science Policies: United States* (Paris: OECD, 1968), provides a comprehensive and accessible guide to the post-war United States science statistics. Recent data appear in the various National Science Foundation statistical series, particularly *Federal Funds for Research, Development, and Other Scientific Activities.*

6. Don K. Price, *The Scientific Estate* (Cambridge: Harvard University Press, 1965), pp. 71-76.

7. Robert C. Wood, "Scientists and Politics: The Rise of an Apolitical Elite" in Robert Gilpin and Christopher Wright, eds., *Scientists and National Policy Making* (New York: Columbia University Press, 1964), pp. 41-72.

8. The actual fiscal year 1966 figures are $12,100,000,000 for extramural research and development and $12,700,000,000 for federal grants-in-aid for state and local governments: National Science Foundation, *Federal Funds for Research, Development and Other Scientific Activities, Fiscal Years 1968, 1969 and 1970,* vol. 18 (Washington: Government Printing Office, 1969), Table C-97, p. 255, and *Economic Report of the President 1970* (Washington: Government Printing Office, 1970), Table C-64, p. 253.

9. Advisory Commission on Intergovernmental Relations, *Fiscal Balance in the American Federal System,* vol. 1 (Washington: ACIR, 1967), p. 153.

10. *Ibid.*

11. Department of Health, Education, and Welfare, *State Data and State Rankings in Health, Education and Welfare* (Washington: Government Printing Office, 1965), Table S 73.

12. Although many can claim to have recognised the potential importance of scientists in shaping a new political system, little theoretical work has been attempted on this topic. A provocative first step can be found in Sanford A. Lakoff, "Scientific Society: Notes Toward a Paradigm," in Paul J. Piccard, ed., *Science and Policy Issues* (Itasca, Ill.: F. E. Peacock, 1969), pp. 51-72.

13. The actual impact these analytical techniques and studies have had on Department of Defense decision-making is an almost unposed and certainly unanswered question. For a positive exposition, see Charles J. Hitch, *Decision-Making for Defense* (Berkeley: University of California, 1967). For a more sceptical empirical assessment, see Harvey Sapolsky, *The Polaris System Development: Bureaucratic and Programmatic Success in Government* (Cambridge, Mass.: Harvard University Press, 1972), Chapters 2 and 4; and "PERT and POLARIS: A Chapter in Organizational Mythology" (forthcoming).

14. See, for example, John Platt, "What We Must Do", *Science,* 156 (Nov. 28, 1969), 1115-1121 and "How Can Science Aid the World Food Crisis?" *Washington Science Trends,* 18, no. 11 (June 19, 1967), 1.

15. The actual disbursement of funds might be slightly different.

16. The figures are for fiscal year 1965. See *Equitable Distribution of R & D Funds by Government Agencies,* Hearings before Subcommittee on Government Research, Committee on Government Operations, Senate, 90th Congress, 1st Session (Washington: Government Printing Office, 1967), Part 3, Exhibit 61 and National Science Foundation, *Geographic Distribution of Federal Funds for Research and Development, Fiscal Year 1965* (Washington: Government Printing

Office, 1967). Initial publication of geographical data appeared in *Obligations for Research and Development and R & D Plant,* Report to the Subcommittee on Science, Research and Development of the Committee on Science and Astronautics, House, 88th Congress, 2nd Session (Washington: Government Printing Office, 1964).

17. Daniel Shimshoni, *Aspects of Scientific Entrepreneurship* (unpublished Ph.D. thesis, Harvard University, 1966); See also Daniel Shimshoni, "The Mobile Scientist in the American Instrument Industry," *Minerva,* 8, no. 1 (January 1970), 59–89; E. B. Roberts, and H. A. Wainer, "New Enterprises on Route 128," *Science Journal,* 4; no. 12 (December 1968), 78–83.

18. For one such vision, see the comments of the governor of Iowa in "Trends of State Government in 1965 as indicated by the Governors' Messages," *State Government* (Spring 1965), p. 130.

19. An analysis of the rationales for the continued public support of science with an assessment of their intellectual validity and political feasibility is contained in Carl Kaysen, *The Higher Learning, the Universities and the Public* (Princeton: Princeton University Press, 1969).

20. President's Science Advisory Committee, Environmental Pollution Panel, *Restoring the Quality of our Environment* (Washington: Government Printing Office, 1965), and Panel on the World Food Supply, *The World Food Problem, A Report* (Washington: Government Printing Office, 1967).

21. There are, unfortunately, no comprehensive studies of state government support of science in the nineteenth century comparable to Dupree, op cit., which examines federal activities in science, and Howard S. Miller, *Dollars for Research: Science and its Patrons in Nineteenth-Century America* (Seattle: University of Washington Press, 1970), which focuses on the role of private philanthropy in the development of science. The most useful and accessible historical studies of state science activities are in geology and agriculture. See Walter B. Hendrickson, "Nineteenth-Century State Geological Surveys: Early Government Support of Science," *Isis,* 52, part III, no. 169 (September 1961) 357–371; H. C. Knoblauch, *et al., State Agricultural Experiment Stations: A History of Research Policy and Procedure,* U.S. Department of Agriculture Miscellaneous Publication 904 (Washington: Government Printing Office, 1962); Gerald D. Nash. "The Conflict Between Pure and Applied Science in the Nineteenth Century Public Policy: the California State Geological Survey 1860–1874," *Isis,* 54, no. 176 (June 1963), 217–228; Peter R. Day, ed., *How Crops Grow: A Century Later,* Bulletin 708, Connecticut Agricultural Experiment Station, New Haven, July, 1969.

22. The National Science Foundation conducted a partial survey (six states) in 1954, and two complete surveys in 1964–65 and 1967–68 of the research and development activities of state agencies. The 1954 survey is reported and discussed in Frederic N. Cleaveland, *Science and State Government* (Chapel Hill: University of North Carolina Press, 1959). See also National Science Foundation, *Scientific Activities in Six State Governments, Summary Report of a Survey Fiscal Year 1954* (Washington: Government Printing Office, 1958) and the individual state studies which were published separately by the Institute for Research in the Social Sciences, University of North Carolina.

23. National Science Foundation. "Research and Development in State Government Agencies, fiscal year 1967 and 1968," *Science Resources Studies Highlights,* NSF 70-13 (June 1, 1970).

24. *Ibid.*

Harvey M. Sapolsky

25. The figure of approximately $200,000,000 for state-financed research and development and research and development plant at state universities and colleges was derived as follows: Table C-3 of NSF 67-16 *R & D Activities of State Government Agencies. FY 1964/65* reports state and local support for research and development in fiscal year 1964 as $154,500,000 and notes (p. 55) that at least 90 per cent of this figure represents state funds. Thus the approximate figure for state-financed research and development would be $143,000,000. The same table reports non-federally financed research and development plant as being $90,700,000 in fiscal year 1964 and notes that two thirds of that figure or about $60,000,000 was state-financed. The estimate of two thirds of the $200,000,000 as being allocated to agricultural research was derived as follows: Table A-3 of NSF 68-22 *Scientific Activities at Universities and Colleges 1964* reports that state governments allocated $117,000,000 in 1964 to agricultural research experiment stations and schools. Table 8 of the same document lists $22,000,000 as the figure for non-federally financed capital expenditures for agricultural experiment stations and colleges in 1964. The assumption was made that all of the $22,000,000 in research and development plant was all state-financed giving a total of state-financed research and development and research and development plant expenditures for 1964 of $139,000,000. This total is verified in Nyle C. Brady, "Organization and Administration of Agricultural Research in the United States," processed by U.S. Department of Agriculture, 1964, which lists state-financed expenditures at experimental stations for 1964 as $133,000,000.

26. These arguments are parallel to and derived in part from the arguments on the limitations of the market system in supporting research activities. See the essays by Harry G. Johnson, and Carl Kaysen, in National Academy of Sciences, *Basic Research and National Goals* (Washington: Government Printing Office, 1965) and Kenneth J. Arrow, "Economic Welfare and the Allocation of Resources for Invention" in National Bureau of Economic Research. *The Rate and Direction of Inventive Activity: Economic and Social Factors* (Princeton: Princeton University Press, 1962), pp. 609–625.

27. This assumption and state programmes are discussed in Boyd R. Keenan, ed., *Science and the University* (New York: Columbia University Press, 1966); Bruce W. Macy, et al., *Impact of Science and Technology on Regional Development*, a report prepared for the Office of Regional Development Planning, U.S. Department of Commerce (Kansas City, Mo.: Midwest Research Institute, 1967) and National Academy of Sciences, National Academy of Engineering, *The Impact of Science and Technology on Regional Economic Development* (Washington: National Academy of Sciences, 1969).

28. *Ibid*, and William H. Gruber, and Donald G. Marquis, eds., *Factors in the Transfer of Technology* (Cambridge: MIT Press, 1969).

29. See Ronald P. Black, and Charles W. Foreman, *Transferability of Research and Development Skills in the Aerospace Industry* (Falls Church, Va.: Analytic Services, Inc., 1965) and E. G. Brown, "Aerospace Studies for the Problems of Men," *State Government*, 39, no. 1 (Winter 1966) and Remarks of Karl G. Harr, Jr., President of the Aerospace Industries Association at the Rotary Club of Philadelphia (Washington: AIA, 1967) and California State Department of Finance, "The Four Aerospace Contracts: A Review of the California Experience," produced January 1966, reprinted Appendix, Volume V of the "Report of the National Commission on Technology, Automation and Eco-

nomic Progress," *Technology and the American Economy* (Washington: Government Printing Office, 1966), and Lawrence Lessing, "Systems Engineering Invades the City," *Fortune*, 77, no. 1 (January 1968), 154 *et seq.*

30. Harold D. Watkins, "Systems Engineering Aids Social Problems," *Aviation Week and Space Technology* 31 (January 1966), 52 *et seq.;* California State Department of Finance, *op. cit.;* Elliot F. Beideman, *State Sponsorship of the Application of Aerospace Industry Systems Analysis for the Solution of Major Public Problems of California* (unpublished Ph.D. dissertation, School of Business Administration, University of Southern California, 1966); Carl F. Stover, "The California Experiment—An Appraisal," National Institute of Public Affairs, 1966 and John S. Gilmore et al., *Defense Systems Resources in the Civil Sector: An Evolving Approach, An Uncertain Market* (Denver: Denver Research Institute, 1967).

31. Typical endorsements of the science advisory system for universal adoption are contained in the *Report of the U.S. Department of Commerce Independent Study Board on the Regional Effects of Government Procurement and Related Policies* (Washington: Government Printing Office, 1967), p. 40; the essays by D. Waldo and C. Nader in C. Nader, and A. B. Zahlen, eds., *Science and Technology in Developing Countries* (Cambridge: Cambridge University Press, 1969), and the addresses of Lee A. DuBridge, Special Assistant to the President for Science and Technology at the 61st Annual National Governors' Conference, Colorado Springs, Colorado, September 3, 1969, and the Western Governors' Conference, Salt Lake City, Utah, March 20, 1970.

32. James R. Killian, Jr., Text of an address before the 55th Annual Governors' Conference, Miami Beach, Florida, July 22, 1963. Monograph, MIT Office of Public Relations.

33. The act and its operations are discussed in Office of State Technical Services, U.S. Department of Commerce Annual Report 1968 (Washington: Government Printing Office, 1969) and D. H. Silva, "State Technical Service—An Emerging Social System," *American Journal of Economics and Sociology*, 28, no. 4 (October 1969), 399–404. The programme was cancelled in 1969 through a combination of Congressional and Administration actions. "Administration Eliminates States Technical Services Funds," *Washington Science Trends* May 5, 1969, p. 1; Andrew Hamilton, "State Technical Services: Congress Swings the Axe," *Science*, 166, no. 3913 (December 26, 1969), 1606–1608.

34. Harvey M. Sapolsky, "Science Advice for State and Local Government," *Science*, 160, no. 3825 (April 19, 1968), 280.

35. Memorandum from H. F. Hersman, National Science Foundation to State Science Advisors and Federal State Coordinators; Steven Dedijer in "A Science Policy for Alaska?" delivered as an address to the 20th Alaska Science Conference, Fairbanks, August 24, 1969, reports a similar finding from a survey which he conducted.

36. The studies were independently prepared by social scientists in each state and will be published separately. A summary and analysis of the major findings of these studies is presented in the text of this paper.

37. Georgia Science and Technology Commission, *Using Science and Technology to Brighten Georgia's Future* (Atlanta: Georgia Science and Technology Commission, n.d.), p. 12.

38. The largest is the Health Research Council of the City of New York,

Harvey M. Sapolsky

whose annual city-appropriated budget of more than $3 million is approximately three times that of any state-sponsored foundation. It should be pointed out that the total budget for the City of New York exceeds $6.5 billion in fiscal year 1970, a budget which is larger than any state except California and New York state.

39. As Circular A-91 of the Bureau of the Budget authorises federal agencies to make grants to state legislatures, the California contract could mark the true beginning of innovative legislation.

40. Dr. Jack Carlson, an official of the Office of Management and Budget, has used the term "Flying Feds" to describe federal officials who have promoted the application of Programme Planning and Budget Systems (PPBS) in state and local governments in the United States.

41. The formation and operation of PSAC is discussed in Robert N. Kreidler, "The President's Science Advisers and National Science Policy," in R. Gilpin, and C. Wright, eds., *Scientists and National Policy Making* (New York: Columbia University Press, 1964), pp. 113-143 and Carl W. Fischer, "Scientists and Statesmen: A Profile of the President's Science Advisory Committee," in S. Lakoff, ed., *Knowledge and Power* (New York: The Free Press, 1966), pp. 315-358. For a report on PSAC's role in national space policy see Enid Bok Schoettle, "The Establishment of NASA," in *Knowledge and Power*, pp. 162-270.

42. Bernice Eiduson, a psychologist who has conducted a series of intensive interviews with a sample of successful scientists, reports that scientists active in federal advisory tasks candidly admit that prestige is their strongest motivation for accepting such positions. See her "Scientists as Advisors and Consultants in Washington," *Bulletin of the Atomic Scientists* October 1966, pp. 26-31, especially p. 25.

43. The Air Force's role in the Oppenheimer case reflects its awareness of the scientists' growing influence in weapons decisions. See Giorgio de Santillana, *Reflections on Men and Ideas* (Cambridge: MIT Press, 1968) pp. 120-126; Philip M. Stern *The Oppenheimer Case* (New York: Harper and Row, 1969).

44. The Air Force's post-war relations with scientists are documented in Thomas A. Sturm, *The USAF Scientific Advisory Board: Its First Twenty Years 1944-1964* (Washington: United States Air Force Historical Division, 1957), and Bruce L. Smith, *The Rand Corporation* (Cambridge: Harvard University Press, 1966) and Nick A. Komons, *A History of the Air Force of Scientific Research* (Arlington, Va.: Office of Aerospace Research, 1966).

45. See Richard H. Blum, and May Lou Funkhouser, "Legislators on Social Scientists and a Social Issue: A Report and Commentary on Some Discussions with Lawmakers about Drug Abuse," *Journal of Applied Behavioral Science*, 1, no. 1, (Jan., Feb., Mar., 1965), pp. 84-112. See also Daniel P. Moynihan, "A Plea to the Candidates: Tell Us What Happened," *Sunday Herald Traveler Magazine*, November 3, 1968, pp. 36-38.

46. In contrast, Carroll argues that the authority of new knowledge undermines and can replace the authority of tradition and law in the society. James D. Carroll, "Science and the City: The Question of Authority," *Science*, 163, no. 3870 (Feb. 28, 1969), 902-911.

47. The concept of "technology fix" is discussed by Alvin M. Weinberg, in "Can Technology Replace Social Engineering?" *The American Behavioral Scientist*, 10, no. 9 (May 1967), 7-10 and placed in perspective in Wallace S. Sayre,

and Bruce L. Smith, "Government, Technology and Social Problems," an occasional paper of the Institute for the Study of Science in Human Affairs, Columbia University, 1969.

48. In 1967 the states employed 20,600 scientists, 34,200 engineers, 41,000 health professionals and 61,900 technicians. Bureau of Labor Statistics, news release, United States Department of Labor 10-360, April 8, 1969.

49. Carl Kaysen, *The Higher Learning, the Universities and the Public* (Princeton: Princeton University Press, 1968), pp. 54-55. For the clear and uncritical statement of this assumption, see National Academy of Sciences, National Academy of Engineering, *The Impact of Science and Technology on Regional Economic Development* (Washington: National Academy of Sciences, 1969).

50. See D. S. Greenberg, "Civilian Technology: NASA Study Finds Little 'Spin-off,'" *Science*, 157, no. 3792 (September 1, 1967), 1016-1018 which summarises John S. Gilmore, et al., *The Channels of Technology Acquisition in Commercial Firms and the NASA Dissemination Program* (Springfield, Va.: Clearinghouse for Federal Scientific and Technical Information 1967). Note also Richard Nelson, et al., *Technology, Growth and Public Policy* (Washington: Brookings, 1967), pp. 83-85 and Samuel I. Doctors, *The Role of Federal Agencies in Technology Transfer* (Cambridge: MIT Press, 1969).

51. Relevant here are studies which analyse the political attitudes of scientists. See especially Yaron Ezrahi, "The Political Resources of Science," a paper delivered at the 1970 meeting of the American Association for the Advancement of Science, Boston, December 30, 1970; Everett C. Ladd, Jr., "Professors and Political Petitions," 163, no. 3874 (March 28, 1969), 1425-1430; Everett C. Ladd, Jr., "American University Teachers and Opposition to the Vietnam War," *Minerva*, 8, no. 4 (October 1970), 542-556; and Robert Gilpin, *American Scientists and Nuclear Weapons Policy* (Princeton University Press, 1962).

20. Federal Technology Policy: Innovation and Problem Solving in State and Local Governments

J. DAVID ROESSNER

In a 1970 report, the Science and Astronautics Committee of the U.S. House of Representatives recommended that "the scientific method and technological research should be increasingly utilized by

Copyright © 1979 by The Regents of the University of California. Reprinted from *Policy Analysis*, vol. 5, no. 2, Spring 1979, pp. 181-200, by permission of The Regents.

regional state, and local organizations in seeking solutions to societal problems."[1] In his 1972 science and technology message, President Richard Nixon referred to the challenge of applying research and development (R & D) resources to public problems and declared that effective use of these resources would require that state and local governments play a central role.[2] Many of the basic policy concerns were set forth in a series of 1972 reports published by the Federal Council for Science and Technology, the Council of State Governments, the Urban Institute, and the National Science Foundation.[3] These reports introduced the term *public technology* and brought a relatively new set of issues to the attention of federal policymakers.

This flurry of interest in public technology helped stimulate individual efforts by federal agencies to provide state and local governments with information about their DOT programs and with mechanisms for obtaining information about R & D activities and findings in specific problem areas. A directory issued in 1975 by the Committee on Domestic Technology Transfer of the former Federal Council for Science and Technology included descriptions of forty-three different federal technology transfer programs, resources, and contact points.[4] The most recent manifestation of federal-level concern with public technology was the provision of the National Science and Technology Policy, Organization, and Priorities Act of 1976 requiring the director of the new White House Office of Science and Technology Policy to establish an intergovernmental advisory panel to identify and define problems at the state, regional, and local levels that science and technology could assist in resolving.

These public technology activities are part of a larger set of federal activities that might be labelled *technology policy* activities, but that assuredly are not the product of a strategically planned and coordinated federal effort. As W. H. Lambright points out, American technology policy and science policy are those allocations of resources and sets of programs that result from a large number of day-to-day decisions made in a variety of locations within the federal establishment.[5] Federal technology policy may be regarded as the sum of federal efforts to change the rate and direction of technological innovation in the American economy, particularly efforts to stimulate or control the development and use of new technology.[6] Public technology policy, from the federal perspective, is that subset of activities directed toward influencing the development and use of new technology by regional, state, and local governments.

The intergovernmental flavor of public technology has implications

Federal Technology Policy

for the context in which public technology programs are operated and policy issues are framed. Federal mechanisms employed to influence the development and use of new products and techniques by state and local governments largely are different varieties of intergovernmental resource transfers: categorical grants, block grants, revenue sharing, and research, development, and demonstration grants. The fragmentation of intergovernmental programs and the complexities of intergovernmental politics—what Allen Schick has called the "intergovernmental thicket"—are thus mirrored in the structure of public technology policymaking.[7] The relatively low priority of intergovernmental relations among domestic problems, the lack of an effective federal coordinating body for either intergovernmental affairs or technology policy, and the weakness in Congress of city and state lobby groups relegate public technology programs to second-class or lower status in most federal agencies.

Coincident with increased interest in public technology in the early 1970s was the Nixon administration's New Federalism: a call for administrative decentralization of federal programs and the devolution of decision-making authority and responsibility from federal agencies to those of state and local governments. As suggested in a 1975 report to the U.S. Office of Management and Budget, the trend toward devolution and decentralization is fueled by factors that reach beyond the ideology of the administration in office.[8] They include a shift in the nature of public problems from equality of opportunity and economic growth to mass transit, crime, housing, pollution, and other problems that have their focus at the subnational level; the increasing administrative burdens of federal assistance programs; and the rising demands being made on state and local services, coupled with their increasing financial instability.[9]

Public technology policy issues, then, rest at the nexus of two broad streams of public concern: the perceived need to stimulate the use of research and new technology by regional, state, and local governments, and the need to assist these same governments as they assume greater responsibilities for the solution of increasingly complex public problems. The problem for federal policymakers is how to develop and implement programs that will at once increase the availability and use of new technologies by state and local governments *and ensure that doing so actually helps solve the complex problems they face.*

The purposes of this paper are to discuss how public technology policy issues can be conceptualized and analyzed, to present the results of empirical research that bears on the resolution of these issues, and to discuss the implications of research findings for federal public technology policy.

Conceptual Approaches to Analysis of Public Technology Issues

Policy analysts frequently say that to identify the problem brings one halfway to the solution. This maxim clearly applies in the case of public technology. Thus, one approach to the analysis of public technology issues is to look more closely at the statement of the problem made immediately above. That statement glibly connects two programmatic goals—to increase the use of new technology (innovation), and to help solve problems—that upon closer examination may not be compatible in many circumstances.

There is no a priori reason to assume that federal programs intended to stimulate the development and use of new products and techniques by state and local governments will, in fact, contribute significantly to those governments' efforts to solve their problems *as they see them;* conversely, federal programs intended to help state and local governments solve their problems need not involve the introduction of new products or techniques. While this would appear to be an obvious distinction, a number of federal programs have failed to clarify their goals along these lines. Moreover, rarely if ever are public technology programs assessed according to problem-solving criteria; more typical criteria include a few success stories, the number of persons requesting information, the number attending seminars, or the number of agencies purchasing or using a new technology.[10] As I demonstrate below, innovation and problem solving at the local level have sometimes been found to be more contradictory than complementary.

An alternative approach breaks the "What's the problem?" question into three more specific questions: Is it a demand problem? (State and local government agencies are inherently unreceptive to new ideas and techniques.) Is it a supply problem? (Private industry, federal agencies, and other suppliers of innovations are unresponsive to the needs of state and local governments.) Is it an infrastructure problem? (The system of communication among state and local governments, and between them and the suppliers of innovations, is inadequate or ineffective.) When stated this way, these questions suggest research as a way to obtain answers, and at least one federal research office has structured its public technology research problem this way.[11]

Federal Technology Policy

A third approach, which I discuss only briefly here, begins with a list of specific types of federal involvement or intervention mechanisms, followed by an assessment of the likely effectiveness of each mechanism. In an early effort of this type, I developed a set of criteria intended to reflect the standard concerns of policymakers (such as cost, effectiveness, risk) and some of the more immediate concerns of the then-current federal administration.[12] I applied these criteria to a list of recommendations for federal action, grouped according to type of action proposed: (1) increase the technical capacity of state and local government personnel; (2) disseminate technical information to state and local jurisdictions; (3) increase the influence of state and local interests in federal science and technology policymaking; (4) encourage technical support to state and local jurisdictions by institutions having scientific and technical resources; (5) support or conduct research and/or demonstration programs. The results showed this to be a useful scheme, though clearly others of this type could prove equally useful.

The conceptual issues do not end here, because attempts to implement any of the above approaches confront a host of problems associated with the applicability of existing research to the analysis of public technology policy. Several recent reviews of the literature potentially applicable to better understanding of technological innovation in state and local government agencies—organizational behavior, economics, diffusion studies, public administration, management science, technology transfer, organizational change—reveal the inadequacy of existing theory in explaining or predicting patterns of adoption, use, and rejection of innovations by state and local governments.[13]

Innovation Versus Problem Solving

The "innovativeness" of organizations or individuals can refer to several different phenomena. From the perspective of diffusion research, innovativeness usually refers to early or frequent adoption of products, processes, or services that are perceived as new by potential adopters.[14] Alternatively, much of the literature on R & D activities in private firms focuses on innovation as the successful development and marketing of new products and industrial processes. Sociologists frequently place innovation within the broader framework of organizational change. Rowe and Boise outline major unresolved definitional issues: (1) whether to include the generation of new ideas as well as their acceptance and implementation; (2) whether to restrict the definitions to organizations that are the first among a set of similar organizations to adopt innovation; (3) whether to require that the introduction of new ideas and so forth be followed by successful implementation; and (4) whether to require that the decision to innovate be made within an organization.[15]

Additional conceptual problems arise because of the lack of a theoretically meaningful typology of innovations. In the absence of such a typology, one cannot say whether early adoption of a "trivial" innovation is more innovative than later adoption of a "significant" one; nor can one say whether the adoption of fifteen innovations of one kind is more innovative than the adoption of five innovations of another.[16] An additional complication occurs if one asks whether an organization that adopts an innovation inappropriate to its goals is more or less innovative than a similar organization that adopts the same innovation for "appropriate" reasons.

These problems spotlight the sterility—from both a theoretical and a practical point of view—of the concept of "innovativeness." What matters is not that an individual or an organization adopts an innovation, but whether the innovation is used, how it is used, and how its adoption and use are linked to the satisfaction of a need or solution of a problem. Existing conceptualizations in the innovation literature, then, lead us to the conclusion that useful research must address the link between the adoption of innovations and the consequences of adoption for organizational or individual problem-solving.

We reach a similar conclusion if we consider the documented experiences of federal programs related to public technology. Through the decade of the 1960s numerous programs attempted to increase the utilization of the results of federally supported R & D.[17] These programs ranged from "spinoff" efforts such as the Technology Utilization program of the National Aeronautics and Space Administration (NASA) to the research utilization activities of social service agencies such as the Office of Education, the Manpower Administration, and the Social and Rehabilitation Service of the U.S. Department of Health, Education, and Welfare.[18] As experience with these programs accumulated, emphasis shifted from passive information dissemination to active efforts such as field offices, research utilization specialists, demonstrations, adaptive engineering, and other forms of technical assistance.

The U.S. Law Enforcement Assistance Administration's Pilot Cities Program was one of these active efforts to provide technical assistance to local governments. As originally conceived, the program called for the establishment in eight cities of a team of researchers and profes-

Federal Technology Policy

sional analysts who would facilitate the introduction and use of law enforcement innovations in the local setting. An evaluation of the program, conducted after more than five years had passed since the first site was funded, concluded that (1) innovation is often unnecessary to improve local law enforcement systems, since what is known about law enforcement and criminal justice far outstrips what is generally practiced, and (2) innovations tend to deal with issues in the periphery of law enforcement/criminal justice needs (teams characteristically had to choose between dealing with a central law enforcement/criminal justice issue or producing a genuine innovation).[19]

The National Science Foundation (NSF) has supported a number of projects intended to demonstrate ways of introducing new technology into cities and states. The California Four Cities Program, jointly sponsored by NSF and NASA, placed aerospace professionals in four California cities for several years to find out if aerospace technology could be used to help solve municipal problems. A study of the program by the U.S. General Accounting Office (GAO) concluded that technically trained persons in city managers' offices could contribute significantly to the solution of city problems, but that aerospace hardware technology applications were practically nonexistent in the four cities. In fact, the technology advisers' primary contributions involved management skills and problem analysis.[20]

This example illustrates from a programmatic perspective the potential conflict between efforts to stimulate innovation in state and local governments and efforts to help solve their problems. This is not to say that innovation and problem solving are invariably at odds; there are too many examples of how innovations in cities have produced cost savings and other evidence of increased productivity for this to be the case.[21] The problems arise when either innovation theorists and researchers or public technology policymakers forget that what matters about innovation is its consequences, not innovation per se.

Demand, Supply, or Infrastructure?

To the extent that the public technology "problem" can be associated with an unwillingness or inability of users—state and local government officials—to adopt and use new products and techniques, it is a *demand* problem, and federal policies should be directed toward the system of incentives that affects local officials' decisions. To the extent that manufacturers and federal agencies produce ideas and techniques and products that are too complex, too expensive, or unrelated to state and local needs and priorities, one can say there is a *supply* problem, and federal policies should be directed toward increasing the responsiveness of innovation suppliers to state and local needs. If users are willing to innovate and suppliers are eager to respond, but the flow of information to suppliers about needs and the flow of new ideas, techniques, and products to users do not occur, one can say there is a communications (*infrastructure*) problem, and federal policies should be directed toward building or strengthening the communications networks that link suppliers with users.

Demand. What evidence exists to confirm or deny the argument that state and local governments are inherently resistant to innovation? William Baumol argues that the gap between public wage levels and public service levels (particularly in cities) will continue to widen inexorably because of conditions endemic to what he calls the "nonprogressive" sector of the economy.[22] He argues that this sector, because it is intrinsically labor-intensive, is largely denied the usual routes to productivity increase—capital accumulation, economies of scale, and technological innovation—which therefore reduces the possibility that wage increases will be offset by increases in productivity. Baumol is not alone in his belief that public organizations are relatively immune to significant productivity increases, especially increases that result from innovative behavior. In making theoretical links between innovativeness and economic sector, economists begin with theories of market efficiency, organization theorists begin with analyses of bureaucratic behavior, and political scientists begin with the ways citizen demands for democratic accountability are translated into the institutional arrangements and legal constraints that form the context of public organizations. Each approach leads from different origins to a common expectation: public organizations are probably less innovative than private organizations.[23] Literature based on the experiences of public officials has tended to support this theoretically based conclusion.[24] But the findings described below, those of more recent, systematic research efforts to improve understanding of innovation in state and local governments, tend to question the conclusions of earlier work.

There is evidence that public organizations operate under a set of incentives different from that of their private counterparts, and that these differences may cause misleading conclusions concerning the innovativeness of public organizations. For example, studies of the diffusion of technologies among industrial firms assume implicitly that firms in a position to adopt innovation have common technical problems, an assumption often unwarranted in the case of city and state agencies.[25] The efficiency with which public agencies provide services (production efficiency) can be viewed as the analog of private industry's profits, but the major incentives that affect public officials may be more numerous and complex than those in private firms.[26] The "performance" of public agencies frequently is judged by criteria such as responsiveness, representativeness, openness, equitable treatment of clients, and accountability, production efficiency is only one of a large number of frequently conflicting organizational goals. Robert Yin has postulated the existence of two innovation processes operating in government organizations: one driven by incentives related to improvements in organizational efficiency or the quality of service output, the other driven by bureaucratic self-interest.[27] Yin argues that bureaucratic interests play a far larger role in public organizations than in private ones. Clearly, it is exceedingly difficult to disentangle the relative influence of efficiency, performance, and bureaucratic self-interest factors in explaining decisions to innovate in public agencies. The point here is that observers of public organizations may be prone to label mistakenly as noninnovative decisions those decisions influenced by factors other than expected performance improvements.

Several institutional features of public organizations act as disincentives for public officials to innovate: (1) The democratic accountability of government agencies to clients, legislative bodies, and higher levels of government means that, relative to private firms, public agencies are less capable of independent action than their private counterparts. Displeased subordinates have multiple, extra-agency routes of appeal. (2) Top leadership changes are both more frequent and more farreaching in public agencies than in private firms. The short tenure of most elected public officials means that political survival is dependent upon production of short-term, highly visible results. Programs to produce these kinds of results must have low risks and quick payoffs, characteristically not the attributes of innovative activities. (3) The client or constituent groups of public agencies tend to be more heterogeneous than those of private firms, particularly in the sense that demographic characteristics such as age, race, education, and health all have political implications. Because the values, interests, and reward structures of public agency constituents vary so much, and because public decisions are so visible, significant changes are difficult to effect. (4) Since public agency outputs are not evaluated in external markets, it is difficult to develop objective performance measures and to specify goals and functions operationally. Goals and objectives consequently lack clarity, which makes developing performance incentives difficult and favors highly visible but superficial change over change that might significantly affect service effectiveness or efficiency in the long run.[28]

Supply. Regardless how receptive state and local governments are to new technology, the rate of technological change in government—the rate at which old or obsolete products and techniques are replaced by newer, improved versions—can be low if the rate of supply of innovations for the state and local government market is low, or if the innovations offered to this market do not match the pattern of demand. Some evidence suggests that both conditions exist, at least in a few specific areas of state and local government services.

In their study of innovation in the fire services, Frohman, Schulman, and Roberts analyzed data on 99 manufacturers and 398 distributors of fire equipment. They found that firms selling primarily to the fire service market are small, financially weak, and unable to obtain outside financing for product development activities.[29] Large firms capable of conducting R & D do market products to the fire services, but this fragmented market is not a major one; these firms tend to orient their product development investment toward more profitable markets.[30] In his study of the law enforcement R & D system Michael Radnor obtained similar results, concluding that "law enforcement cannot, by itself, be a large enough market to make anything other than very small companies profitable unless they deal in other markets as well."[31]

J. F. Blair's careful analysis of data from a sample of nearly 200 heads of firms in the Delaware Valley (Philadelphia area) of Pennsylvania revealed a number of features of industrial participation in, and attitudes toward, the municipal market.[32] The impact of the local/municipal government market in the Delaware Valley appeared to be small relative to other markets: municipalities represented an insignificant or minor share of business for firms in the valley. Responding firms also indicated that their major marketing efforts were not directed toward municipal markets; Blair suggests that this limits the exposure of municipalities to innovative technologies.[33] Additional interviews with sixteen of the most active local venture capitalists revealed that few of them recognized "municipalities and local governments as a promising market for ventures based upon technological innovation."[34]

This gloomy picture is mitigated somewhat by the fact that it does not hold uniformly across all government service areas. Feller, Menzel, and Engel's study of the diffusion of highway and transportation

Federal Technology Policy

innovations among states showed, at least with respect to impact absorption devices, a high level of innovative behavior among the relevant firms and vigorous marketing to the state highway departments.[35] Feller and Menzel's more recent study of diffusion at the municipal level showed that firms do exercise independent influences on the rate and patterns of diffusion of new technologies among cities, and that these influences include the location of firms in different regions, firm resources, and marketing networks.[36]

The federal government is another potential source of innovations intended for use by state and local governments. Some domestic agency R & D programs were created to provide results useful to state and local government practitioners (such as teachers, vocational rehabilitation workers, water pollution control agents), and the growing number of technology transfer and research utilization programs and activities reflects increased efforts to get the products of this federal R & D used.[37] Unfortunately, few systematic studies of these efforts have been done that would enable conclusions to be drawn about how responsive these R & D programs are to the needs of their state and local government clients. Scattered case analyses and a recent Rand Corporation study of federal demonstration projects do not offer grounds for optimism, however.[38] Possible reasons for the lack of evaluations of these kinds of programs include the difficulty in developing program effectiveness criteria, the cost of conducting surveys of state and local government users, and agency concern that the results might be embarrassing.

There does appear, then, to be a "supply" problem, but it is due more to the fragmented state and local market than to the unresponsiveness of the private firms to state and local needs. In other words, if the market were more attractive, private manufacturers apparently would respond with more aggressive marketing efforts, larger investments in R & D, and a portfolio of projects more responsive to the state and local market. Impressionistic evidence suggests that federal agency suppliers, for whom market profitability is not an immediate concern, have difficulty accurately identifying state and local government needs, using those needs to plan R & D programs, and stimulating the widespread use of federal R & D products.

Infrastructure. Suppliers and users are necessary but not sufficient ingredients in any system that provides innovations responsive to state and local government needs. Needs must be communicated to suppliers; products must be advertised; regulations and standards must be set, communicated, and enforced; and evaluative information about new products and techniques must be shared among users.

J. David Roessner

Where "users" are state and local government agencies, key components of the infrastructures that perform these functions include public interest groups (the International City Management Association), professional associations (the American Public Works Association), trade associations (the National Association of Wholesaler-Distributors), standard-setters and product evaluators (the National Bureau of Standards, the National Fire Protection Association) and the myriad distributors of industrial products. Except in scattered cases, little is known about how these kinds of organizations act to facilitate or inhibit the adoption and use of new technologies by state and local agencies.

Professional associations appear to be a key component of the communication networks that operate horizontally among similarly-placed state and local officials in different cities and states, and vertically among functional area counterparts at the local, state, regional, and federal levels. Feller, Menzel, and Engel observe that state highway and transportation departments are part of a strong, tightly integrated network that includes organizations such as the American Association of State Highway and Transportation Officials and the Transportation Research Board of the National Academy of Sciences, but that analogous organizations in the air pollution field either do not exist or do not perform intermediary roles as effectively.[39]

Similar contrasts appear to exist at the local government level. Limited evidence suggests that an effective "grapevine" operates among fire chiefs and other fire officials, who rapidly pass around information about the existence of new products and techniques in the fire services.[40] This informal network also functions, but not as effectively, to transmit information about how well these new products or techniques actually work. The perceived importance of rules, regulations, and laws enacted at different levels within the federal system varies widely across the fire, traffic, air pollution, and solid waste services.[41] Differences also exist among the four functional areas with respect to their perceptions of the importance of information from different sources and its adequacy in assessing new technical innovations.[42]

Summary and Implications for Public Technology Policy

A demand, supply, infrastructure approach to the analysis of public technology policy issues suggests that there are problems, and hence opportunities for improvement, in all three areas. State and local government agencies evidently respond favorably, and rapidly, when exposed to a new technology or technique that addresses a priority problem *as they define it* and that is accompanied by *credible* information about implementation problems, performance, and consequences. The unique institutional context of public organizations means that problems and priorities vary widely across cities and states that on the surface appear similar: the social and political demography of citizens differ, laws and regulations differ, and traditions concerning the appropriate role of government and citizen-government interaction differ. The highly localized nature of problem definition and priority-setting also means that innovations require a long period of incubation before they become widely accepted; in some cases this is due to the lack of an effective or credible source of evaluative information, in others to the lack of an effective communications network among practitioners, in others to the fact that innovation requires a large amount of adaptation to local conditions.

The institutional context of public organizations also has implications for the rate of supply of innovations to the public market. Public accountability requirements in purchasing decisions, coupled with the large diversity of local needs, creates a fragmented market in many state and local government service areas. Private firms do not regard this market as profitable relative to other markets, and the low profit margins that result from limited markets and competitive bidding requirements do not encourage investment in new product development. The situation summarized here suggests that federal public technology policy should emphasize certain strategies.

Whenever a federal agency's mandate and political climate permit, it should avoid attempting to develop and foster the utilization of particular "solutions"—technologies, techniques, or systems—and work toward building the analytic and evaluative capabilities of its state or local government clientele. Given the difficulty of specifying common needs of state and local governments, communicating these to federal agencies, incorporating them in R & D and technical assistance programs, and disseminating the results of R & D projects, it would appear to be far more efficient and effective to bolster the ability of local decision makers to define their problems in technical terms, identify alternatives that may or may not include innovative solutions, and evaluate alternatives themselves or have the expertise to buy useful evaluative information from the outside. This type of strategy would have the additional benefit of sharpening the nature of state and local government demand for new products, possibly resulting in some aggregation of local markets.

Recent research and experience concerning the innovative behavior of state and local officials suggests that *the most effective way to improve these officials' technical capability is to utilize what Yin calls "natural" points of entry into the incentive system in which they live.*[43] Natural points of entry are those groups and activities toward which state and local officials look for rewards and cues for action. They include professional communications networks, particularly public interest and clientele groups such as the International City Management Association and the National Association of State Purchasing Officials, and colleagial networks made up of acquaintances in neighboring jurisdictions or in jurisdictions looked to for leadership in the technological or problem area involved. Other natural points of entry are the tests used in government personnel systems as bases for promotion, and the curricula used in education and training programs for public management positions. Ensuring that the latest accepted analytic methods and management techniques are part of the curricula of the schools that produce public managers, and introducing the latest accepted ideas, practices, and techniques into civil service examinations for higher positions in local government departments should have salutary effects on local officials' awareness and use of the best available techniques and information. Similarly, some professional associations could be stimulated through federal action to upgrade their efforts to evaluate innovations and communicate the results to their clients, or to develop and disseminate training and educational materials that utilize current, best-practice techniques. Finally, federal grants could be used more extensively than at present to help upgrade the scientific and technical capabilities of state legislatures through the support of in-house analytic staffs.[44] It is worth pointing out, however, that heavy federal involvement in professional associations and public interest groups can lead to loss of these groups' independence, and hence of their credibility with their clients.

If a federal agency is faced with the problem of disseminating specific technologies or techniques to state and local governments, some strategies may be more effective than others. First, to the extent possible, *agencies should focus on the process by which they select R & D projects.* Analyses of federal demonstration and civilian R & D programs reveal that project failures are frequently associated with the absence of any prior assessment of the demand for the product or technique at an early state in R & D planning.[45] Market analysis evidently needs to become a far more common practice in federal civilian R & D agencies than is now the case. Second, and relatedly, *agencies need to be aware of the strengths, weaknesses, and operation of the innovation system or systems in which their primary client groups exist.* For example, decisions by the Law Enforcement Assistance Administration to support the development

Federal Technology Policy

and demonstration of new, lightweight body armor requires information about the roles and behavior of the International Association of Chiefs of Police, police unions, standards agencies, body armor manufacturers, and distributors. Any reasonably complete market analysis of the likely demand for lightweight armor would require information about the behavior and interaction of each of these involved parties with respect to new products.

These suggested strategies may not be looked upon with favor by federal agencies. They tend to be long term in their effects (such as changing curricula in schools of public management, changing questions on civil service examinations), and most of the benefits of building state and local capacities are intangible relative to measures such as the number of pieces of information requested or of technologies in use. Perhaps most problematic from the federal agency perspective is that increased analytic capability at the state and local levels may mean increased rejection of the technologies offered by federal agencies.

Current trends suggest that the locus of control for federal agency decisions on what R & D projects will be done and for whom, how they will be packaged, and how they will be disseminated will shift away from the agency and closer to the intended client: greater influence will be manifested by professional associations, advisory groups, and public interest groups. The Cooperative Extension Service of the U.S. Department of Agriculture and the Federal Highway Administration in the U.S. Department of Transportation represent agencies whose locus of control has shifted considerably from the federal official to the state or local client. The benefits, costs, and peculiarities of such arrangements need to be explored carefully as American public technology policy follows the trend of devolution of responsibility that began with the New Federalism.

Notes

1. U.S., Congress, House, Committee on Science and Astronautics, *Toward a Science Policy for the United States,* Committee Print, 91st Cong., 2d sess., October 15, 1970.

2. President's Message to the Congress, March 16, 1972, reprinted as "Appendix" in J. M. Logsdon, "Toward a New Federal Policy for Technology: The Outline Emerges," Program of Policy Studies in Science and Technology, Staff Discussion Paper 408 (Washington, D.C.: George Washington University, 1972).

3. Federal Council for Science and Technology, *Public Technology: A Tool for Solving National Problems* (Washington, D.C.: U.S. Government Printing Office, 1972); Council of State Governments, *Power to the States—Mobilizing*

J. David Roessner

Public Technology (Lexington, Ky., 1972); The Urban Institute, *The Struggle to Bring Technology to Cities* (Washington, D.C., 1971); *Action Now, Partnerships—Putting Technology to Work,* Report of the National Action Conference on Intergovernmental Science and Technology Policy, Harrisburg, Pa., June 21-23, 1972 (Harrisburg: Pennsylvania Office of Science and Technology, 1972).

4. National Science Foundation, *Directory of Federal Technology Transfer,* NSF 75-402 (Washington, D.C., 1975).

5. W. H. Lambright, *Governing Science and Technology* (New York: Oxford University Press, 1976), p. 187.

6. It is worth noting that federal antitrust, tax, patent, and economic policy may have as much or more impact on the rate and direction of technological innovation as programs intended to apply direct influences. See *Technological Innovation and Federal Government Policy,* Research and Analyses of the Office of National R & D Assessment, NSF 76-9 (Washington, D.C.: National Science Foundation, 1976).

7. Allen Schick, "The Intergovernmental Thicket: The Questions Still Are Better Than the Answers," *Public Administration Review,* Special Issue, December 1975, pp. 717-22.

8. U.S., Executive Office of the President, *Strengthening Public Management in the Intergovernmental System,* prepared for the Office of Management and Budget by the Study Committee on Policy Management Assistance, NSF 75-600 (Washington, D.C.: National Science Foundation, 1975).

9. Ibid., pp. 1-5.

10. National Science Foundation, *Federal Technology Transfer: An Analysis of Current Program Characteristics and Practices,* Report to the Committee on Domestic Technology Transfer, Federal Council for Science and Technology NSF 76-400 (Washington, D.C.: National Science Foundation, 1976).

11. National Science Foundation, Office of National R & D Assessment, *Official Program Plan for Support of Extramural Research,* FY 1975 (Washington, D.C., 1975).

12. J. D. Roessner, "Policy Options for Public Technology," in *Serving Social Objectives Via Technological Innovation: Possible Near-Term Federal Policy Options* (Washington, D.C.: National Science Foundation, Office of National R & D Assessment, 1973).

13. J. D. Roessner, "Innovation in Public Organizations" (Paper presented at the National Conference on Public Administration, Syracuse, N.Y., May 1974); E. M. Rogers, "Innovation in Organizations: New Research Approaches" (Paper presented at the 1975 Annual Meeting of the American Political Science Association, San Francisco, September 2-5, 1975); Paul Berman and M. W. McLaughlin, *Federal Programs Supporting Educational Change* R-1589/1-HEW (Santa Monica, Calif.: Rand Corporation, 1974), vol. 1, *A Model of Educational Change;* E. R. House, *The Politics of Educational Innovation* (Berkeley, Calif.: McCutchan, 1974); K. E. Warner, "The Need for Some Innovative Concepts of Innovation: An Examination of Research on the Diffusion of Innovation," *Policy Sciences,* 5 (1974), 433-451; G. W. Downs, Jr., and L. B. Mohr, "Conceptual Issues in the Study of Innovation, *Administrative Science Quarterly,* 21 (December 1976), 700-714.

14. See E. M. Rogers and F. F. Shoemaker, *Communication of Innovations* (New York: Free Press, 1971).

15. L. A. Rowe and W. B. Boise, "Organizational Innovation: Current Research and Evolving Concepts," *Public Administration Review*, 34 (May/June 1974), 284-293.

16. See Downs and Mohr, "Study of Innovation"; and Robert Eyestone, "Confusion, Diffusion, and Innovation," *American Political Science Review* 71 (1977), 441-448.

17. S. I. Doctors, *The Role of Federal Agencies in Technology Transfer* (Cambridge, Mass.: MIT Press, 1969).

18. R. G. Havelock and D. S. Lingwood, *R & D Utilization Stretegies and Functions: An Analytical Comparison of Four Systems* (Ann Arbor: University of Michigan, Institute for Social Research, 1973).

19. C. A. Murray and R. E. Krug, *The National Evaluation of the Pilot Cities Program* (Washington, D.C.: American Institutes for Research, 1975), pp. 178-179. It should be noted that the "failure" of the Pilot Cities Program was due largely to implementation problems rather than to flaws in the basic design of the program. The finding that innovation and problem solving were frequently incompatible was an important finding, but does not account for the failure of the program.

20. Comptroller General of the United States, *Technology Transfer and Innovation Can Help Cities Identify Problems and Solutions*, PSAD-75-110 (Washington, D.C., 1975).

21. National Commission on Productivity, *Third Annual Report* (Washington, D.C., 1974); E. K. Hamilton, "Productivity: The New York City Approach," *Public Administration Review*, 32 (November/December 1972), 784-795.

22. W. J. Baumol, "Macroeconomics of Unbalanced Growth: The Anatomy of Urban Crisis," *American Economic Review*, 57 (June 1967), 415-426.

23. For an elaboration of these arguments, see J. D. Roessner, "Incentives to Innovate in Public and Private Organizations," *Administration and Society*, 9 (November 1977), 341-365.

24. For examples of this literature, see T. W. Costello, "Change in Municipal Government: A View from the Inside," in L. W. Rowe and W. B. Boise, eds., *Organizational and Managerial Innovation: A Reader* (Pacific Palisades, Calif.: Goodyear, 1973); F. O'R. Hayes, "Innovation in State and Local Government," in F. O'R. Hayes and J. E. Rasmussen, eds., *Centers for Innovation in the Cities and States* (San Francisco: San Francisco Press, 1972); and Urban Institute, *Technology to Cities*.

25. K. E. Warner, "Concepts of Innovation"; H. P. Utech and Ingrid Utech, *The Communication of Innovations between Local Government Departments*, Pilot study for the National Science Foundation Office of National R & D Assessment (Washington, D.C., 1974); Irwin Feller and D. C. Menzel, *Diffusion of Innovations in Municipal Governments*, Final report on National Science Foundation grant DA-44350 (University Park, Pa.: Pennsylvania State University, Center for the Study of Science Policy, 1976); Irwin Feller, D. C. Menzel, and Alfred Engel, *Diffusion of Technology among State Mission-Oriented Agencies*, Final report on National Science Foundation grant DA-39596 (University Park, Pa.: Pennsylvania State University, Center for the Study of Science Policy, 1974).

26. For a full discussion, see J. D. Roessner, "Designing Public Organiza-

tions for Innovative Behavior" (Paper presented at the Thirty-fourth Annual Meeting of the Academy of Management, Seattle, Wash., August 20, 1974); and idem, "Incentives to Innovate."

27. R. K. Yin et al., *A Review of Case Studies of Technological Innovations in State and Local Services* (Santa Monica, Calif.: Rand Corporation, 1976).

28. I made these points in my paper "Designing Public Organizations for Innovative Behavior," pp. 12-15.

29. A. L. Frohman, Jr., Marc Schulman, and E. B. Roberts, *Factors Affecting Innovation in the Fire Services,* Report to the National Bureau of Standards, contract 1-35905, March 1972.

30. Ibid., p. 121.

31. Michael Radnor, *Studies and Action Programs on the Law Enforcement Equipment R & D System: Evaluative Study of the Equipment Systems Improvement Program,* Report to National Institute for Law Enforcement and Criminal Justice, grant no. 74-NI-99-0004-G (Evanston, Ill.: Northwestern University, 1975), vol. 2, p. 298.

32. J. F. Blair, Jr., *Industry, Innovation and the Municipal Market,* Final report on U.S. Department of Commerce grant F-C3431 (Philadelphia, Pa.: Franklin Institute Research Laboratories, 1973).

33. Ibid., pp. 11-25.

34. Ibid., pp. 11-86.

35. Feller, Menzel, and Engel, *Diffusion of Technology.*

36. Feller and Menzel, *Diffusion of Innovations,* p. 277.

37. See National Science Foundation, *Directory of Federal Technology Transfer.*

38. See W. S. Baer et al., *Analysis of Federally Funded Demonstration Projects: Final Report,* prepared for the Experimental Technology Incentives Program, U.S. Department of Commerce contract no. 4-35959 (Santa Monica, Calif.: Rand Corporation, 1976); and N. B. McEachron et al., *Management of Federal R & D for Commercialization,* prepared for Experimental Technology Incentives Program (Menlo Park, Calif.: SRI International, 1978).

39. Feller, Menzel, and Engel, *Diffusion of Technology.*

40. Utech and Utech, *Communication of Innovations.*

41. Feller and Menzel, *Diffusion of Innovations,* p. 185.

42. Ibid., p. 177.

43. R. K. Yin, "R & D Utilization by Local Services: Problems and Proposals for Further Research," mimeographed (Santa Monica, Calif.: Rand Corporation, 1976).

44. See D. C. Menzel, R. S. Friedman, and Irwin Feller, *Development of a Science and Technology Capability in State Legislatures: Analysis and Recommendations,* Final report on National Foundation grant GT-34868 (University Park, Pa.: Pennsylvania State University, Center for the Study of Science Policy, 1973). The National Conference of State Legislatures has an active office of science and technology that provides scientific and technical support to state legislatures.

45. See Baer et al., *Federally Funded Demonstration Projects;* and McEachron et al., *Management of Federal R & D.*

I. Citizen Participation

Having discussed the policy-setting roles of various governmental entities regarding technology, we now turn to the role for direct citizen participation. Technological issues present problems for participation in that nonprofessionals must understand them sufficiently to make informed choices. Participation also offers promise of technological development that is more responsive, more attuned to human needs, and more appropriate.

In Chapter 21, James D. Carroll offers a response to the disenchantment with technology—what he calls "participatory technology," that is, inclusion of the public in the processes of developing, implementing, and regulating technology. Technology is inextricably intertwined with politics—decisions represent and affect deeply held social values. Indeed, the realization of social goals or values, such as decent housing for all, often depends on technology. Development and control of technology is thus itself a political (value-laden) act. Carroll explores three routes for citizen participation in policy processes concerning technology: litigation, technology assessment, and ad hoc activities.

In Chapter 22, Hazel Henderson describes citizen movements as a valuable social feedback mechanism. They bring forth information, the basic currency of all economic and political decisions, to modify perceptions of problems and issues. In turn, they thereby modify institutions, values, and technological developments. Their critiques of conventional wisdom (potentially abetted by new technologies such as two-way cable television) open new social, cultural, and technological alternatives.

21. Participatory Technology

JAMES D. CARROLL

In recent decades the idea of the alienation and estrangement of man from society has emerged as one of the dominant ideas of contemporary social thought. While interpretations of the concept of social alienation vary, Etzioni *(1)* has expressed the core of the idea as "the unresponsiveness of the world to the actor, which subjects him to forces he neither comprehends nor guides... Alienation... is not only a feeling of resentment and disaffection but also an expression of the objective conditions which subject a person to forces beyond his understanding and control."

There is considerable speculative and observational testimony and some empirical evidence *(2)* that the scope and complexities of science and technology are contributing to the development of social alienation in contemporary society. Keniston *(3)*, for example, suggests that technology and its effects have been a factor in the alienation of many young people. At the same time he notes that the attitude of many young people toward technology is ambivalent because a revolt against the effects of technology must inevitably exploit the technology it opposes. In a different vein, de Jouvenel *(4)* has testified to the adverse psychological impact of scientific and technological complexities on sustaining general confidence in one's judgment. "Because science saps such individual confidence, we have a problem, which I feel we can meet but which it would be imprudent to deny." In a more general observation Mesthene *(5)* recently has referred to "the antitechnology spirit that is abroad in the land."

Participatory Technology

In this article I analyze the incipient emergence of participatory technology as a countervailing force to technological alienation in

Copyright © 1971 by the American Association for the Advancement of Science. Reprinted from *Science*, Vol. 171, February 19, 1971, pp. 647–652, by permission.

Participatory Technology

contemporary society. I interpret participatory technology as one limited aspect of a more general search for ways of making technology more responsive to the felt needs of the individual and of society. The term *participatory technology* refers to the inclusion of people in the social and technical processes of developing, implementing, and regulating a technology, directly and through agents under their control, when the people included assert that their interests will be substantially affected by the technology and when they advance a claim to a legitimate and substantial participatory role in its development or redevelopment and implementation. The basic notion underlying the concept is that participation in the public development, use, and regulation of technology is one way in which individuals and groups can increase their understanding of technological processes and develop opportunities to influence such processes in appropriate cases. Participatory technology is not an entirely new social phenomenon, but the evidence reviewed below suggests that its scope and impact may be increasing in contemporary society.

I first analyze several facts of which people are becoming increasingly aware that suggest why participatory technology is emerging as a trend, and I then analyze different forms of this trend. Finally, I evaluate some of its implications.

Underlying Realizations

One primary reason for the emergence of participatory technology is the realization that technology often embodies and expresses political value choices that, in their operations and effects, are binding on individuals and groups, whether such choices have been made in political forums or elsewhere. In the language of contemporary political science, by "political value choices" I mean choices that result in the authoritative allocation of values and benefits in society. In its most significant forms politics culminates in the determination and expression of social norms and values in the form of public law, public order, and governmental action. To an indeterminate extent, technological processes in contemporary society have become the equivalent of a form of law—that is, an authoritative or binding expression of social norms and values from which the individual or a group may have no immediate recourse. What is at issue in the case of the computer and privacy, the supersonic transport and noise levels, highway development and the city, the antiballistic missile and national security, and the car and pollution is the authoritative allocation of social values and benefits in technological form.

The second realization is a correlative of the first. Technological processes frequently are the de facto locus of political choice. They are often political processes in which issues are posed and resolved in technical terms. In the absence of appropriately structured political processes for identifying and debating the value choices implicit in what appear to be technical alternatives, technical processes become, by default, the locus of political value decisions. In the context of a concern for the environment, technical questions of waste disposal systems involve value choices. In the context of a concern for urban development, technical questions of highway location and development involve value choices. In the context of a concern for privacy, technical questions of data collection and retrieval involve value choices. Technological processes often embody significant value questions that are difficult to identify and resolve in public forums because the processes are technically complex and occur in administrative organizations to which citizens do not have easy access.

Third, there is the realization that the public order of industrial society is not particularly well structured for identifying, publicizing, and resolving in public forums political questions implicit in technological processes. The public order of industrial society is founded on, and perpetuates, values, compromises, and perceptions that are being rendered obsolete by transformation of the social and political conditions from which they were derived. The public order of industrial society preeminently expresses perceptions of material need and the values of economic growth—perceptions and values rooted in the experience of material want and economic insecurity of past generations. Because of the development of powerful technologies of production, and because of other factors, these perceptions and values, as embedded and expressed in public institutions and processes, do not encompass the total area of concern, which is expanding to include the quality of the environment, race, urban development, population growth, educational opportunity, the direction of technology, and other matters. Established means of structuring and expressing political concern themselves often border on obsolescence, because they are often based on geographical and functional jurisdictions that are unrelated to the issues on which the public must take action. If these jurisdictions were otherwise defined—for example, were defined to include an entire metropolitan area—they might provide the structure for more effective representation of diverse views and might facilitate public action through bargaining and tradeoffs.

Today, in the face of population growth and technological com-

Participatory Technology

plexity, legislative bodies, except in unusual cases such as that of the antiballistic missile, delegate to administrative agencies the responsibility for regulating, developing, and controlling technology. The general objectives of these administrative agencies involve mixed questions of value and technique, and the agencies resolve such questions in terms of their bearing on realization of the general objectives. Often the general objectives further the interests of individuals and groups allied with a particular agency. To the Department of Defense the question of the desirability of developing, maintaining, and transporting chemical and biological agents is primarily a matter of national defense policy. It is not primarily a question of the humaneness of such agents, or of their ultimate effects on the environment, or of their value or threat to man in contexts other than that of national defense.

By default, the responsibility for scrutinizing mixed questions of technology and value from the perspective of societal well-being often passes to special-interest groups and to individuals who may or may not be in a position, or be well equipped, to learn of and to influence such decisions. This is one aspect of the more general phenomenon of the devolution of authority from public representatives and administrators to "private" groups and individuals in contemporary society.

Fourth, there is the realization that, in contemporary society, political action directed toward the achievement of political value objectives, such as the production of 2.6 million housing units a year, often depends on the ability to translate the desired objective into technical tasks. Marcuse (6) observes that "the historical achievement of science and technology has rendered possible the *translation of values* into technical tasks—the materialization of values. Consequently, what is at stake is the redefinition of values in *technical terms,* as elements in the technological process. The new ends, as technical ends, would then operate in the project and in the construction of the machinery, and not only in its utilization [emphasis in the original]."

To a considerable extent, the achievement of more effective processes of education, housing, delivery of health care, postal service, public safety, and urban development depends on the political and technological capacity of contemporary society to agree on, and to translate, value objectives into technological acts. Traditional legislative declarations of intent are not sufficient. The establishment of a right to a decent home in a suitable environment requires more than a legislative act declaring that such a right exists. It also depends on the development of technical capability to translate the right into reality.

This does not mean that, in the formulation of political objectives, a

James D. Carroll

technological, problem-oriented mode of thought must replace humanistic, intuitive, moral, and other modes of thought. It means that other modes of thought often depend for realization in public life and action on their expression in technical form, and that the development and control of that form is itself a political value-oriented act.

Fifth, there is the realization that the status enjoyed by technology as an agent for both bringing about and legitimatizing social change contributes to the growth of participatory technology. There is a tendency, stressed by Ellul (7), Rickover (8), and others, for contemporary man to accept change in technological form as inevitable and irresistible. In some cases, new technologies probably are accepted because of the specific results they produce for the individual, such as the mobility that, under some conditions, is made possible by the automobile. But there seems to be an additional social, psychological, and economic element at work—what Ellul calls "technological anaesthesia"—that generates acceptance of technological innovation irrespective of the particular effects that may result. Many people seem willing to use cars in urban areas even though such use may contribute little to mobility and may adversely affect the environment and health. It seems paradoxical but true that, while some changes in institutions and behavior are strongly resisted, other changes often are readily accepted when a technological element in the situation is the agent of change.

Participatory technology is one limited way of raising questions about the specific technological forms in terms of which social change is brought about. It is directed toward the development of processes and forums that are consistent with the expectations and values of the participatory individuals, who may resort to them in the absence of other means of making their views known. In participatory technology, however, as in other participatory processes, the opportunity to be heard is not synonymous with the right to be obeyed.

I here analyze three kinds of activities to illustrate some of the empirical referents of the concept of participatory technology.

Litigation

The first is the citizen lawsuit, directed toward the control and guidance of technology. As Sax (9) indicates, "The citizen-initiated lawsuit is . . . principally an effort to open the decision-making process to a wider constituency and to force decision-making into a more open and responsive forum. . . . [The] courts are sought out as an

Participatory Technology

instrumentality whereby complaining citizens can obtain access to a more appropriate forum for decision-making."

The courts, of course, rely heavily on adversary proceedings, various forms of which have been suggested (10) as appropriate for handling scientific and technological issues involving the public interest. Not only can litigation restrict the use of technology, it can also lead to the modification and redevelopment of existing technology and stimulate the development of new technology to satisfy social values expressed in the form of legal norms, such as a right to privacy.

The legal response to cases involving technology has taken two forms. The first is an extension of those aspects of the legal doctrine of standing which determine who has a right to be heard in court on particular issues involving activities undertaken or regulated by public agencies. The second is a search by legal scholars, practicing lawyers, and judges for systems of conceptual correspondence in the terms of which scientific and technological developments and activities can be conceptualized and evaluated as changes in social values and norms that may warrant a legal response. The appropriate role of law in the regulation of genetic experimentation is an example.

An extension of the doctrine of standing has occurred in several recent cases involving technology, although the extension is not limited to such cases. In the words of the United States Supreme Court (11), "The question of standing is related only to whether the dispute sought to be adjudicated will be presented in an adversary context and in a form historically viewed as capable of judicial resolution." The basic question is "whether the interest sought to be protected by the complainant is arguably within the zone of interests to be protected or regulated by the statute or constitutional guarantee in question" (12). The question of standing is a question not of whether a party should win or lose but of whether he should be heard.

The current extension of the doctrine is sometimes called the "private attorney general" concept. Under this concept a private citizen is allowed to present a case as an advocate of the public interest. A leading case is Scenic Hudson Preservation Conference *v.* Federal Power Commission (13), decided by the Second Circuit of the United States Court of Appeals on 29 December 1965. On 9 March 1965 the Federal Power Commission granted a license to Consolidated Edison Company to construct a pumped storage hydroelectric project on the west side of the Hudson River at Storm King Mountain in Cornwall, New York. A pumped storage plant generates electric energy for use during peak load periods by means of hydroelectric units driven by water from a headwater pool or reservoir. The Storm King Project, as proposed by Consolidated Edison, would have required the placement of overhead transmission lines on towers 100 to 150 feet (30 to 45 meters) high. The towers would have required a path some 125 feet wide through Westchester and Putnam counties from Cornwall to the Consolidated Edison's facilities in New York City—a distance of 25 miles (40 kilometers). The petitioners were conservation and other groups and municipalities who claimed that the project, as designed by Consolidated Edison and as approved by the Federal Power Commission, would destroy the character of the land and the beauty of the area.

The Federal Power Commission argued, among other things, that the petitioners did not have standing to obtain judicial review of the legality of the license because they "make no claim of any personal economic injury resulting from the Commission's action."

The Court of Appeals held that the petitioners were entitled to raise the issue of the legality of the license and the licensing procedure even though they might not have a personal economic interest in the question. The court reasoned that a citizen has an interest in actions that affect the nature of the environment, and that this interest is arguably within the zone of interests that are or should be protected by law. On the merits of the case, the court held that the Federal Power Commission was required to give full consideration to alternative plans for the generation of peak-load electricity, including a plan proposed by one of the petitioners for the use of gas turbines.

The Scenic Hudson case is significant because it set a precedent for the enlargement of the opportunity of citizens, acting as citizens and not as private parties, to secure judicial review of the actions of public agencies, and of actions of the interests these agencies often regulate, in cases involving technology as well as other matters. The decision supports the proposition that, in certain cases, citizens will be recognized in court as advocates of a public interest, on the grounds that, as members of the public, they have been or may be injured by the actions complained of. They need not claim that they have been or will be injured economically or otherwise as private persons (*14*).

The development of the "private attorney general" concept does not mean that substantive changes will automatically occur in the constitutional, statutory, and common law doctrines that regulate rights and duties pertaining to the development and use of science and technology—analysts such as Patterson (*15*), Frampton (*16*), Cowan (*17*), Miller (*18*), Cavers (*19*), Mayo and Jones (*20*), Korn (*21*), Green (*22*), Ferry (*23*), Wheeler (*24*), and others (*25*)—indicates the difficulties of developing systems of conceptual correspondence between scientific and technological developments and legal concepts and doctrines. Scientific, technological, and legal systems often further different values and serve different purposes, and the reconciliation of conflicts in these values and purposes is only in part a juridical task. The "private attorney general" concept, however, does invite more active judicial scrutiny of such conflicts and may contribute to substantive changes in legal doctrine in the future (26) in areas such as the computer and privacy; air and water supply and pollution; noise control; medical, genetic, and psychological experimentation; drug testing and use; nuclear energy and radiation; food purity and pesticides; and the control and handling of chemical and biological weapons.

While the legal form of citizen participation in the control and development of technology has severe limitations because it tends to be (i) reactive rather than anticipatory, (ii) controlled by restrictive rules of evidence, and (iii) subject to dilatory tactics, litigation has proven, over time, to be a significant element in the efforts of individuals and groups to influence the processes and institutions that affect them.

Technology Assessment

A second form of participatory technology comes within the scope of existing and proposed processes of "technology assessment." While the concept of technology assessment can be interpreted to include the kinds of legal action I have discussed (27), the term usually is used to refer to activities that are somewhat more anticipatory in nature and broader in scope.

To some extent "technology assessment" is a new label for an old activity—the attempt to comprehend, and to make informed decisions about, the implications of technological development. The movement to formalize and improve this activity in a public context was initiated in 1967 by Senator Edmund Muskie (28) in the Senate and by Representative Emilio Q. Daddario (29) in the House of Representatives. This movement has successfully directed attention to some limitations in the way technological questions are currently considered in the American system of politics and government.

"Technology assessment" was defined in the bill introduced by Daddario in the House of Representatives on 7 March 1967 as a "method for identifying, assessing, publicizing, and dealing with the implications and effects of applied research and technology." The bill asserted that there is a need for improved methods of "identifying the potentials of applied research and technology and promoting ways and means to accomplish their transfer into practical use, and identifying the undesirable by-products and side effects of such applied research and technology in advance of their crystallization, and informing the public of their potential danger in order that appropriate steps may be taken to eliminate or minimize them."

The strengths and weaknesses of various forms of existing and proposed technology assessment are extensively analyzed in the hearings conducted by the Muskie (*30*) and Daddario (*31*) subcommittees; in the studies undertaken for the Daddario subcommittee by the National Academy of Sciences (*32*), The National Academy of Engineering (*33*), and the Science Policy Research Division of the Legislative Reference Service (*34*); and in related analyses, such as those made by the Program of Policy Studies in Science and Technology of George Washington University (*35*).

In these hearings and reports, citizen participation in technology assessment is both described and advocated. The analysis by Coates (*36*) of 15 case histories of technology assessments identifies one case that involved direct citizen participation—the examination of consumer products undertaken by the National Commission on Product Safety, which was established by Congress on 20 November 1967. In 1968 and 1969, the commission investigated the safety of such products as toys and children's furniture, architectural glass, power mowers, power tools, glass bottles, and aerosol cans. Citizens testified before the commission and directed the commission's attention to various incidents and problems. Coates observes that citizens participated in this particular assessment because the experience of members of the public with various products was itself part of the subject matter of the inquiry. There was no direct citizen participation in the other assessments examined by Coates, but the subject matter of several of the assessment processes suggests that some form of citizen contribution, either direct or through representative intermediaries, would have been appropriate. This is true, for example, of the assessments of environmental noise, and of future public transportation systems of advanced type.

In his written testimony submitted to the Daddario subcommittee, Mayo (*37*) stresses the importance, in assessment processes, of direct participation or representation of persons affected by a technology. He emphasizes the fact that technology assessment has a dimension beyond the identification and analysis of the impacts of technology. This is the dimension of evaluation of the social desirability or undesirability of such impacts. Since different segments of the public may view the impacts in various ways, as beneficial or detrimental, comprehensive evaluation is difficult without direct inputs from such segments. While special-interest groups can be relied on to express their views, they cannot safely be regarded as representative of the views of all major segments of the public that may be concerned.

Of the various hearings and reports generated by the Daddario subcommittee, the report of the technology assessment panel of the National Academy of Sciences places the greatest emphasis on citizen participation and representation. This panel asserts that legislative authorization and appropriation processes are inadequate as technology assessment processes because legislative processes frequently consider only the contending views of well-organized interest groups and often do not direct attention to long-range consequences. The panel further argues that, while technology assessment occurs in industry and in government agencies, with few exceptions the basic questions considered concern the probable economic and institutional effects of a technology on those who are deciding whether to exploit it. Existing processes fail to give adequate weight to "the full spectrum of human needs" because not enough spokesmen for diverse needs have access to the appropriate decision-making processes.

In the judgment of the panel, extensive citizen participation and representation in the assessment process is necessary both for practical reasons and for reasons of democratic theory. There are two practical reasons. First, citizen participation in the early stages of the development of a technology may help to avoid belated citizen opposition to a technological development after heavy costs have been incurred. Second, "objective evaluation" is impossible unless the diverse views of interested parties have been considered. On the level of political theory, the panel suggests that, in a democratic framework, it is necessary to consider the views of those who will be affected by a particular course of action.

The National Academy of Sciences panel explicitly acknowledges that technology assessment in some of its aspects is a political process because it involves questions of value (32, p. 83): "We can hope to raise the level of political discourse; we must not seek to eliminate it." The panel concludes (32, pp. 84 and 87) that there is a "need to accompany any new assessment mechanism with surrogate representatives or ombudsmen to speak on behalf of interests too weak or diffuse to generate effective spokesmen of their own.... Means must also be devised for alerting suitable representatives of interested groups to the fact that a decision potentially affecting them is about to be made.... *Whatever structure is chosen, it should provide well-defined channels through which citizens' groups, private associations, or surrogate representatives can make their views known.... It is particularly important to couple improved assessment with improved methods of representing weak and poorly organized interest groups"* [emphasis in the original].

As the National Academy of Sciences report states, and as Folk (*38*) stresses, to be effective technology assessment must function as part of the political process. What is at issue is the distribution and exercise of a form of decision-making power over technology. New technology assessment processes and structures probably would open decision-making processes to a wider constituency than now exists, and might change the distribution of power over some decisions involving technology. At the very least, new processes and structures might make it difficult for those accustomed to making technological decisions to do so without the knowledge of many other concerned people. It is doubtful that new assessment processes would be regarded as neutral either by those who now dominate technological decision-making processes or by those who might disagree with the results. Even though every effort were made to analyze questions of value as dispassionately as possible, or to exclude such questions entirely from assessment processes, dissatisfied parties almost certainly would attack the results and seek to offset them by other forms of political action.

Persuasion, bargains, and trade-offs in values are at the heart of political processes. Whether effective assessment can or should attempt to avoid these processes is questionable. Because technology assessment is to some extent a political process, the participation or representation of citizens may be not only desirable from the perspective of democratic theory but also necessary in political practice. Even such participation may not assure the effectiveness of the process in a larger political context.

Ad Hoc Activity

A third form of participatory technology encompasses a variety of ad hoc activities of individuals and groups beyond the scope of structured processes of litigation and assessment. This form includes activist intellectualism of the sort undertaken by Carson (*39*). Nader (*40*), and Commoner (*41*); quasi-official action of the kind undertaken by Congressman R. D. McCarthy concerning chemical and biological warfare (*42*); political and informational activities (*43*) of the sort undertaken by such groups as the Citizens' League Against the Sonic Boom, the Scientists' Institute for Public Information, the Sierra Club, Friends of the Earth, and Zero Population Growth; and

Participatory Technology

sporadic activities of loose coalitions of individuals and groups energized by particular situations and issues.

Rather than attempt to survey such ad hoc activities, I here briefly describe and analyze an example of abortive participation that occurred in 1967 and 1968 in the initial efforts to develop a new town on the site of Fort Lincoln in Washington, D.C. (*44*). In some ways the Fort Lincoln example is typical of problems that often arise in processes of citizen participation in urban development. In other ways the case is distinctive because the primary purpose of the Fort Lincoln project was to demonstrate on a national basis the potentials of technological and administrative innovation for urban development.

On 30 August 1967, President Johnson publicly requested several members of his administration and of the government of the District of Columbia to begin at once to develop a new community on the site of Fort Lincoln, which consists of 345 acres of nearly vacant land in the northeast section of Washington, D.C. The President explained the purpose of the project as the development of a community that would demonstrate the potentials of administrative and technological innovation in urban development. The Fort Lincoln project was conceptualized as the leading project in a national program to develop "new towns intown" on federally owned land in various cities throughout the country.

On 25 January 1968, Edward J. Logue, who had achieved national recognition as an urban development administrator in New Haven and Boston, was retained as principal development consultant for Fort Lincoln. In the following 10 months, Logue and his associates developed an ambitious and innovative plan (*45*) that was based on, among other things, a thorough analysis (*46*) of the potentials for technological innovation in the development of Fort Lincoln and on a proposal (*47*) for an innovative educational system for the new community.

Fort Lincoln was a federal urban renewal project. Some form of citizen participation in urban renewal projects is required by law. Logue and the government officials involved in the Fort Lincoln project had had extensive experience with citizen participation in other urban development projects, including a model cities project in Washington, D.C. In developing the plans for Fort Lincoln, they made extensive efforts to fashion a participatory structure that would be acceptable to the citizens of the northeast section of Washington. For the most part they failed. Political activists in the area perceived the technical planning process as the locus of political opportunity and choice concerning such questions as the number of low-income families to be housed on the site. Although these activists disagreed over who could speak for the citizens, they agreed that the residents of the area should be granted funds to hire professionals to participate with and for them in the technical planning and development processes. At one point the Department of Housing and Urban Development offered to grant money for this purpose to the council that represented the citizens, but for various reasons the council rejected the offer.

The Nixon Administration suspended development of Fort Lincoln in September 1969, pending further study. One analyst (48) has argued that the project was suspended because neither federal nor local officials believed that the development plan was either technologically or politically feasible. Other analysts (49) have suggested that the project was suspended because members of the Nixon Administration regarded it as a personal undertaking of President Johnson's and as an example of the overly ambitious social engineering activities of "the Great Society."

The struggle over citizen participation diminished support for the project in the neighborhood and among its potential supporters in other areas of the city. No strong political constituency favored the project. The Nixon Administration could and did suspend it without antagonizing any strong or vocal interest group.

Fort Lincoln is one example of the extent to which technical planning and development processes can become the locus of political conflict when these processes are perceived as the de facto locus of political choice. It is also an example of some of the difficulties that can arise in the course of efforts to reconcile the dictates of administrative and technological reasoning with the dictates of the political thinking of participating individuals in particular situations.

Problems

Like many other participatory processes, participatory technology raises questions about the adequacy of the theory and practice of representative government.

According to traditional theories of American public life, citizens should express their demands for public action to their political and governmental representatives. Conflicting demands should be reconciled by persons elected or appointed to policy-making positions in which they are publicly accountable for their actions. Administrative and technical processes are not, in theory, the appropriate locus for the exercise of political influence and the reconciliation of political

Participatory Technology

conflicts, because these processes are not usually structured as open political forums, and because most administrators and technical people are not directly accountable to electorates.

This theory of government is a prescriptive rather than a descriptive one. It does not correspond well with the realities of the exercise of political power in and through administrative and technical activities. Among other things, increases in population, the expansion of the public sector, and the increase in technological complexity have changed the number and, to some extent, the nature of demands and possibilities for governmental action in recent decades. While legislative bodies and individual elected officials continue to respond to some of these demands, many other demands are considered and resolved in administrative processes of limited visibility. The very act of translating most legislation into specific processes usually involves an exercise of political choice. Furthermore, agencies often invite demands upon themselves as a way of expanding the scope of their support and powers.

The politicalization of administration in this century, especially in response to the activities of interest groups, is a widely recognized phenomenon (50).

Participatory technology is an attempt to influence public agencies directly, and, through them, the quasi-public and private interests they often influence and regulate. Like other participatory processes, participatory technology in some of its forms circumvents traditional processes of expressing demands through elected representatives and of relying on representatives to take appropriate action.

The hazards of participatory technology are many. On the one hand it can be used by administrative and technical people in a manipulative way to generate the illusion of citizen support of a particular course of action. On the other hand it can degenerate into forums for the exercise of obstructionist, veto-power techniques and paralyze public action. It can generate an overload of demands that agencies are not equipped to handle. It can be used as an instrument by an aggressive minority to capture decision-making processes and to impose minority views on a larger community. It can simply shift the locus for the exercise of "the tyranny of small decisions" (51) from one group to another or merely enlarge the core group that exercises control. Finally, it can lead to the dominance of technological know-nothing over the judgments of qualified individuals who are legally responsible for, are dedicated to, and understand processes of public action.

At the same time, as Spiegel and Mittenthal (52) observe, "Citizen participation can occur in partnerships with a governmental unit as well as against it. Its nature can be cooperative and integrative or conflicting and oppositional..." Participatory technology, if appropriately structured, can contribute to decision-making processes that take into account alternative points of view, and can help an agency perform its functions in a more effective and open manner. It can provide a means by which the individual who feels powerless in the face of technological complexity can find a forum for the expression of his views.

The basic questions are these: In what cases is citizen participation in technological processes warranted, and according to what rationale? How should participation be structured and conducted? How much weight should participation be given in decision-making processes?

To provide a priori answers to these questions is impossible because of the variety of situations to which they apply. For this reason it is recommended that public agencies, scientific and technical associations, and individual members of the scientific, technological and political communities undertake analyses of these questions in the various situations for which they have responsibility or to which they have access. No single activity by a particular organization such as the National Academy of Sciences can meet the need. The analysis must be as broad-based as the activities to which these questions apply.

At the same time, the men responsible for policy making in foundations should consider the establishment of an experimental center for responsive technology. Such a center would analyze, on a continuing basis, the question of the ways in which public participation in technological decisions involving a public interest can be structured, and would support such participation in appropriate cases. The center might also support the education of proponents of technology, who would be qualified to recognize alternative conceptions of the public interest in technological matters and to present these conceptions to decision-making bodies.

Summary

The hunger to participate that exists today in various segments of the American public is in part a response to what some people perceive as an unresponsiveness of institutions and processes to the felt needs of the individual and of society. It is also, in part, an expression of a desire for a redistribution of power in American public life.

Technology is one of the major determinants of the nature of public as well as private life in contemporary society. Participatory technology is an attempt on the part of diverse individuals and groups to influence technological processes through participation in existing or new public processes by which technology is or can be developed, controlled, and implemented. Like other processes of direct citizen participation in governmental decision making, it raises many questions about the adequacy of existing theories and practices of representative government. These questions cannot be answered on an a priori basis. Members of the educational, scientific, technical, and governmental communities should analyze these questions in an effort to develop answers that are appropriate to the particular situations for which they are responsible and with which they are concerned.

Notes

1. A. Etzioni, *The Active Society* (New York: Free Press, 1968), pp. 617–622.
2. The best review is E. Chaszer, *Science and Technology in the Theories of Social and Political Alienation* (Washington, D.C.: George Washington University, 1969). See also V. C. Ferkiss, *Technological Man: The Myth and the Reality* (New York: Braziller, 1959), and H. M. Sapolsky, *Science,* 162 (1968), 427.
3. K. Keniston, *The Uncommitted: Alienated Youth in American Society* (New York: Harcourt, Brace & World, 1968); and *Young Radicals* (New York: Harcourt, Brace & World, 1968). See also T. Roszak, *The Making of a Counter Culture* (New York: Doubleday, 1969).
4. B. de Jouvenel, in *Science and Society,* E. Vavoulis and A. Colver, eds. (San Francisco: Holden-Day, 1966), p. 85.
5. E. Mesthene, *Technology Assessment* (hearings before the Subcommittee on Science, Research, and Development of the House Committee on Science and Astronautics. 91st Congress, 1st Session) (Washington, D.C.: Government Printing Office, 1969), p. 246.
6. H. Marcuse, *One Dimensional Man* (Boston: Beacon, 1954), p. 232.
7. J. Ellul, *The Technological Society* (New York: Knopf, 1964).
8. H. Rickover, *American Behavioral Scientist,* 8 (1965), 3.
9. J. Sax, *Annals of the American Academy of Political and Social Sciences,* 389 (1970), 72.
10. J. Conant, *Science and Common Sense* (New Haven: Yale University Press, 1961); J. Killian, in H. Woolf, ed., *Science as a Cultural Force* (Baltimore: Johns Hopkins Press, 1964); A. Kantrowitz, *Science,* 156 (1967), 763. H. Wheeler, *Center Magazine,* 2 (1969), 59; H. Green, in R. C. Kasper, ed., *Technology Assessment: The Proceedings of a Seminar Series* (Washington, D.C.: George Washington University, 1969). See also H. W. Jones, ed., *Law and the Social Role of Science* (New York: Rockefeller University Press, 1966); L. Mayo, *Scientific Method, Adversarial System, and Technology Assessment* (Washington, D.C.: George Washington University, 1970).
11. Flast v. Cohen. *United States Supreme Court Rep. No. 392* (1968), 83.

12. Association of Data Processing Service Organizations v. Camp, *United States Law Week,* 38 (1970), 4194.

13. U.S. Court of Appeals, Second Circuit, *Federal Reporter No. 354* (1965), 608; certiorari denied, *United States Supreme Court Rep. No. 384* (1966), 941.

14. In several recent cases the Scenic Hudson doctrine has been applied to matters such as highway location, the displacement of people by urban renewal projects, the protection of navigable waters, and the protection of lumber preserves. See *Cornell Law Review,* 55 (1970), 761. The proposed Environment Protection Act of 1970, introduced in the U.S. Senate in early 1970 by Senators Hart and McGovern, would clarify and extend the right of private citizens to bring antipollution suits against government agencies, industries, and private citizens.

15. E. Patterson, *Law in a Scientific Age* (New York: Columbia University Press, 1963).

16. G. Frampton, *Michigan Law Review,* 63 (1965), 1423.

17. T. Cowan, *George Washington Law Review,* 33 (1964), 3.

18. A. Miller, ibid., p. 17.

19. D. Cavers, *Michigan Law Review,* 63 (1965), 1325.

20. L. Mayo and E. Jones, *George Washington Law Review,* 33 (1964), 318.

21. H. Korn, *Law and the Determination of Facts Involving Science and Technology* (New York: Columbia University Law School, 1965).

22. H. Green, *Bulletin of the Atomic Scientists,* 23 (1967), 12.

23. W. Ferry, *Saturday Review,* 51 (1968), 50.

24. H. Wheeler, *Center Magazine,* 2 (March 1969), 59.

25. See *Vanderbilt Law Review,* 17 (1963), 1; *Report of a Conference on Law and Science* (London: David Davies Memorial Institute, 1964); *George Washington Law Review,* 33 (1964), 1; *Michigan Law Review,* 63 (1965), 1325; *Case Western Reserve Law Review,* 19 (1967), 1; *George Washington Law Review,* 36 (1968), 1033; *University of California Los Angeles Law Review,* 15 (1968), 267; *Cornell Law Review,* 55 (1970), 663.

26. Suits initiated in recent years to affect the control of technology through new applications of, or substantive changes in, legal doctrines include actions to ban the use of pesticides; to prevent airlines from using jets that pollute the air at the Newark, New Jersey, airport; to enjoin offshore drilling; to order a paper company to provide air pollution controls at a pulp mill; and to prevent a gas company from extending pipelines across a wooded tract. Several of these and similar suits are discussed in J. W. Moorman, "Outline for the Practicing Environmental Lawyer" (Washington, D.C.: Center for the Study of Responsive Law, 1969); L. J. Carter, *Science,* 166 (1969), 1487; ibid., p. 1601.

27. See, for example, B. M. Portnoy, *Cornell Law Review,* 55 (1970), 861.

28. 90th Congress, 1st Session, "Creation of a Select Committee on Technology and the Human Environment," *Senate Resolution 68* (January 25, 1967).

29. 90th Congress, 1st Session, "Technology Assessment Board," *House of Representatives Bill 6698* (March 7, 1967). See also "Technology Assessment" (statement of Emilio Q. Daddario, chairman, Subcommittee on Science, Research, and Development of the House Committee on Science and Astronautics, 90th Congress, 1st Session) (Washington, D.C.: Government Printing Office, 1968).

Participatory Technology

30. *Establish a Select Committee on Technology and the Human Environment* (hearings before the Subcommittee on Intergovernmental Relations of the Senate Committee on Government Operations, 90th Congress, 1st Session) (Washington, D.C.: Government Printing Office, 1967).

31. *Technology Assessment Seminar* (proceedings before the Subcommittee on Science, Research, and Development of the House Committee on Science and Astronautics, 90th Congress, 1st Session) (Washington, D.C.: Government Printing Office, 1967); *Technology Assessment* (hearings before the Subcommittee on Science, Research, and Development of the House Committee on Science and Astronautics, 91st Congress, 1st Session) (Washington, D.C.: Government Printing Office, 1969).

32. National Academy of Sciences, *Technology: Processes of Assessment and Choice* (Washington, D.C.: Government Printing Office, 1969).

33. National Academy of Engineering, *A Study of Technology Assessment* (Washington, D.C.: Government Printing Office, 1969).

34. Science Policy Research Division, Legislative Reference Service, *Technical Information for Congress* (Washington, D.C.: Government Printing Office, 1969). See also National Academy of Public Administration, *A Technology Assessment System for the Executive Branch* (Washington, D.C.: Government Printing Office, 1970).

35. These analyses are described in *Report: 1967-1968* and *Report: 1968-1969* (Washington, D.C.: George Washington University, 1970).

36. V. Coates, "Examples of Technology Assessments for the Federal Government" (Washington, D.C.: Government Printing Office, 1970).

37. L. Mayo, *Technology Assessment* (hearings before the Subcommittee on Science, Research, and Development of the House Committee on Science and Astronautics, 91st Congress, 1st Session) (Washington, D.C.: Government Printing Office, 1969), pp. 83-102.

38. H. Folk, paper presented at the Boston meeting of the AAAS (December 1969).

39. R. Carson, *Silent Spring* (Boston: Houghton Mifflin, 1962).

40. R. Nader, *Unsafe at Any Speed* (New York: Grossman, 1965).

41. B. Commoner, *Science and Survival* (New York: Viking, 1966).

42. See R. D. McCarthy, *The Ultimate Folly* (New York: Vintage, 1969).

43. See G. DeBell, ed., *The Environmental Handbook* (New York: Ballantine, 1970); J. G. Mitchell and C. L. Stallings, eds., *Ecotactics* (New York: Pocket Books, Simon & Schuster, 1970); R. Rienow and L. T. Rienow, *Moment in the Sun* (New York: Ballantine, 1967); W. A. Shurcliff, *SST and Sonic Boom Handbook* (New York: Ballantine, 1970).

44. The account given here is derived from a longer study: J. D. Carroll and J. Zuccotti, "The Siege of Fort Lincoln, circa 1969: A Study in Nonparticipatory Technology," paper presented at the Eastern Regional Conference on Science, Technology, and Public Programs, Boston, 1970. See also M. Derthick, *New Towns In-Town* (Washington, D.C.: Urban Institute, 1970), and "Fort Lincoln," *The Public Interest,* no. 20 (1970), 3.

45. *Fort Lincoln New Town Final Planning Report* (District of Columbia Redevelopment Land Agency, National Capital Planning Commission, and Government of the District of Columbia, Washington, D.C., 1969).

46. D. A. Crane, A. H. Keyes, F. D. Lethbridge, D. H. Condon, *Technologies Study: The Application of Technological Innovation in the Development of a New*

Hazel Henderson

Community (District of Columbia Redevelopment Land Agency, National Capital Planning Commission, and Government of the District of Columbia, Washington, D.C., 1968).

47. M. Fantini and M. A. Young, "Design for a New and Relevant System of Education for Fort Lincoln New Town," (New York: 1968).

48. M. Derthick, *New Towns In-Town* (Washington, D.C.: Urban Institute, 1970).

49. J. D. Carroll and J. Zuccotti, "The Siege of Fort Lincoln, circa 1969: A Study in Nonparticipatory Technology," paper presented at the Eastern Regional Conference on Science, Technology, and Public Programs, Boston, 1970.)

50. See T. Lowi, *The End of Liberalism* (New York: Norton, 1969); see also J. C. Charlesworth, ed., *Theory and Practice of Public Administration* (Philadelphia: American Academy of Political and Social Sciences, 1968).

51. A. S. Kahn, *Kyklos-International Review of Social Sciences,* 19 (1966), 23.

52. H. Spiegel and S. Mittenthal, in H. Spiegel, ed., *Citizen Participation in Urban Development* (Washington, D.C.: National Training Laboratories Institute for Applied Behavioral Science, 1968), vol. 1, pp. 12-13. See also P. Davidoff, *Journal of American Institute of Planners,* 31 (1965), 331. E. M. Burke, ibid., 34 (1968), 287. S. R. Arnstein, ibid., 35 (1969), 216. A. Altshuler, *Community Control* (New York: Pegasus, 1970).

22. Information and the New Movements for Citizen Participation

HAZEL HENDERSON

America's emerging social values might well be termed "post-industrial" and are espoused by a growing number of more educated, politically influential citizens who are disenchanted with many existing institutions and priorities but, for the most part, still believe that their objectives can be reached by restructuring business and government machinery through constitutional means. They include environmentalists, militant consumers, students and young people, middle and upper-income housewives newly activated by the con-

Originally appeared in vol. 412 of *THE ANNALS of The American Academy of Political and Social Science* (March 1974, pp. 34-43). © 1974, by The American Academy of Political and Social Science. All rights reserved. Reprinted, with additional material, in *Creating Alternative Futures* by Hazel Henderson (Berkley Windhover Books), pp. 277-295. Copyright © 1978 by Princeton Center for Alternative Futures, Inc.

Information and Citizen Participation

sciousness of the women's rights movement or the boredom of suburban life, the public-interest lawyers, scientists, engineers, doctors, social workers and other politicized professionals, the joiners of political organizations such as Common Cause, the activist stockholders and the various crusaders for "corporate responsibility."

The new "post-industrial" values of such groups are to a great extent needs described by the humanistic psychologists Abraham Maslow, Erich Fromm and others as transcending the goals of security and survival and are therefore less materialistic, often untranslatable into economic terms, and thus beyond the scope of the market economy and its concept of "homoeconomicus." As I tried to alert business leaders at the White House Conference on the Industrial World Ahead in 1972, they constitute a new type of "consumer demand," not for products as much as for life-styles, and include yearnings which Maslow referred to as "meta-needs": for meaning and purpose in life, a closer sense of community and cooperation, greater participation in social decision making, a general desire for social justice, more individual opportunities for self-development and more options for defining social roles within a more esthetic and healthful environment.

Ironically, these new values attest to the material successes of our current business system and represent a validation of a prosaic theory of traditional economics which holds that the more plentiful goods become, the less they are valued. For example, to the new "post-industrial" consumers the automobile is no longer prized as enhancing social status, sexual prowess or even individual mobility, which has been eroded by increasing traffic congestion. Rather it is seen as one component of a mode of transportation forced upon them by the particular set of social and spatial arrangements dictated by an interlocking group of powerful economic forces embodied by the auto, oil, highway and rubber industries. Such consumers have begun to view the automobile as the instrument of this monolithic system of vested interests and client group dependencies, which has produced an enormous array of social problems and costs: decaying, abandoned inner cities, an overburdened law-enforcement system, an appalling toll of deaths and injuries, some 60% of all our air pollution and the sacrifice of millions of acres of arable land to a highway system that is the most costly public-works project undertaken by any culture since the building of the pyramids and the Great Wall of China!

It has become expedient of late for business spokespeople to excoriate the views held by these new consumers. At best, they are seen as esoteric, at worst, un-American: but certainly a luxury not affordable by the average American family, let alone those living in poverty.

Hazel Henderson

And yet it must be acknowledged that these views are increasingly validated by the realities of environmental degradation, unemployment, continued poverty in spite of a climbing Gross National Product, and other visible evidence of the shortcoming of current social and economic arrangements. At the same time, some of these "post-industrial" values are surprisingly congruent with values being expressed by the poor and less privileged. Such groups—whether welfare recipients or public employees, less powerful labor unions or modest homeowners and taxpayers—seem to share the same demand for greater participation in the decisions affecting their lives and disaffection with large bureaucracies of both business and government. Environmentalists found themselves agreeing with labor and minorities that human service programs, which also tend to be environmentally benign, should be expanded rather than cut. There was also agreement that a federal minimum income program is more needed than ever, because it would create purchasing power for instant spending on unmet basic needs, such as food and clothing and, by affording the poor greater mobility to seek opportunities in uncrowded areas, would help relieve the overburdened biosystems of the cities. As another example, to recast the environmentalists' disenchantment with our automobile-dominated transportation system, we may note the very different but equally vocal objections of the poor. Over 20% of all American families do not own an automobile. For the poor, many of them inner-city residents, the cost of even an old model is prohibitive. This decreases their mobility and narrows their job opportunities, while the decline in mass transit and increased spatial sprawl permitted by wide automobile use worsen the situation. The highway building spree is too often experienced by poor and minority groups as the callous cutting of roads for white suburbanites through black and poor neighborhoods, permitting even more of the cities' remaining middle and upper-income taxpayers to flee urban problems for greeener pastures.

Therefore, although it is possible to dismiss these "post-industrial" consumers as irrelevant—and indeed they may well be less of a market for consumer goods—they nevertheless represent a new and different challenge of vital concern to business and politics as usual. Even though they are no longer willing to perform the heroic feats of consumption which have heretofore been successfully urged upon them by massive marketing barrages, their opinion-leadership roles and trend-setting life-styles will continue to influence traditional consumer tastes. This influence has been felt in the new anarchism and casualness in clothing fashions, the popularity of bicycling, the trends

Information and Citizen Participation

away from ostentatious overconsumption toward more psychologically rewarding leisure and life-styles, and reflecting the astounding growth of encounter groups and other activities associated with the human potential movement. In addition, the "meta-needs" of the "post-industrial" consumers will express themselves in increasingly skillful political activism and advocacy as they continue to find in their more holistic concepts greater congruity between their own goals and the aspirations of the less privileged. Furthermore, their growing confrontations with corporations over their "middle-class" issues, such as the environment and peace, have led them to discover the role of profit-maximizing theories in environmental pollution and of the military-industrial complex in defense expenditures and war. These insights, together with their awareness of their own privilege and their acceptance of guilt and concern for social injustice, are leading to the kind of convergence with other socioeconomic group interests so much in evidence in the movement for corporate responsibility.

Many of the corporate campaigns have been equally concerned with peace, equal opportunity in employment, pollution, the effects of foreign operations, safety and the broadest spectrum of social effects of corporate activities. Typical was Campaign GM, which simultaneously sought representation on General Motors' board of directors for minorities, women, consumers and environmental concerns. The same convergence is evident in the Washington-based Urban Environment Conference, composed of labor unions and a cross section of environmental groups pledged to stand united in the face of a growing number of corporations that attempt to prevent implementation of pollution-control laws by raising fears of unemployment, plant shutdowns or even relocating in more "favorable" states or in other countries. Both labor unions and environmentalists view such tactics as more often power plays and bluffing or the result of poor management than bona fide cases of corporate hardship.

Similarly, environmentalists and unions have worked together to reduce in-plant pollution and the ravages of such occupational diseases as black lung, or in fighting the wholesale destruction of small farms, open lands and streams through the excesses of strip-mining. This convergence is also visible in the comprehensive manifestos and social critiques of theoreticians, whether in the movements for civil rights, peace, women's liberation, or environmental and consumer protection. Most of these critiques tend to explain war, racism, sexism and all forms of social and environmental exploitation as being interrelated and stemming from current patterns of power and distribution and their roots in prevailing economic and cultural assumptions.

Hazel Henderson

This growing understanding of the political nature of economic distribution has naturally focused on the dominant economic institution of our time: the corporation and its political as well as economic role.

A consistent theme underlying the activities of United States citizen movements has been one of alienation from prevailing perceptions of reality. While the commercial mass media have projected subtle, but compelling, images of the kind of split-level suburban lifestyles conducive to the needs of a mass-consumption economy, the citizen movements, whether for peace, consumer and environmental protection or social equality, have marched to a different drummer. They have focused on the unresearched, often suppressed information concerning the dis-economies, dis-services and dis-amenities which we are now learning constitute the other side of the coin of industrial and technological development. They have risen naturally as social feedback mechanisms in response to the increasingly visible second-order consequences of our uncoordinated economic and technological activities and fragmented, tunnel-vision social decision making.

Citizen Movements as Social Feedback Mechanisms

It is not surprising that citizen organizations focus their efforts on modifying the goals and structures of those institutions, such as large corporations, which they believe have grown powerful enough to avoid normal social and political constraints. Not only corporations, but also government agencies, legislative bodies, labor unions and religious organizations tend to institutionalize past needs and perceptions and are ill-designed to perceive new needs and to respond to new conditions. Some measure of institutional lag is inevitable and necessary to avoid disruptive and rapid oscillations in societies. However, in conditions of rapid technological and social change, new complexities and interdependencies which are characteristic of the United States today, this institutional lag often becomes clearly counterproductive—for example, the current imbalance in the United States transportation system caused by the entrenchment of a powerful array of public and private institutions with vested interests in promoting automobile use.

Many citizen leaders realize that all institutional structures are, by definition, designed to screen out any information they perceive as unwanted or irrelevant so as to better concentrate on the purposes for which they were organized—hence, their capacity for selecting, concealing, distorting and impounding information and the resulting shortcomings of their planning and goal-setting processes. Indeed, organizational theorist Bertram Gross claims that it is impossible to

Information and Citizen Participation

measure the performance of any system independent of its structure.[1] For example, the United States Congress and its committee structure impedes information flow by slicing reality into fragments which fit its somewhat arbitrary scheme of organization. A case in point was the clash with the executive branch of government over the Congress's inability to reintegrate the welter of fragmented information needed to ascertain the total size and shape of the federal budget. Academia exhibits the same paralyzing lack of communication among disciplines and fields of research. The Executive Branch, itself, suffers the same myopia; scores of single-purpose agencies pursue their narrow goals, often addressing long forgotten problems.

A clue to the shortcomings of humanly designed structures may be found in nature. As ecologist Gregory Bateson has noted, it is rare to find ecological or biological systems which are activated by a specific need or which seek to maximize single variables. Meanwhile, information at the interfaces between many of our social problems is sparse because our society is ill-equipped to perceive, let alone research, these overlooked areas of interplay.

We have seen that all of this uncoordinated institutional activity in the United States today is based on the Cartesian view of the world which has held sway in our minds for three centuries. This has led to the growth of narrow-purpose structures and reductionism in our academic disciplines, and has in turn over-rewarded analysis, while discouraging synthesis; sustained property rights, while ignoring amenity rights; fostered unrealistic mental dichotomies, such as those between public and private goods and services; and over-rewarded competitive activities, while ignoring the equally vital role of cooperation in maintaining the cohesion and viability of the society as a whole. This tunnel-vision has clouded all our perceptions and caused us to focus on objects and entities, while ignoring their fields of interplay. It is precisely in these fields of interplay that public interest citizen movements have naturally sprung up and taken root. They have begun this vital task of filling in the information gaps on the effects of all this uncoordinated activity and painstakingly documenting our growing social costs—what British political economist E. J. Mishan calls the "bads" that inevitably come with all the goods.

We are now witnessing the collapse of policies based on this Cartesian world view. The new view now being developed by futurists is holistic, modeled on the concepts of general systems theory and analogies drawn from biological and ecological disciplines. It assumes that parts of systems can be understood and analyzed only within the contexts of ever larger macrosystems. Our current problems—whether we designate them as social, economic or environmental crises—are all part of the larger crisis of our myopic perceptions. These myopic perceptions, which were adequate for a quieter, slower moving age, no longer provide us with enough lead time to correct our course. Not only are our fragmented social structures inappropriate for describing, or dealing with, macroproblems, but the very goals of these subsystems are antithetical.

This must lead us to question a very basic assumption underlying both our economic and political systems: first, the assumption in our economic system that the aggregated goals and activities of microeconomic units will somehow add up to the public welfare. There is increasing evidence in our mounting social and environmental costs that the very opposite may be true. For example, the basic thesis of economist K. W. Kapp, expressed in his book *The Social Costs of Private Enterprise*, first published in 1950, holds that maximization of net income by microeconomic units—individuals and firms—is likely to reduce the income or utility of other economic units and of society at large. Public interest groups intuitively have understood this proposition; for example, corporations and special-interest groups can be subjectively successful or profitable by concealing or ignoring information on the external costs to society imposed by their activities.

Secondly, the same questionable assumption underlies our political system; we should have new doubts whether the competing special interests and their jostling can ever add up to even an approximation of the public interest. As discussed earlier, in many other disciplinary contexts, such as general systems theory, biology and ecology, it is considered almost axiomatic that optimizing subsystem goals is antithetical to optimizing the macrosystem of which they are a part.

The rise of the public-interest citizen movements is an expression of such new awareness—nowhere more pronounced, perhaps, than in the environmental movement in which such general systems views have become the dominant organizing principle. Similarly, as their new research enters our economic and social decision making, it diversifies the range of options under consideration, redressing the instabilities and errors caused by single-purpose-dominated policy making.

The Role of Information in Modifying Institutions and Values

Vastly increased information flows may prove to be our best hope for irrigating our impacted social system and modifying its structures, easing some of them into oblivion, while deflecting the course and redefining the goals of others. Information is, of course, the basic

Information and Citizen Participation

currency of all economic and political decision making. The quality and quantity of information and the way it is structured, presented and amplified control all of our resource allocations. In fact, information programs and directs energy—although, as yet, we do not know the equations representing this process.

Citizen leaders know that political and social conflicts are fought with information; this understanding is clearly illustrated in the diverse information-gathering operations of Ralph Nader's many public-interest research groups. Other organizations share the same appreciation for the power of information: the Council on Economic Priorities and the National Council of Churches' Corporate Information Center research the social, rather than the economic, performance of corporations and disseminate their findings to the press, money managers and stockholders. Information is also the weapon of the new advocacy professional groups, such as the activist doctors of Health-Political Action Committee, the Scientists Institute for Public Information, the Union of Concerned Scientists, the Center for Science in the Public Interest and the Union of Radical Political Economists, as well as the legions of radicalized lawyers, sociologists, librarians, psychologists, architects, planners and even management consultants.

The strategies and tactics of these groups are based on the following, shared assumptions that new or restructured information, when deployed and amplified, can:

—Alter human perceptions of reality

—Create changes in personal values, preferences and goals, which are later reflected in new collective and institutional goals

—Explode the boundaries of academic disciplines by creating cognitive dissonances and conflicts, often leading to gradual paradigm shifts

—Successfully challenge the rationality and legitimacy of resource allocations and decisions of governmental and private institutions

—Strengthen the power of consumers and citizens to perceive and protect their own interests and to understand how individual interests coincide more frequently when viewed within ever larger system contexts until, when finally viewed in planetary and ecological contexts, they literally become identical

—Short-circuit hierarchical, pyramidal and bureaucratic control

—Illuminate the intricate chains of causality and interdependence in complex societies and their reciprocal exchanges with equally complex host ecosystems

For example, urban dwellers must now understand the interdependencies involved in provisioning a modern city and managing its wastes if they are to be capable of making rational political judgments: suburbanites must learn how dependent their way of life is on such factors as the political mood of Arab nations, the oil needs of Japan, the viability of central cities or even the continued willingness of women to spend large amounts of time as unpaid chauffeurs. Finally, by spreading cognitive dissonance among citizens, information can release individuals from their subservience to prevailing cultural norms and create new opportunities for insight and learning.

The Alternate Media Movement

Having developed remarkable skills in gathering and restructuring information, public-interest groups also have mastered, to a significant degree, the modulation and amplification systems available in a technologically advanced society: from mass media to the informal mail and telephone networks of trust through which they channel their new information so that it acquires political and economic potency. For example, in the late 1960s activists and demonstrators began to master the staging of media events and guerrilla theater to manipulate drama-hungry news editors and reporters. They learned the commanding of media time and space through means as varied as challenging broadcasters' licenses, picketing outside newspapers and demanding hot-line, talk-in shows for the discussion of local issues and the redress of grievances; they could end-run politicians, administrators and corporations. These activities are based on the insight that in a complex, technologically advanced society free speech is an empty platitude if one is denied access to the mass-media amplification apparatus, which represents the vital nervous system of such a body politic.

In addition to opening up access to existing commercial mass-media channels, citizen groups developed their own alternate or underground media catering to the new consciousness of minority groups, women, environmentalists and communards. In the mid-1960s cost and lack of sophistication still limited underground media to the mimeograph machine and the printing press. Thereafter, underground radio began to develop as students learned how to move in on university-operated stations, and the transition to television became a gleam in every media activist's eye.

One of the most important influences during these developments was that of a New York-based group, Radical Software, which—through its magazine of the same name—linked alternate media

Information and Citizen Participation

people working all over the country and in Canada and helped them share new skills, such as videotaping with portable video cameras, splicing, editing and producing TV programs which reflected their own radical views and lifestyles.

At the same time alienated minorities of all kinds began to appreciate the potential for political power in the control of media. The battles to obtain conventional broadcasting licenses, as well as the new cable TV franchises, began. Media activists saw in cable television the possibilities for linking up communities in two-way exchanges and dialogues, covering school board and planning meetings, articulating hitherto unnoticed grievances and resolving conflicts—not to mention the longer range potential of such systems for generating revenues by providing library and data services, education, medical diagnosis and even shopping assistance. Local battles over the regulation of cable television franchises resulted in conditions being imposed upon would-be cable operators to provide channels set aside for public access.

Open Channel, the New York group founded in 1970, set out to show how the public could organize to use their newly won access channels, teaching local community groups, PTAs and voluntary service organizations how to make professional quality videotape programs interpreting their goals and activities for showing to New York's cable television audience. Other models have emerged for the use of television and print media in resolving conflicts, profiling value changes and locating possible new consensus areas, such as Choices '76, a series of media town meetings developed by another media innovator, Michael McManus, in 1973 for New York's Regional Plan Association.

Through the efforts of such innovators, media theorists—such as Marshall McLuhan, Amitai Etzioni, Chandler H. Stevens, Jerome Barron, John Culkin, Everett Parker and Donald Dunn—and activists—such as Nicholas Johnson, Peggy Charron, Albert Kramer and others; counter-advertisers Glenn Pearcy and Roger Hickey of the Public Media Center and public interest opinion researchers, citizens are gaining insight into the decision role of mass media in shaping our images and culture. They are learning also that television and computers need not necessarily be monolithic, centralizing forces, but can be used in decentralizing and coordinating modes. Instead of central data banks designed to find out about people, they can also function as random-access systems for gathering data on what people think and for funneling these opinions into political and economic decision centers.

Hazel Henderson

Resource One, a radical computer group in California, developed a random-access computer network to link citizen-action groups which share its data base on resources available for fighting consumer, environmental or social equity battles. Such citizen information networks can draw on many other key data, such as the computerized files maintained by the nonprofit Citizens Research Foundation of Princeton, which cross-references all political campaign contributions of over $500, or files maintained by the League of Conservation Voters of voting profiles of every United States congressman. Similar is the New York Regional Federal Trade Commission program, mentioned earlier, which does not even require activation by a complaining citizen, on the theory that many of the most victimized people in society are too unaware or apathetic to fight for their own rights.

The falling of the last bastion against full citizen participation is, as discussed, embodied in the concept of the electronic referendum which, of course, is technically possible today. The family television set could provide the citizen with information inputs on policy options and choices, with the telephone serving as the output device whereby the votes on issues could be instantly recorded at the appropriate legislative matrix. This scenario causes nightmares for the political scientist who is cognizant of both the role of vote horse-trading between politicians as an indispensable device for conflict resolution and the role of the representative form of government in damping the daily passions of the electorate.

However, one must note that a process similar to vote trading also takes place at the grass-roots level among citizen groups and opinion leaders—in garnering support for their issues and in the forming of ad hoc coalitions—which serves an equally important conflict-resolving purpose and which would still function in referenda situations. Such electronic referenda proposals only highlight more sharply the ever increasing need for information on the part of citizens in a modern democracy. Only if citizenship became a full-time job could the citizen in a complete industrial society possibly master the mountains of information necessary to make wise choices on such a daily plethora of issues.

Indeed, Edwin Parker and Donald Dunn call for the setting up of a computer-cable television information utility to be available in most United States homes by 1985.[2] Another proposal, put forward by Governor Ella T. Grasso of Connecticut when she was a Congresswoman, would make all communications by letter or phone from constituents to their federal legislators free of charge. Meanwhile, the best hope for broader access to continuing education would seem to

Information and Citizen Participation

be the model of Britain's Open University, which is conducted, with minimum barriers to enrollment, over nationwide television. For, to leave educational innovation to our schools and universities is probably as ill-advised as it might have been to put buggy-whip manufacturers in charge of development of the automobile.

Are Citizens Movements an Index of Societal Adaptability?

The new urges on the part of citizens to develop their own research capabilities and to communicate more frequently with each other and the public at large also may be seen as the validation of the thesis presented by Wilbert E. Moore and Melvin Tumin in their illuminating paper "Some Social Functions of Ignorance."[3] They point out that ignorance serves the social purposes of, among other things: preserving privileged position, reinforcing traditional values and preserving social stereotypes. Such social purposes are seen today as having little redeeming worth in a society racked by the adaptation requirements of technological change. Indeed, as Gregory Bateson has noted, adaptability is a resource as surely as is coal or oil, and we need to develop a new economics of adaptability in order to understand how humans use, conserve or waste their precious stock of this commodity.

In fact, do the new citizen cadres in our midst represent a vital part of our society's stock of adaptability; if so, how can we conserve and employ their energies most productively? Gunnar Myrdal has pointed out that citizens organized for their own purposes can often serve the same functions as costly, regulatory bureaucracies.[4]

However, these citizen organizations and their research operations are often subject to severe resource limitations, even though they have proved their competence and validated their research by a much more rigorous process than that by which academic research is validated. As Ralph Nader has pointed out, any one of these independently produced studies is immediately pounced on by hundreds of paid experts in the employ of institutions which may feel themselves threatened by such information. No expense or effort is spared to fault or discredit such studies, as the Council on Economic Priorities discovered when it published its studies in 1971 and 1972 of the United States pulp and paper industry and the investor-owned electric utility industry.

Nevertheless, these studies were found worthy of publication by the Massachusetts Institute of Technology Press. The special values of these citizen-research groups is that their innovative and unorthodox approach provides priceless, if often painful, critiques of conventional wisdom and infuses the body politic with a rich new yeast of social, cultural and technological alternatives. For example, it was the determination of environmentalists which forced the consideration of research and development alternatives, such as solar energy, geothermal steam, magnetohydrodynamics and fuel cells, into the current debate on energy policies. It was the Scientists Institute for Public Information which challenged the Atomic Energy Commission's emphasis on the breeder-reactor program, and it fell to the Project on Corporate Responsibility and the National Affiliation of Concerned Business Students to develop new curricula on social performance criteria for corporate management for our university's schools of business administration. Many groups, such as the Council on Municipal Performance, already use sophisticated computer systems for their research which compares the efficiency of municipal governments in such areas as housing, crime control and waste management.

As Mancur Olsen has pointed out in *The Logic of Collective Action*, the classic economic problems of free goods and free riders enter into the prognosis for sustaining the voluntary, nonprofit activities of citizens' groups: the work and effort the citizen participant invests in protecting common property or amenity rights automatically accrues to all, regardless of whether they shared the opportunity costs incurred by their more public-spirited neighbors.[5] Often, the result bears out the old proverb: "everybody's business is nobody's business." Therefore, society provides little motivation for the disinterested champion of the public interest and, in fact, often imposes severe financial and other penalties on do-gooders. Luckily, there are less obvious motivations at work, including those described by humanistic psychologist Abraham Maslow as ascending hierarchical needs beyond mere survival and security: knowledge, an esthetic environment, social justice, leisure and greater participation in decisions affecting one's life. These transcendent needs seem to characterize many citizen activists in this and other advanced economies in which survival is no longer a widespread preoccupation.

Value changes—based on perceptual changes, in turn based on new information—are, then, the mainspring of the new citizen movements.

Public-interest groups are now also aware of how heavily economics relies on the false assumption of adequate information availability, both in the marketplace and in illuminating such social choices. In fact, they are now beginning to end-run the economists by forcing

Information and Citizen Participation

their information into such decisions and not only successfully affecting the outcomes, but actually altering the value preferences which economists accept as immutable, given data.

An equally disturbing phenomenon for economists used to the idea that increasing information will decrease uncertainty is that the increased flow of formerly uncollected or suppressed information provided by citizen groups actually increases uncertainty, because it tends to constitute information on what is not known and what must still be researched. For example, the environmental-impact statements required of federal projects under Section 102 of the National Environmental Policy Act of 1969, more often than not contain this kind of information, which provides useful questions rather than answers.

A case in point which demonstrates the citizen groups' interest in raising the right questions involved the Congressional Office of Technology Assessment. A coalition of citizens organizations drew up criteria for opening the technology assessment process to wide public participation and scrutiny, recommending the concepts of adversary science and dialectical cost-benefit analysis; for they understood that technology assessment is essentially a normative process involving value conflicts and, therefore, cannot be left to the technologists.

Suffice it to say that the information struggle with all the communication and structural roadblocks our Cartesian perceptions have created, will continue. Public-interest groups will continue their efforts to manipulate these fragmented concepts of rationality, because their second-order consequences will continue to activate their concern. The war of symbols between the new and the old consciousness will also continue as advertisers destroy word meaning and debase language currencies. Love is now a soft drink and a cosmetic; ecology has similarly been bankrupted of its content. Bureaucrats and politicians add to the Orwellian confusion: war is peace, bombing is protective reaction and full-employment targets shift when we do not meet them.

Public-interest groups retaliate by seeking to destroy the narrow meanings of words, such as profit, efficiency, utility and progress. When the vast value and paradigmatic shifts we are now experiencing reach some new equilibrium, we may also have discovered some higher-system-level discourse which may permit us to discuss our macroproblems in a common language. Until then, we can only hope that, in all the static and confusion of our free marketplace of ideas, the best information will win and will be allowed to be the basis for action.

Hazel Henderson

Notes

1. Bertram Gross, "Social Indicators," in Raymond A. Bauer, ed., *The State of the Nation: Social Systems Accounting* (Cambridge, Mass.: M.I.T. Press, 1966), pp. 36-48.

2. Edwin B. Parker and Donald A. Dunn, "Information Technology: Its Social Potential," *Science,* June 30, 1972, 1392-99.

3. Wilbert E. Moore and Melvin M. Tumin, "Some Social Functions of Ignorance," *American Sociological Review,* 14 (February 1949), 787-95.

4. Gunnar Myrdal, *Beyond the Welfare State* (New Haven, Conn.: Yale University Press, 1960).

5. Mancur Olson, *The Logic of Collective Action* (Cambridge, Mass.: Harvard University Press, 1965).

J. Public Choice

In this last section, we try to pull together (at least in a loose fashion) the issues raised concerning the *process* of addressing technology-intensive policy issues. We have heard about the problems and the potentials of technology; we have considered its political and economic workings at the international, federal, state, and local levels; and we have opened the inquiry into the possibility of broadened participation in policy making concerning technology.

In Chapter 23, Michael S. Baram presents a model of the interaction between forms of social control and the course of technological innovation from the application of resources to the production of effects on society. This illuminates opportunities to improve what Edward Wenk, Jr. has called the "social management of technology." Baram concludes that a citizenry that expresses a diversity of interests through a variety of channels is the most effective mode for promoting the accountability of policy makers.

Mark R. Berg's analysis in Chapter 24 addresses one means to social control of technology, a version of policy analysis called technology assessment (TA). TA is a systematic study of the effects on society of technological change, with particular emphasis on the unintended, indirect, and delayed impacts. Berg places TA in a properly political context. He notes the difficult juxtaposition of scientific, analytical credibility (TA's source of power) and the influences on TA to fit with the institutional structures and processes that use it. The political ramifications of, and on, TA run deep. TA is inherently value-laden. Determining who the assessors are (a new coterie of experts) matters. Even the selection of methodology has significant value implications. We must yet come to grips with how to employ TA as an ingredient in the wise social management of technology.

We close this book with Duncan MacRae's discussion of interrelationships between science and practicing democracy in Chapter 25. MacRae takes issue with the "engineering model" of technology policy whereby experts address technical means, not ends, to inform decision makers. He thinks that this leaves the citizen and even the politician unenlightened and incapable of judging the technical considerations. MacRae explores several ways to disseminate technical information. Two counterposed strategies are to return to the notion of the well-informed citizen (infuse technical knowledge into the citizenry) or to develop representative science (infuse a range of values into the scientists and technologists). The consideration of how to use expertise raises deeper concerns. How should preferences be shaped by a democracy? How should they be confined? He offers policy analysis as a promising way to draw technical competence into the policy realm.

23. Technology Assessment and Social Control

MICHAEL S. BARAM

The emerging concepts of corporate responsibility and technology assessment are, to a considerable extent, responses to problems arising from technological developments and their applications by industry and government. These problems appear in the relatively discrete sectors of consumer protection and occupational safety and in the diffuse sectors of community quality of life and the national and international environments.

Consumer Protection

As products have become more sophisticated and defects in them less easily detected by the consumer, the common-law principle of caveat emptor, "let the buyer beware" has been largely abandoned by the courts, and the principle of strict corporate liability has been frequently adopted (1). Federal and state legislation and regulatory agencies for consumer protection have multiplied with this shifting of responsibility. Nevertheless, common law, legislation, and regulation pertaining to product safety have been largely ineffective (1, p. 2):

> ... federal authority to curb hazards in consumer products is virtually non-existent ... legislation consists of a series of isolated acts treating specific hazards in narrow product categories. ... Despite its humanitarian adaptations to meet the challenge of product-caused injuries, the common law puts no reliable restraint upon product hazards.

As a result, Ralph Nader and other crusaders have mobilized citizens against specific technological developments embodied in hazardous

Copyright © 1973 by the American Association for the Advancement of Science. Reprinted from *Science*, Vol. 180, May 4, 1973, pp. 465–472, with permission.

products and processes—such as the Corvair and various food additives.

The 92nd Congress enacted the Consumer Product Safety Act, thereby creating an independent commission with the authority to develop mandatory safety standards for many product categories and to carry out related functions to protect consumers (2). However, regulation of automobiles, drugs, boats, foods, and other product categories is excluded and left to existing programs. The commission is expected to maintain the regulatory agency tradition of reliance on industrial testing and reports; and "Except for the availability of [commission] information and the opportunity for litigants to argue the fact of compliance or noncompliance with mandatory Government standards, the law is expected to have little effect on products liability litigation" (3). It is too early to determine whether or not the law will bring about an effective regulatory program.

Occupational Health and Safety

The incidence of harm to workers, the difficulties of employee recovery under the common law, and the inability of the judicial system to internalize such "costs" sufficiently to bring about a preventive approach by corporate management are among the factors that led to workmen's compensation laws and insurance programs (4), and agency standards for occupational hazards. The National Labor Relations Act (5), and most recently the Occupational Safety and Health Act (6) have provided frameworks for decision-making on automation and hazardous technological developments. Nevertheless, high injury rates persist in several industrial sectors (7) as old and new technology continues to create lethal environments for employees—for example, "The National Academy of Sciences reports a study showing that the life-span of radiologists is five years shorter than the national average . . ." (8, p. 13).

The introduction of new automation technology has traditionally brought about strong union opposition because of impacts on job security (9). Now, impacts on employee health provide new bases for opposition. As a result, some new, highly automated plants have been shut down—Rio Tinto's lead processing plant in the United Kingdom and General Motors' Vega plant in Lordstown, Ohio, have recently suspended operations until the economic and the physical and mental health effects of new automation technology on employees could be determined and diminished (10).

Technology Assessment and Social Control

Community Quality of Life

The impacts of industrial and government technology on health, land use, esthetics, and other aspects of community quality of life (*11*) have finally aroused organized citizen oppositon. Government transportation and energy programs are now persistently opposed by local communities. Corporations that have traditionally provided the economic base for communities are now increasingly confronted by litigants seeking compensatory damages, restraining orders, and injunctions; by newly aggressive local officials responding to citizen complaints and invoking long-dormant police powers against noise, smoke, and other nuisances; and by state and federal officials enforcing air and water quality programs. Despite judicial reluctance to enjoin ongoing industrial activity that concurrently provides local economic benefits and environmental degradation (*12*), the expanding enforcement of public nuisance and pollution control laws has recently brought about a number of plant closures (*13*).

Nevertheless, the economic objectives of states and local communities and the fear of job losses and other dislocations that would arise from project or plant shutdowns will continue to determine the pace at which community quality of life is rehabilitated and environmental degradation controlled (*14*). The complex task of resource management must be undertaken by state and local governments. How else to reconcile the objectives of economic and social opportunity—housing, economic development, transportation, and so on—with enhanced community quality of life—open space, recreation, esthetically pleasing surroundings, population stability? The reconciliation of such diverse objectives will not be possible until the consequences of technology can be systematically assessed, until rational siting and land use guidelines have been established, and until state and regional planning find a viable political structure.

National Environmental Quality

Ehrlich, Commoner, and other early crusaders may have been critically received, but nations are now embarking on serious, more effective pollution control programs. In the United States, the new water pollution control program has been designed to achieve use of the "best practicable" pollution control technology by 1977, the "best available" technology by 1983, and a national "no pollution discharge" goal by 1985 (*15*). The air quality program provides authority for federal control over new stationary sources of air pollution, over automotive emissions, and over all sources of air pollutants hazardous to human health (*16*). New legislation has established federal authority to limit the noise emissions of numerous corporate products (*17*); and laws to tighten up control over pesticides and hazardous materials have again been enacted (*18*).

The national commitment now authorizes control over most forms of pollution caused by technological processes, ensuring more rigorous analysis, regulation, enforcement, and citizen participation. Nevertheless, many technology-created pollution problems remain—the management and disposal of radioactive waste, toxic materials, sludge, and solid waste. In addition, new technologies such as weather modification and marine resource extraction are now being developed and experimentally applied, and they will undoubtedly create new problems and new legislation in our already "law-ridden society" (*19*, p. 32). The pattern is obvious and disturbing: the development of a technological advance, insistence upon its application by interest groups in industry and government, utilization, the appearance of environmental problems, legislation, regulation, and extensive litigation to control environmental impacts (*20*).

Assumptions

These problems of consumers, employees, communities, and nations are the results of the processes we use to develop, apply, and regulate our technology—of our methods of social control. Social control is, in turn, the result of complex interactions of underlying political, economic, and cultural forces.

What is to be done? We can continue to grapple with the problems as they crystallize, using the established and ineffective patterns of post hoc legislation, regulation, and litigation. On the other hand, we can boldly attempt to alter the underlying forces or causes, and their interactions, but this calls for information we do not have and demands an acknowledgement that the forces at work in different political systems are yielding substantially similar problems (*21*).

The most feasible strategy appears to be one of intervening in those decision-making processes of the public and private sectors that bring about technological applications; such intervention would take the form of introducing new frameworks for planning and decision-making. The development and use of coherent frameworks for technology assessment and utilization could meet many of the demands for corporate and governmental responsibility. Clearly, the

Technology Assessment and Social Control

use of such frameworks will affect the underlying social forces not directly confronted and will entail considerable reliance on established legal and regulatory procedures (22, 23).

The task of developing frameworks for technology assessment and utilization must be undertaken in full recognition of several realities.

1. Application of any such framework to a particular technological advance will yield differences in opinion and information from professionals, as well as from concerned citizens.

2. Continuing research, monitoring experiments, and changing designs will not necessarily resolve such differences, but will generally reveal the trans-scientific nature of decisions to be made about the further development and utilization of a specific technological advance: for example, the decisions will ultimately involve value-based consideration of the probable harm of the advance and the scope, magnitude, and acceptability of that harm (24).

3. Receptors—consumers, employees, and citizens generally—will find elitist decision-making and compensatory solutions to possible harmful effects inadequate, and they will actively seek to participate in the planning, design, and implementation stages of the technology application process.

4. A multiplicity of inadequate decision frameworks for technology assessment and utilization already exist and are employed by, for example, Congress, regulatory agency officials, corporate management, insurance rate-setters, courts, and organized citizen's groups.

Given this statement of the problem and these assumptions, it appears that the task is to somehow "get it all together"—to develop an understanding of how technology interacts with society and its institutions of social control; to demonstrate that citizens, corporations, and public institutions are all interrelated in specific patterns and thereby share responsibility for rational planning and decision-making; and to shape a common conceptual framework that can be readily applied by each decision-maker, in order that the different results can be compared meaningfully and used to choose knowledgeably among alternatives.

Developing a Coherent Framework

Technology is dependent upon processes that occur in four interrelated contexts: basic research, applied research, the development of prototypes for testing or experimentation, and ongoing production and utilization. Although it is difficult to pinpoint the path of any specific development, it is clear that most technology (in the form of processes, products, or techniques) in use today was brought about by the interactions of people and findings in these four contexts (25).

Within each context different levels and kinds of resources, or inputs are required—for example, manpower, funds, time, facilities, education, and materials—but large social and economic commitments and irreversible commitments of natural resources are usually made only when the development and experimentation phase is undertaken. These large commitments lend an inevitability to the technological advance, because few courts and federal agencies have been willing to halt major socioeconomic commitments, irrespective of hazards to individuals or society (26).

The technology that emerges subsequently brings about social and environmental effects, or outputs—direct and indirect, primary and secondary, beneficial and detrimental, measurable and unmeasurable. Whether one uses nuclear power or the snowmobile as an example of current applications of technology, several classes of effects are apparent. These include effects on health (mental and physical, somatic and genetic), economy (individual and corporate, local and national, international), environment (pollution, disruptions of ecosystems), resources (availability of materials, land, and waters for competing uses), values (changes that are ultimately reflected in new law and policy), and socio-political institutions and processes (structural and substantive changes). As these and other effects are aggregated, they determine the quality of life.

We have no quantifiable information on many of these effects; nor can we accurately predict potential effects, their synergism, or the intervention of exogenous forces such as population migration or natural disasters. We do not have devices sophisticated enough to monitor and assess many of these effects, nor do we have articulated goals or indices to measure progress toward such goals (23). Decisions on goals, indices, and effects are now, and will probably always remain, transscientific.

But we have learned one thing well—that impacts and amenities which are unmeasurable or unquantifiable are nevertheless real and should be as integral to decision-making as quantifiable technical and economic considerations. At the federal level, this has been clearly expressed in the National Environmental Policy Act (NEPA) of 1969 (27), which requires that "un-quantified environmental amenities and values" be considered along with technological and economic or quantitative inputs to public agency decision-making on projects, permits, contracts, and other major actions when such actions are likely to result in significant environmental impacts. Agencies are now struggl-

Technology Assessment and Social Control

ing with this new requirement as they develop environmental impact assessments, which are subsequently exposed to the public for review before agency action. Public response to over 3,000 impact statements during the past 2 years has ranged from acquiescence, to intervention in agency proceedings, to political pressure, to extensive litigation (28).

Following this brief discussion of inputs to and outputs of the process of technological advance, a simple model can be developed which relates a specific technological development to resources (inputs) and effects (outputs) (Figure 1).

The implementation of each program will depend on a variety of decision-makers in both public and private sectors and at varying jurisdictional levels—local, state, regional, and federal. These decision-makers function as controls on any program in essentially two ways (Figure 2): (i) by *controlling resources* (for example, public and private sources of manpower and funds for research and develop-

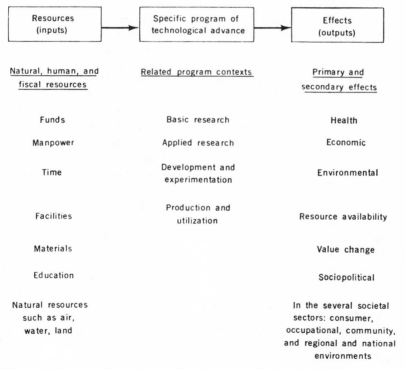

Figure 1. Resources (inputs) and effects (outputs) of technological developments.

457

Michael S. Baram

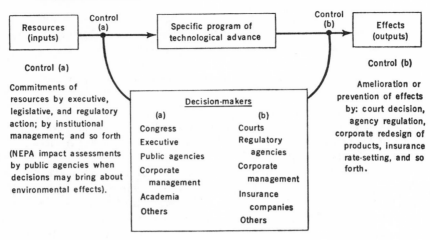

Figure 2. Decision-makers.

ment; land use and natural resource authorities; federal and state legislatures, whose enactments may be essential to the availability of other program resources; and educators, who determine training programs) and (ii) by *controlling the detrimental effects* (for example, the courts by means of preliminary or permanent injunctions or awards of compensatory damages; federal agencies, such as the Food and Drug Administration and the Environmental Protection Agency, and their state counterparts by engaging in standard-setting, regulation, and enforcement; and program managers, corporate management, and insurance rate-setters by bringing about program or product redesign to abate or ameliorate specific effects).

To further develop this model, some of the major influences on decision-makers who control technological developments must be determined. These influences (Figure 3) include information on: (i) resource availability; (ii) technical and economic feasibility; (iii) actual and potential effects; and (iv) operational-institutional values, which are comprised of the common law, legislation, economic and social policy, institutional management policies, and other "given" values that have been recognized and accepted by decision-makers as of the time any specific decision is made regarding further program development. These include diverse and often conflicting laws and policies—for example, NEPA (to foster the conservation and rational use of resources) and the oil depletion allowance (to foster rapid exploitation of resources).

To complete this general model, the social dynamics of any pro-

Technology Assessment and Social Control

gram of technological advance must be considered further—specifically, the responses of individual citizens and organized interest groups to perceived resource commitments and program effects (Figure 4). These responses can be manifested through institutional procedures for changing the laws and policies that influence decision-makers—a lengthy process requiring extensive aggregation of voters or shareholders and generally undertaken in order to influence future decisions, not the particular decision that provoked the response.

Responses can also be manifested through formal, adversarial procedures to challenge decision-making—for example, injured consumers can go to court and disturbed environmentalists can intervene in agency proceedings or seek judicial review of agency decisions. Finally, a variety of informal procedures can be employed to feed back responses to decision-makers—such as demonstrations, employee absenteeism, product boycotts, consumer choice, or quasipolitical campaigns. The environmental and consumer protection movements serve as vivid examples of these new pressures on decision-makers, pressures new only in their intensity.

Citizens responding to perceived detrimental effects or resource misuse comprise a diverse group of consumers, shareholders, unions, crusaders, and citizens' organizations, ranging from those with national objectives (for example, the Sierra Club) to those with local or

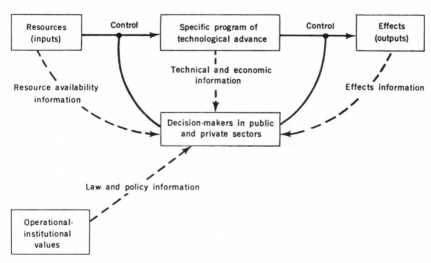

Figure 3. Information flows to decision-makers.

459

Michael S. Baram

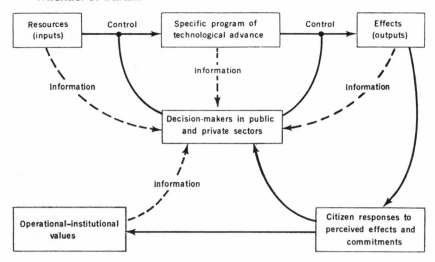

Figure 4. Summary of influences on decision-making.

self-interest objectives (for example, labor unions, airport neighbors). The responses manifested through institutional, formal-adversarial, or informal procedures for exerting pressure on decision-makers may, in time, become so widespread or aggregated that they will be incorporated into the common law or form the basis for new management policy or legislation and, as such, become part of the matrix of operational-institutional values. This has already occurred to a considerable extent with regard to environmental and consumer protection responses.

Although the sector of society that responds adversely to the effects of a specific technological development does not normally constitute a democratic majority in its early stages, the issues raised by such responses deserve serious consideration, and the procedures for eliciting such responses are being strengthened by the courts and legislatures. First, the responses represent new perceptions, new "pieces of the truth" that were either unknown to, ignored, or lightly considered by decision-makers. Second, they represent market and political influence that can be magnified by use of the media. Third, although they may be ignored at first, these responses will continue to appear in various forms and may bring about delays that are more costly after a program has been started (the utilities and the Atomic Energy Commission, for example, are now finding this out as they attempt to further the nuclear power program: plant construction and operation are running more than 2 years behind schedule, with greatly increased costs, because of extensive litigation and hearings (29), which

Technology Assessment and Social Control

resulted from an earlier failure to consider citizens' concern about thermal and radioactive waste disposal, reactor safety, and related ecological and health issues. Fourth, such responses are based on real concerns, will often find larger public support, and eventually could result in stringent legislation or judicial findings that decision-makers would have to live with (30).

Finally, a citizenry that expresses a diversity of interests is the most effective mode of promoting the accountability of decision-makers to the full social context in which they operate. Too often, decision-makers in all institutions have failed to inform the public about the bases and risks of decisions, thereby precluding feedback of larger social issues and humanistic concerns in their effort to promote institutional or self-interest objectives (31). But the benefits of an informed and responsive public have now been adequately demonstrated. Cars will be cleaner by 1975; the Army Corps of Engineers will not continue to dam rivers and spend public funds without more rigorous analysis of impacts and needs; the Food and Drug Administration will begin informing the public of the chemical contents and quality control criteria of specific consumer products they regulate; maximum permissible exposures of workers and the public to power-plant radiation have been falling. These are some of the recent "accountability" benefits that are being derived from public pressure.

Decision-making in both public and private institutions supporting technological programs and applications is becoming more complicated and less efficient, in the institutional, short-term sense; but long-term efficiencies, in terms of larger social interests such as public health, can be expected. In more pragmatic economic and political terms, it has become increasingly apparent that it is in the long-term self-interest of decision-makers and their institutions to be open and responsive to the interests of the public. As David Rockefeller has defined the issue for the private sector (32):

> The question really comes down to this: Will business leaders seize the initiative to make necessary changes and take on new responsibilities voluntarily, or will they wait until these are thrust upon them by law? Some adjustments are inevitable ... there may have to be new laws to force consideration of the quality-of-life dimension so that more socially responsive firms will not suffer a competitive disadvantage. It is up to the businessman to make common cause with other reformers ... to initiate necessary reforms that will make it possible for business to continue to function in a new climate as a constructive force in our society.

In the public sector, opposition to projects and the failing credibility of programs have prompted several agencies to increase citizen participation in program planning and design—beyond the environmental impact statement requirements of NEPA (33).

The model I have presented (Figure 4) does not provide any answers, but it can be used for several purposes: to widen the perceptions of planners, designers, and decision-makers responsible for specific technological advances and applications; to depict the interrelationships of resources, effects, decision-makers, institutions, and citizens; to develop policy, management, or program alternatives in the corporate, congressional, and public agency sectors that support and regulate technological development and utilization; and to assess, with public participation, the impacts of technological developments before they are utilized. Above all, the model articulates an accounting system, or framework, for decision-making that is dynamic and that can be used by all of the decision-makers, irrespective of their interests. The model has also proved helpful in the development of curricula and research: by making possible the ordering and integration of diverse perspectives and events and by providing an understanding of the patterns of technological development, application, and impacts, as well as social responses to technology. This understanding extends to technology in general, as well as to developments in such specific area as mariculture, housing, and bioengineering (34).

Reforms in Process

A number of recent legal developments can be related directly to the model, particularly to the sector designated "citizen responses to perceived effects and commitments" of technology. For citizen responses to be responsible, the flow of information to the public about effects and commitments—actual and potential—must be coherent and balanced, and it must present alternatives with their uncertainties in comparable terms. For citizen responses to be meaningful, the processes of planning, design, and decision-making must be accessible to citizens and open to their concerns.

For example, NEPA requires federal agencies to assess environmental impacts before "major actions" are taken. These actions range from the Atomic Energy Commission's approval of a construction license for a nuclear plant to be built by a utility, to the funding of increments of the highway program by the Department of Transportation, to authorization by the Department of Agriculture for the use of herbicides and pesticides. The responsibility for assessment is broad and must include full consideration of five issues (35):

Technology Assessment and Social Control

1. Potential environmental impacts
2. Unavoidable adverse impacts
3. Irreversible commitments of resources
4. Short-term use considerations versus long-term resource needs
5. Alternatives to the proposed action

Draft and final impact assessments are made available to other governmental officials and to the public for review and further development under guidelines established by the Council on Environmental Quality (*36, 37*). Although NEPA does not provide veto power to any official, even if the project poses real environmental hazards, the act does provide new information to the public—by exposing the extent to which environmental effects are being considered by the agency—and provides an enlarged record for judicial review of agency decisions. Obvious deficiencies in an agency's procedure, the scope of its statement, or the content of its statement will, on the basis of experience since NEPA was enacted, result in citizen intervention in agency processes, political opposition, and litigation. Many projects proposed and assessed have been delayed, and, in some cases, projects have been abandoned. Other projects have proceeded after being redesigned to ameliorate those effects on the environment that generated controversy (*23*, pp. 221–267).

Most projects involve applications of existing technology, but a few involve the development of new technologies—for example, the Department of Transportation's air cushion vehicle, the Atomic Energy Commission's liquid metal fast breeder reactor, cloud seeding experiments of the National Science Foundation and the National Oceanographic and Atmospheric Administration, and the use of polyvinylchloride containers, to be approved by the Internal Revenue Service, for alcoholic beverages (*38*).

NEPA does not expressly require consideration of social, health, or economic impacts or of secondary effects such as subsequent population migration and land development. These considerations are frequently ignored or treated in cursory fashion, even though they are integral to comprehensive assessment of project impacts and decision-making. NEPA does not impose assessment and exposure processes on industry or the private sector, but, whenever a utility, corporation, or other private institution is the applicant or intended beneficiary of federal agency funds, license, or other "major action," its proposal is subject to the NEPA process. There have been suggestions that NEPA be extended directly to the private sector, but as yet these have not been seriously considered at the federal level. However, variants of NEPA have been adopted by several states, and more states are expected to follow suit (*39*). Because of state and local control of land use, state versions of NEPA have the potential for directly affecting land development activities in the private sector. This potential has been realized in California, where the state supreme court has determined that the state's Environmental Quality Act requires county boards of supervisors to conduct environmental assessments before issuing building permits for housing projects and other land developments to the private sector (*40*). Similar requirements may apply to the private sector in Massachusetts, where the new environmental assessment requirements are imposed on "political subdivisions" as well as on state agencies and officials (*41*).

Therefore, the model can be further developed by adding environmental impact assessments by public decision-makers at the point where resources are to be committed to certain types of projects that apply "old" technology, as well as to certain activities that will involve the further advance or application of new technology. Concomitantly, the flow of information to citizens has been enhanced.

The development of impact statements is a meaningless exercise unless they are actually used in decision-making (*42*). It is difficult to use impact statements because of the diversity and the essentially unquantifiable nature of the new factors they present—since most agency decision-making depends on quantification of technical and economic factors (*37*). The use of impact statements in the last stage of a project, such as the awarding of construction contracts, is deceptive. The earlier stages of planning and design may not have included assessment, thereby precluding citizen inputs at a time when more important changes in project plans and alternatives could have been accomplished. In other words, effective use of impact assessment techniques and citizen feedback can be more readily achieved in the earlier, less tangible stages of a project—precisely when most agencies prefer to plan and design without public intervention. Hopefully, litigation and subsequent judicial review will impose the NEPA framework earlier in agency processes (*43*).

Further difficulties with the NEPA process have become apparent. There is an inherent conflict in the requirement that the agency proponent of a project assess it and discuss alternatives. After all, the agency has already selected an alternative and has undertaken the impact assessment essentially to justify its choice. Subsequent discussion of alternatives is too often a superficial process of setting up "straw alternatives" for facile criticism. Clearly, independent review of all the alternatives, including the proposed agency action, would be desirable. However, independent review would also require the structuring

Technology Assessment and Social Control

of new agency procedures and independent institutions for assessment (*44*).

Finally, the problem of dealing with unquantifiable impacts remains. The assignment of values and weights to environmental and social amenities may either be arbitrary or intentionally designed to produce decisions that had been predetermined by agency officials.

Despite these difficulties and the numerous conflicts and increased costs that now attend agency programs, NEPA is slowly forcing wiser environmental practices, more sensitive agency bureaucracies, and more effective roles for citizens. It is possible that the NEPA process could eventually provide the basis, not for conflict in the courtroom or at agency hearings, but for negotiation in good faith between interested parties over points of dispute as revealed by the environmental assessment (*45*). The resolution of labor-management conflicts under the National Labor Relations Board provides useful experience that should be reviewed for possible application to the NEPA context.

A major extension of NEPA practices to the assessment of new technology may have been accomplished with the passage of the Technology Assessment Act of 1972 (*46*). This law established within the legislative branch an Office of Technology Assessment (OTA) to "... provide early indications of the probable beneficial and adverse impacts of the applications of technology and to develop other coordinate information which may assist the Congress...." The office is required to undertake several tasks (*46*. sect. 3):

1) identifying existing or probable impacts of technology or technological programs;
2) where possible, ascertaining cause and effect relationships;
3) identifying alternative technological methods of implementing specific programs;
4) identifying alternative programs for achieving requisite goals;
5) estimating and comparing the impacts of alternative methods and programs;
6) presenting findings of completed analyses to the appropriate legislative authorities;
7) identifying areas where additional research or data collection is required...
8) undertaking... additional associated activities....

Assessments to be carried out "... shall be made available to the initiating... or other appropriate committees of the Congress... [and] may be made available to the public..." (*46*).

The law does not distinguish between technological developments in the public agency and private sectors and presumably includes technology being developed with private funds. Although provided with the authority to subpoena witnesses, OTA "... shall not, itself, operate any laboratories, pilot plants, or test facilities." The broad language of the assessment requirements and the way in which assessments are used by Congress effectively preclude a substantial replication of the litigation and other conflicts that have characterized the NEPA experience.

Political conditions will inevitably determine the initiation of OTA studies and their use by congressional committees, and it appears that the public will, in general, be unable to secure judicial review to promote accountability of OTA and Congress.

The burden of formulating guidelines to describe when OTA should be called upon by Congress and prescribing procedures for providing information to the public clearly lies with the OTA board and advisory council. Above all, it appears essential that OTA develop and articulate a coherent framework for all technology assessments to be undertaken. Such a framework would prevent OTA assessments from becoming skillfully contrived, ad hoc case studies, which would be essentially closed to the introduction of important information from citizens and interest groups. OTA therefore has the additional burden of laying out a framework that will replace the multiple, partial models employed by different interests, that will promote inputs from interdisciplinary and humanistic sources, and that will clearly present, in a replicable format, the quantifiable and unquantifiable costs and benefits of new technological developments and applications.

Procedures to enhance the flow of balanced information on technological developments to the public will inevitably face the problem of information manipulation and secrecy practices (47).

> The public's need for information is especially great in the field of science and technology, for the growth of specialized scientific knowledge threatens to outstrip our collective ability to control its effects on our lives.

Secrecy on the part of public agencies and the executive branch is still common practice to protect decision-making processes from public criticism, despite the 1967 Freedom of Information Act (48). However, sustained public pressures for the release of non-classified information have made such secrecy more controversial and somewhat more difficult to justify. The recent passage of the Federal Advisory Committee Act may bring about the diminution of another important form of secrecy in the public sector—agency advisory committee proceedings and recommendations, which are used in setting standards and other decision processes (*49*).

The common law of trade secrets is similarly invoked to protect corporate information—presumably from the competition (the common law basis for the concept) (*50*), but increasingly from the public and government. The Environmental Protection Agency has been unable to secure information on the quantities of polychlorinated biphenyls (PCB's) made and sold by the one American manufacturer, despite evidence that PCB's are now part of the international pollution problem (*51*). In other industrial technology sectors, however, congressional legislation has provided the government with access to information and procedures normally cloaked by trade secrecy. For example, section 206(c) of the Clean Air Act (*16*) provides that the Environmental Protection Agency may:

> ... enter at reasonable times, any plant or other establishment of such [auto engine] manufacturer, for the purpose of conducting tests of vehicles or engines in the hands of the manufacturer or... to inspect... records, files, papers, processes, controls, and facilities used by such manufacturer in conducting tests... [regarding motor vehicle and engine compliance with EPA regulations].

A similar section in the 1972 Water Pollution Control Act (*15*, sect. 308) also provides the Environmental Protection Agency access to secret information held by water polluters. It appears that Congress is now aware of trade secrecy as an obstruction to pollution control and is willing to begin limiting the antisocial uses of secrecy to some extent.

Finally, trade secrecy, in its present forms, will certainly obstruct the development of meaningful "corporate social audits" that David Rockefeller and other industrial leaders have called for. Legal sanctions for corporate secrecy obviously must be challenged if corporate responsibility and technology assessment are to be realized.

Beyond secrecy lies the problem of corporate advertising for new products and technological processes. Here, too, developments in the courts and regulatory agencies indicate that better information must be provided the public. The rapid evolution of the "Fairness Doctrine" now means that radio and television broadcast licensees must make reasonable and fair presentations of the contrasting sides of a controversial issue, once such issue has been raised (usually by advertising) on licensee broadcast time. As expressed in a recent law review note: "This obligation is incurred even at the licensee's expense if no sponsorship is available... [although] the licensee has discretion to determine how the contrasting sides will be presented and who will be the spokesman" (52, p. 109).

The doctrine has been applied by federal courts to cases of product advertising (cigarettes, large-engine automobiles, and high-test gasolines) in which it was felt that only one side of a controversial issue—the effect of such products on public health—was being presented by Federal Communications Commission (FCC) licensees in the form of advertisements. In the case of cigarettes, Banzhaff v. FCC (53), the court noted that its ruling for equal time for countercommercials or presentations promoted the first amendment policy of fostering the widest possible debate and dissemination of information on matters of public importance. In the case of commercials for automobiles and high-octane gasolines, the court noted, "When . . . the hazards to health implicit in air pollution are enlarged and aggravated by such products, then the parallel with cigarette advertising is exact . . ." (54) and ignored possible impacts on advertising and licensees as it sent the case back to the FCC for redetermination.

The idea that broadcast licensees should present balanced information on advertised but controversial technological processes or products is now a reality. Once again, the flow of information to the public, as indicated on the model (Figure 5), is being enhanced and new corporate attitudes and advertising practices should follow. (The NEPA, OTA, secrecy, and "Fairness" developments can now be depicted on the model.)

How will this enhanced flow of information be used by citizens responding to the effects of technology? What will be the nature and forms of the resulting new pressures on decision-makers?

On the model, the broad arrow from citizens to decision-makers represents not a flow of information, but adversarial processes in courts and agency proceedings. For decision-makers to learn from an endless series of adversarial processes is a slow, costly, and painful task that benefits only lawyers. The task facing the public sector and corporate decision-makers who are responsible for applications of technology is to transform this relationship from an adversarial one to one of joint decision-making and negotiation of differences in good faith among all interested parties—in short, to establish an ongoing dialogue and joint effort at assessing and planning the uses of technology (55). This effort will require new institutional management procedures, the development of more sophisticated assessment techniques, the articulation of assumptions by decision-makers, an opening up of project or program planning and design stages, and, ultimately, structural and substantive changes in the political system.

Technology Assessment and Social Control

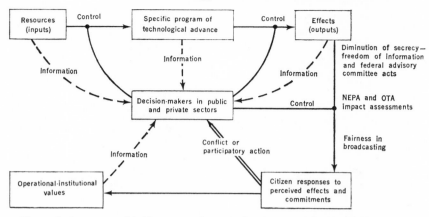

Figure 5. Summary of influences and recent developments for decision-making.

"Who speaks for the public?" will become a central issue—one that the federal agencies and the courts are now grappling with in the context of NEPA (56). Perhaps technology itself may provide some assistance here. Citizen-feedback technology exists, has been used experimentally, and has demonstrated a remarkable potential for both informing citizens and eliciting opinions and information useful for decision-making (57). The enhanced process orientation that could result from applications of the recommended model, improved information flow, and new citizen-feedback techniques would ensure continuing recognition in decision-making of the pervasive social impact of technology.

Can these numerous, fragmented developments in technology and in our legal and political systems be integrated into a coherent framework for the social control of technology? It has been noted that (58, p. 729):

> ... two major intellectual developments of the 17th century occurred almost simultaneously in law and science. The first was the drive for systematic arrangements and presentation of existing knowledge into scientifically organized categories... the second... was the concern with degrees of certainty or... probability.... By the end of the 17th century... traditional views... had been upset and new methods of determining truth and investigating the natural world had replaced those that had been accepted for centuries... there was a strong movement toward arranging both concepts and data into some rational ordering that could be easily communicated and fitted into the materials of other fields so that a universal knowledge might emerge... traditions of legal history and legal argumentation that assume the law's autonomous march through history are seriously in need of correction....

469

Michael S. Baram

It is now time to replicate this experience, develop a coherent framework for the social control of technology, and ensure that forthcoming processes of technology assessment and utilization will be systematic and humane. (59)

Notes

1. *Final Report of National Commission on Product Safety* (Washington, D.C.: Government Printing Office, 1970), pp. 73-79.
2. Public Law 92-573 (1972).
3. *U.S. Law Week,* 41, no. 16 (1972), 1061.
4. See, for example, J. Sweet, in *Legal Aspects of Architecture, Engineering and the Construction Process* (St. Paul, Minn.: West, 1970), section 30.07, pp. 634-637.
5. 29 U.S. Code 151.
6. 29 U.S. Code 651. See *Job Safety and Health Act of 1970* (Washington, D.C.: Bureau of National Affairs, 1971) for collection of relevant materials.
7. D. Cordtz, *Fortune,* November 1972, p. 112.
8. As discussed by F. Grad, in *Environmental Law* (New York: Bender, 1971), pp. 1-115.
9. *Harvard Law Review,* 84 (1971), 1822.
10. Coverage in the media has been extensive. See the 1971 and 1972 issues of *London Observer* and *New York Times*—for example, *New York Times* (March 7, 1972), p. 17.
11. *Man's Health and the Environment* (Washington, D.C.: Department of Health, Education and Welfare, 1970), pp. 97-125.
12. Boomer v. Atlantic Cement Co., 26 New York 2nd ser. 219, 257 New Eng. 2nd ser. 870 (1970) provides a classic example of judicial caution.
13. See "Economic Dislocation Early Warning System Reports" of the Environmental Protection Agency, Washington, D.C. (mimeographed).
14. Note, for example, the numerous requests for variances from air pollution control requirements by industry and chambers of commerce that are now being processed and granted.
15. Public Law 92-500 (1972).
16. 42 U.S. Code 1857, as amended by Public Law 90-148 (1967), Public Law 91-604 (1970), and Public Law 92-157 (1971).
17. Public Law 92-574 (1972).
18. Public Law 92-516 (1972).
19. *Legal Systems for Environment Protection,* legislative study no. 4 (Rome: U.N. Food and Agriculture Organization, 1972), pp. 23-32.
20. Congressional recognition of the relationship between technological advance and environmental deterioration is expressed in Title I, Section 101(a) of Public Law 91-190 (1970).
21. C. S. Russell and H. H. Landsberg, *Science,* 172 (1971), 1307; M. I. Goldman, *Science,* 170 (1970), 37.
22. The Council on Environmental Quality has partially defined the task (23, p. 343): "The contemporary world is to a great extent determined by technology.... The scale and speed of technological change may well have outstripped the ability of our institutions to control and shape the human

Technology Assessment and Social Control

environment.... It is important to understand the emerging technologies of the future and their implications for the environment and our way of life.... Predicting what and how new technologies will shape the future is a difficult task.... Even more difficult than predicting future technological developments is assessing what the full impact of any particular technology will be.... Despite the difficulties of assessing technology, it is essential that it be done.... We must develop the institutional mechanisms capable of making technology assessments...." Implicit in the council's proposal is the need for new methods to be employed in the development of assessments and the need for assurance that such assessments will indeed be used in decision-making in relevant public and private institutions.

23. Council on Environmental Quality, *Environmental Quality: Third Annual Report* (Washington, D.C.: Government Printing Office, 1972).

24. A. Weinberg, *Science,* 177 (1972), 27.

25. See, for example, *Technology in Retrospect and Critical Events in Science* (Washington, D.C.: National Science Foundation, 1968).

26. B. Portnoy, *Cornell Law Review,* 55 (1970), 861.

27. 42 U.S.Code 4321.

28. See *102 Monitor,* the monthly report of the Council on Environmental Quality, for listings of environmental impact assessments and periodic reviews of litigation related to NEPA. Also see (*23*), pp. 221–267, for a comprehensive survey of NEPA implementation.

29. No data available at this time. Statement based on conversations with professionals familiar with nuclear power program.

30. See, for example, Calvert Cliffs Coordinating Committee *v.* Atomic Energy Commission, 449 Fed. Rep., 2nd ser. 1109 (D.C. Cir. Ct., 1971).

31. As Senator Sam Ervin (D-N.C.) has said: "When the people do not know what their government is doing, those who govern are not accountable for their actions—and accountability is basic to the democratic system. By using devices of secrecy, the government attains the power to 'manage' the news and through it to manipulate public opinion. Such power is not consonant with a nation of free men ... and the ramifications of a growing policy of governmental secrecy is extremely dangerous to our liberty" (*The Nation,* November 8, 1971, p. 456).

32. Boston *Globe,* May 5, 1972, p. 17.

33. See *Congressional Record* October 5, 1972, p. 517059, regarding the Corps of Engineers and *Policy and Procedure Memorandum 90-4* (Washington, D.C.: Department of Transportation, 1972) regarding the federal highway program.

34. The model is being used in the presentation of "Law and the Social Control of Science and Technology" and "Legal Aspects of Environmental Quality," two graduate courses, and in several research projects at M.I.T. by the author.

35. Public Law 91-190 (1970), sect. 102 (2) (c).

36. *Federal Register,* 36 (April 23, 1971), 7724.

37. Council on Environmental Quality, "Memorandum for Agency and General Counsel Liaison on NEPA Matters," May 16, 1972 (mimeographed).

38. See the *102 Monitor* of the Council on Environmental Quality for abstracts of draft and final impact assessments, some of which grapple with new technological developments.

39. See *102 Monitor,* 1, no. 6 (July 1971), 1, for action by six jurisdictions. Since this review, Massachusetts has adopted its version of NEPA: Chap. 781 of Massachusetts Acts of 1972, amending Chap. 30 of Massachusetts General Laws. Connecticut is now considering similar action.

40. Friends of Mammoth *v.* Mono County, 4 Environ. Rep. Cases 1593, Calif. S. Ct. (1972).

41. Chap. 30, Massachusetts General Laws, sect. 62.

42. In Calvert Cliffs Coordinating Committee *v.* AEC (30), the court's ruling included discussion of the "balancing process" that agencies must undertake in project decision-making to comply fully with NEPA, in addition to procedural compliance in the development of impact assessment: "The sort of consideration of environmental values which NEPA compels is clarified in Section 102(2) (A) and (B). In general, all agencies must use a 'systematic, inderdisciplinary approach' to environmental planning and evaluation 'in decision-making which may have an impact on man's environment.' In order to include all possible environmental factors in the decisional equation, agencies must 'identify and develop methods and procedures ... which will insure that presently unquantified environmental amenities and values be given appropriate consideration in decision-making along with economic and technical considerations.' 'Environmental amenities' will often be in conflict with 'economic and technical considerations.' To 'consider' the former 'along with' the latter must involve a balancing process. In some instances environmental costs may outweigh economic and technical benefits and in other instances they may not. But NEPA mandates a rather finely tuned and 'systematic' balancing analysis in each instance."

43. See, for example, Stop H-3 Association *v.* Volpe, 4 Environ. Rep. Cases 1684 (1972), where the U.S. District Court for Hawaii held that highway project design work and further test borings be enjoined until an impact assessment has been developed and used, since such work "would increase the stake which ... agencies already have in the [project]" and reduce any subsequent consideration of alternatives.

44. M. Baram and G. Barney, *Technology Review,* 73, no. 7 (1971), 48.

45. The "Leopold Matrix" is a useful mechanism for promoting rational discussion and systemic resolution of project impacts by the proponents and opponents of a project in a nonadversarial setting. The matrix disaggregates impacts, calls for designation of probability of magnitude and significance of each impact, and can be completed by each of the interested parties in a project controversy. Comparative analysis of the results reveals important areas of difference of opinion and enables the parties to consider a variety of strategies for reducing these differences, such as design change or the need for concurrent projects to offset specific impacts. For example, waste water from a housing project may be one of the bases for community opposition, yet state and federal funds and programs may be available to reduce the problem. See *A Procedure for Evaluating Environmental Impact,* circular no. 645 (Washington, D.C.: U.S. Geological Survey, 1971). Also see P. Bereano (unpublished manuscript, 1971) for application of the "Leopold Matrix" to technology assessment.

46. Public Law 92-484 (1972). For the text of the bill and relevant background, see U.S. Senate, Committee on Rules and Administration, subcommittee on computer services, *Office of Technology Assessment for the Congress* (92nd Congr., 2nd sess., 2 March 1972).

47. Soucie v. David, 2 Environ. Rep. Cases 1626 (D.C. Cir. Ct., 1971).
48. 5 U.S. Code 552.
49. Public Law 92-463 (1972) and Executive Order 11686 (1972). Also see U.S. House of Representatives, Committee on Government Operations, *Advisory Committees* (92nd Congr., 2nd sess., November 4, 1971).
50. M. Baram, *Harvard Business Review*, 46, no. 6 (1968), 66.
51. *Chlorinated Hydrocarbons in the Marine Environment* (Washington, D.C.: National Academy of Sciences, 1971), p. 17: "Recommendation: Removal of obstacles to public access to chemical production data. Among the causes contributing to the lack of available data on the chlorinated hydrocarbons is a legal structure that allows manufactures of a given material, when there are no more than two producers, the right to hold their production figures as privileged information.* The panel recognizes the economic rationale that deters the release of production figures by such manufacturers and understands that our government is charged by law with the protection of their proprietary interest. Indeed, we approve the principle that governmental action should not artificially affect competition. However, we also feel that there are times when it is not in the public interest for government to maintain as privileged data that are necessary for research into the state of our environment and for an assessment of its condition. In that regard, we recognize the possibility that it is not always competitive concerns alone that determine the less than candid posture assumed by industry concerning production figures. We recommend that the laws relating to the registration of chemical substances and to the release of production figures by the Department of Commerce and the Bureau of the Census be reexamined and revised in the light of existing evidence of environmental deterioration. The protection afforded manufacturers by government is an artificial obstacle to effective environmental management, particularly with reference to the polychlorinated hydrocarbons. In view of other impediments—technological, methodological, and financial—such protection is clearly inappropriate."
*For example, the Monsanto Chemical Company has refused to release its production figures for PCB's, although requested to do so by many scientists and government officials."
52. B. Wiggins, *Natural Resources Journal,* 12, no. 1 (1972), 108.
53. 405 Fed. Rep. 2nd ser. 1082 (1968); certiorari denied, 396 Supreme Ct. 824 (1969).
54. Friends of the Earth v. Federal Communications Commission, 2 Environ. Rep. Cases 1900 (D.C. Cir. Ct., 1971).
55. Of course, the achievement of a consensus is not sufficient to ensure responsible decisions: there must also be an integration of technical perspectives on long-term material and individual needs, which may have been ignored by the parties to the consensus. Such needs are usually too remote (for example, teratogenic effects) or hidden (for example, ground water depletion) to be accorded full consideration by project proponents and citizen adversaries.
56. See Sierra Club v. Morton, 45 Supreme Ct. 727 (1972), wherein the Supreme Court provided the latest answer to when "... a party has a sufficient state in an otherwise justifiable controversy to obtain judicial resolution of that controversy...." The court noted that injury other than economic harm is sufficient to bring a person within the zone of standing; that merely because an injury is widely shared by the public does not preclude an indi-

vidual from asserting it as a basis for personal standing; that injury sufficient for standing can include esthetic, conservational, and recreational injury, as well as economic and health injury. But the court noted that "... broadening the categories of injury that may be alleged in support of standing is a different matter from abandoning the requirement that the party seeking review must have himself suffered the injury..." and that "... a party seeking review must allege facts showing that he is himself adversely affected..." in order to prevent litigation by those "... who seek to do no more than vindicate their value preferences through the judicial process."

57. T. Sheridan, "Technology for Group Dialogue and Social Choice," M.I.T. report to the National Science Foundation on grant GT-16. "Citizen Feedback and Opinion Formulation," 1971; and D. Ducsik, N. Lemmelshtrich, M. Goldsmith, E. Jochem, "Class Exercise Simulating Community Participation in Decision-making on Large Projects: Radiation Case Study" (unpublished manuscript, 1972).

58. B. Shapiro, *Stanford Law Review*, 21 (1969), 727.

59. I wish to thank Dennis W. Ducsik, a doctoral candidate in the department of civil engineering at M.I.T. who is pursuing an interdisciplinary program in environmental resource management and technology assessment, for his help as a research assistant in the project and in the development of this article. I also wish to acknowledge the support of the National Endowment for the Humanities (grant no. EO-5809-71-265).

24. The Politics of Technology Assessment

MARK R. BERG

Technology Assessment and Public Policy: Towards the Success of a Merger

During the past several decades a string of "systems science" approaches have been applied to problems of public policy. Sometimes this merger between government and the scientific/technical community has been successful, other times it has not. Advocates and critics alike can find examples of systems analyses, operations research, cost/benefit analyses, decision analyses and a host of other "systems science" approaches and techniques which have been successfully used, abused or ignored when applied to areas of public policy.[1,2]

Reprinted with permission from *Journal of the International Society for Technology Assessment*, vol. 1, December 1975, pp. 21–32.

The Politics of Technology Assessment

However, very little empirical work has been done to systematically document the track record of the "systems sciences" in terms of utilization of results. The theoretical literature which deals with the failures in the science/government merger tends to emphasize, as the source of the problem, either characteristics of the government sector, the scientific/technical sector, or basic incompatibilities between the two sectors.[3]

Technology assessment (TA) can be viewed as one of the more recent attempts to apply the "rational, systematic" approaches to an area of public policy—in this case, to the management and regulation of technology. As described by one of its most commonly used definitions, technology assessment is a class of policy studies which systematically examines the effects on society that may occur when a technology is introduced, extended, or modified with special emphasis on those consequences that are unintended, indirect and delayed.[4] The central theme here, and in much of the TA literature, is one of TA as a nonpolitical, evenhanded, information source. This, of course, is the scientific/technical side of the partnership speaking. Not surprisingly, it underemphasizes the perspective of the political sector which tends to view information and knowledge as sources of power and the political process as the working out of power relationships among competing interest groups and constituencies.[5,6,7]

The point to be made here is not that one perspective is necessarily more correct or appropriate than the other, simply different. The scientific/technical perspective tends to be more rigorous, systematic, neutral and empirical while the political, or societal governance, perspective tends to be less rigorous, more politicized, and more ideological. Thomas Kuhn's valuable work on the concept of paradigms would suggest that these two perspectives, or paradigms, would differ in terms of relevant values and purposes, acceptable data, methodologies, guarantors of knowledge, language, etc.[8] This in turn, along with the experiences of the earlier systems science approaches, would suggest that *the success of the scientific/political partnership as proposed by TA cannot be taken for granted.*

A central thesis of this paper is that the TA community must be consciously aware of, and responsive to, the often conflicting perspectives, goals and processes of the two different paradigms TA is attempting to bridge. Without this self-consciousness TA runs the risk of nonutilization, misutilization, unthinking drift, elitism, cooptation, and even outright corruption.

As a contribution to this needed self-consciousness I examine, below, a number of elements potentially relevant to the success of the TA movement as it operates in a political environment. I begin by examining TA's sources of power within that environment, and what sectors of society, as advocates and adversaries, are likely to use and abuse the power which may accrue to an effective TA capability. The reciprocal relation between TA and the political sector is then examined, wherein the form and the content of each sector is shown to affect the form and content of the other sector. From this are then drawn a number of implications and alternatives which TA practitioners will have to face up to in their attempts to cope with, and operate effectively in, a political environment.

The Politics of Technology Assessment

Perhaps a valid social indicator of the degree of breakdown in value consensus and the increase in value conflict in society is the number of books and articles published with the title, "The Politics of...." This increased politicization of society and the increasing demands being made on our political processes and institutions are the result of developing trends of the last decades such as increasing education and affluence, and rapid social and technological change, as well as rising pressures from economic, population, resource and environmental problems. Technology assessment as an intervention and policy-oriented response to at least some of these pressures cannot be isolated from politics on the basis of its having "objective" empirical and analytic content nor as a nonpartisan management tool. *All the way from the sources of demand for technology assessment to its final outputs and effects, TA is neither value free, nor value neutral, nor nonpolitical.* For when assessors and politicians begin to make the inevitable choice about what problems to study, what variables to attend to, what parties at interest to consider, and what alternatives to legitimate, they make value choices. And when the processes and outputs of technology assessment begin to influence outcomes and policies, TA becomes a political activity. For at that point technology assessment is influencing answers to the basic political questions of "who gets what, when and how."[9] To be less than this is to leave TA as an ineffective endeavor, of little serious value, and incapable of justifying its claims for scarce resources and intellectual attention.

TA And Its Sources of Power

Technology and control of technology are modern-day sources of power, just as military might and ownership of land and capital have been in the past.[10,11] Thus if technology assessment represents an embryonic source of information about and control over technology,

The Politics of Technology Assessment

it too is a new source of power. Assuming that, in time, TA will be able to muster the intellectual and methodological muscle needed to achieve its purposes, it is instructive to examine those aspects of TA which serve to generate its societal power or political muscle.

At a somewhat abstract level one might say that power accrues to TA because of its access to and production of information and knowledge which in turn serve to reduce uncertainty and risk. In a sense, then, TA's power comes from its ability to provide knowledgeable and believable imagery about the future.[12] The believability and trust factor should be emphasized. *If TA cannot maintain its analytic credibility or if it becomes associated with a partisan political bias, it could easily lose its political power-base*—thus the paradoxical conclusion that TA's political muscle may hinge on its ability to maintain a commonweal perspective and an apolitical, evenhanded public image.

At a more operational level the sources of TA's power are several. To a significant degree TA's power resides in its need to set priorities for subjects of study. "Knowledge is power," and thus the choice of what knowledge shall be generated and available for use by affected parties is also power. In other words, TA has the ability to change the content and context of national debate. It can influence opinion and generate ideas in "good currency."[13,14] That is, in its role as a source of "objective" expert knowledge TA can both influence and legitimize public policy decisions by having the ear of both the policy maker and the public.[15] Finally, TA cannot only alter the context of public debate, but by raising questions and issues on its own, it can serve as an early warning device which influences the national agenda.

It should be stressed that to suggest that technology assessment is a source of political power is not the same as arguing that it is an overwhelmingly strong power. In some cases it may be, in others it may not. The effects and clout of technology assessments will probably be highly issue and project specific. Outcomes will flow from the interactions between many other actors and information sources beside TA's, and in view of the pluralistic nature of society and the many problematic aspects of TA, that is how it should be. Given this pluralistic perspective it will clarify the politics of technology assessment to ask where the demand for TA is coming from. That is, who in the political arena hopes to gain from the development of a technology assessment capability?

The Demand for TA

Technology assessment advocates exist in a number of sectors of society and have a number of different and often mixed motivations for their support. For example, there are what might be called the *"rational policy advocates,"* otherwise known as the *"good government reformers."* The dominant underlying assumption here is that better and more open information will reduce the power of special interests and allow for the avoidance or reduction of negative effects through better planning and policy decisions. From a slightly different perspective another name for this group would be the *"technological fix advocates."* These come in two basic varieties. First is the rapidly dying breed of technological optimists who hope and/or believe that with a few strategically placed patches and detours the 300-year march of progress can continue down the throughway to heaven on earth. The second variety might be called the "technological pragmatists" who believe that technological fixes will be needed, but that each technological change will create its own chain of residue problems.[16] This group, while not having totally abandoned the concept of progress, is less sanguine about our abilities to cope with our problems and the mixed blessings of technology and science.

At the far end of the technological optimism pessimism scale are the "technological pessimists" or "neo-luddites." This group not only sees technology as the source of almost every major modern-day societal problem, but also feels that applying complex technological fixes to technologically induced problems will only exacerbate the problem. While many within this group make up the back to the land movement nineteenth century style, others offer a somewhat more positive and promising program under titles such as "soft" technology, humanistic technology, people's technology, etc. It is towards this group that critics of TA most accurately direct their accusations of "technology harrassment and arrestment."

This taxonomy of technology assessment advocates can be no more than description of ideal types. Technology has become one of the most emotional, vexing, and uncertain symbols of our society. Its immense threats and promises have generated a love/hate relationship which affects us all.[17] When technology, as a single symbol of our time, can embody both the threat of our destruction and the promise of health and abundance, it should not be surprising to find individuals approaching the concept of technology assessment with both mixed motives and objectives.

The kinds of advocates described above can be found in varying proportions in business, government, academia, the "counterculture" —in fact, in all sectors of society. Despite the different nuances coming from the various advocates and sectors, the central theme of their demand for technology assessment is the need for man to take control

The Politics of Technology Assessment

of technology and use it for the ends of man. While this is as it should be, one must also be aware that support for and interest in TA are also generated at a much more mundane and pragmatic level as well. For example, over the last decade a cadre of "experts" has been created who, to a significant degree, depend on (or look forward to) technology assessment as the basis for their teaching, research, and consulting careers. Alongside the altruistic motives which certainly drive this group is the fact that they have become a special interest group in society whose power, fame, and fortune hinge on increasing the demand for technology assessment and other futures-related activities. Closely related are the bureaucrats and administrators whose domains of responsibility include technology assessment. Their bureaucratic and political power, career advancement, and sense of self worth are now, to a significant degree, tied to the fortunes of TA and to building up a demand and constituency for their services.[18,19,20] Examples of this role can now be found in NSF, the new Office of Technology Assessment, other federal agencies, and a host of universities, think tanks and consulting firms. *Taken together the bureaucrats and "experts" may represent a significant positive feedback loop in the growth of, and demand for, technology assessment.*

Still further demand for TA is generated by politicians and other public sector decision makers. While this group may advocate TA on its merits as a tool for improved public policy decision making, they can also be expected to see and make use of its other potentials. For example, to be associated with technology assessment, or science and technology policy in general, may provide a useful power base on a range of important issues. Congresspersons with assignments on the "right" committees thus gain influence on particular policy decisions as their opinions take on the added legitimacy associated with especially knowledgeable sources.

Looking to the darker sides of political motivation, it will not be surprising to see particular assessment projects being supported as stalling tactics so that action on unpopular or controversial issues can be postponed until after an important election, or as a way of buying time for the building up or tearing down of relevant political coalitions. Similarly, one can expect to see particular assessments supported (and then influenced) by politicians, agencies, and business leaders in hopes of finding support and legitimacy for decisions actually made on other grounds.[21,22] A strategically timed assessment may prove to be a helpful scapegoat for a politician faced with conflicting constituent pressures and a "damned if you do, damned if you don't" decision. It should be expected, as well, that individual assessment teams will provide readily expendable scapegoats for decision makers whose assessment-based decisions prove to be wrong or unpopular. Still other assessments will be supported by relatively weak actors in search of a third party to be used in attacking powerful opponents, thereby reducing their own risk of direct retaliation.[23] These motives and their consequences for TA are further discussed later. (For a glimpse of some of these motives in operation, see S. Ebbin's insightful case history of the Jamaica Bay/Kennedy Airport technology assessment.)[24]

TA and Structural Power

Having examined the sources of TA's power and the sectors in society which advocate its use, the question still remains as to how that power is operationalized. In other words, how will power relationships and authority be altered by the introduction of TA into the structure and processes of the decision network? The questions of who in society will shape and have access to TA-based information and power, and who will have the resources and skills needed to use it effectively are not trivial or academic. In an information-based society, such as ours, they are fundamental political questions.

As a new form of societal information and analysis, technology assessment will be partially shaped by and, in turn, partially shape the institutional structures and procedures in which it will be used. Institutional reforms, such as changes in communication patterns, expectations, behavioral norms, and reward systems, will result and be necessary if TA is to be carried out and used effectively.[25] However, at this point in the development of TA it is, perhaps, still an open question as to whether such reforms and institutional innovations will facilitate democratic control of technology or whether, instead, they will reinforce the potential for antidemocratic control by a technocratic elite.

As suggested above, technology assessment does not enter into a vacuum, but instead affects and is affected by current power relations. These political constraints on TA are exemplified by the controls and institutional format imposed on the Office of Technology Assessment. For example, the informational and nonpolitical nature of OTA have been stressed, while at the same time great care was involved in designing partisan balance among the Technology Assessment Board members who will set policy for OTA. It will be this politically motivated group which will be setting constraints on the outputs of OTA (not to mention the additional control which can be exerted by the congressional appropriations committees). Still further

The Politics of Technology Assessment

constraints and compromises are inherent in the potential conflicts and animosity between OTA and other agencies and committee staffs with interests in similar areas. These constraints and pressures created by the political nexus which surrounds OTA will show their operational effects in terms of subjects to be studied, perspectives to be emphasized, and conclusions to be drawn (or not drawn).

In addition to OTA, the National Science Foundation has been a major actor in creating and shaping the technology assessment movement. While NSF is one of the least political of the federal agencies, it nonetheless exists in a political environment and must therefore respond to manifest and potential pressure from clients, agencies and elected officials who comprise its coalition of supporters. While the record of NSF is quite good in terms of the autonomy allowed its sponsored researchers, political criteria have played a part, and thus shaped, its decisions on what projects to solicit and accept. This is not a criticism, it is merely another example of the ways in which the current political structure, process and balance of power has shaped and will shape the form and content of technology assessment.

Importantly, the inverse proposition is also valid. That is, technology assessment has shaped and will shape the political process and structure. For example, it is design, not accident, which has put a technology assessment capability under the control of Congress. The fact that the capability is congressional should provide Congress with additional leverage and expertise on policy matters vis-à-vis the executive branch and private sector. Similarly, the fact that OTA is under congressional control is probably a reasonable guarantee of assessments which take a more commonweal perspective, rather than the more partisan or special interest perspective which might be expected from a bureaucratic agency or narrow interest group. Such a commonweal perspective should provide poorly funded public interest groups with improved leverage vis-à-vis typically well-funded special interests.[26] For example, the cost of a well-funded TA could represent a major portion of, or even overwhelm, the budget of a local government or citizens action group, whereas it would only represent a small portion of the advertising budget a large industry could use to counter its potential impact.

A reasonable model for the type of shift in power and leverage which might be expected through the institutionalization of technology assessment is the shift which has resulted from the environmental impact statements required by the National Environmental Policy Act. While technology assessments are not, at this point, required of new, large-scale technologies, the fact that the capability exists (or can exist), in both the public and private sectors, may nonetheless insure that TA will become an integral element in decision processes. To not use the capability, in a timely manner, could leave corporate and public decision makers open to the charge and consequences of nonfeasance.

It should be stressed that such shifts in leverage are not at all certain, but merely potential. They will be resisted by those who stand to lose. In many cases the inherent structural power and economic leverage of large industries, such as oil or automobiles, will simply overwhelm any effects a TA study might produce by increasing public information or participation. Critics of TA will be quick to label it as "technology harrassment and arrestment" which can only serve to retard the dynamic growth of modern technology. Similarly, Congress will need much more than just better information if it is to accrue the added power which could flow from improved decision-making capabilities.

Thus, the exact way in which TA may affect current power balances, if at all, is simply not yet clear. A major uncertainty rests with the assessors themselves and how broadly they will define their constituencies and variables of interest. For example, will assessment studies successfully communicate the issues and alternatives to the general public, or will they be oriented to just "top level" decision makers? Will TA further move our system towards centralized control on the basis of recommendations for monitoring and regulation? Similarly, it is not yet clear whose issues and decision criteria will be stressed, e.g., the individual vs. the "public interest," big business vs. small, local government vs. national, etc. TA has the potential to expand legitimate decision criteria beyond the traditional questions of short-term technical and economic feasibility. In theory, at least, technology assessment should aid in attempts to consider a wider range of issues and perspectives, such as environmental impacts, mental and physical health, effects on productivity, inflation, and unemployment, questions of distributive justice, and even the rights of future generations. How far TA can and will go on these and other issues is not clear. However, the farther it goes, the more technology assessment will reverberate throughout the political environment.

The Political Implications of Methodology

One of the causal factors which will be operating to affect the balances of power discussed above will be the methodologies used in doing technology assessments. TA has not yet developed any over-

The Politics of Technology Assessment

whelming consensus with respect to appropriate methodologies. In view of this it is important to recognize that *choice of methodology is not just a technical matter, but has significant political implications as well.* There is no such thing as a universal methodology which can operate meaningfully within all world-views, which can emphasize all values equally, or which can be applied to all problems with equal success. Instead there are many different methodologies, each of which tends to define problems in different ways, point toward different solutions, emphasize different values, and employ different people in its application.[27] Each of these outcomes has potential political content, and thus, the choice of methodology has political content as well.

The point can be readily illustrated with some examples. For instance, consider the difference between the exploratory (or tracing) methodologies which forecast from the present forward, and the normative methodologies which start from desirable alternative futures and work backward toward present decisions which may get us where we want to be. A technology assessment which uses only the former, more conservative, method tends to tell the public, "here is the future," or at best, "here are the futures you can have." A TA which uses both methodologies, however, tends to first ask the public, "what futures do you want?" and then goes on to examine the ways to get there. It should not go unnoticed that the extrapolation of the present as used by the exploratory methods is relatively easy to mask as apolitical, whereas the more goal-oriented stance of the normative methods is much more difficult to mask as apolitical technical analysis. For those assessment teams who choose to operate under an apolitical, neutral model of TA, this methodological distinction may represent a critical political constraint on the development of TA.[28]

A less subtle and philosophic example would be an assessment which uses a cost/benefit methodology as its only form of economic analysis. While this may be a useful technique, cost/benefit analysis does not adequately handle the question of who gets the benefits and who bears the costs. Such an assessment could do little for the cause of distributive justice. A similar issue arises when one realizes that the decision as to what constitutes a "representative sample" for survey research is both a technical issue and an ideological issue. Two hundred years ago, for example, one might have argued that a representative opinion poll should be based on only white, male property holders. In an even less democratic society the very idea of a public opinion poll would have little value.

One final example would be the choice of whether to use a complex, mathematical, computer model or a more participatory gaming simulation to describe and explore the techno-social system under study. Of course, both might be useful. But the point is that one tends to limit public participation (although not totally), and the other encourages it and attempts to use it constructively. *Public participation is an issue which will bedevil TA for some time to come. As a methodological problem it cannot be resolved solely within the traditional scientific/technical terms of "truth," efficiency, control, etc. Of equal importance will be typically political criteria such as justice, advocacy, an informed citizenry, consensus, and coalition building.*

TA and the Power of the Expert

In my attempts to underscore the political implications of technology assessment I have run the risk of overstating the potential power and influence of a mature TA capability. While keeping this risk in mind, it is important to extent the argument to one last area, that is, the potential power and conflict of interest of those who do assessment studies. I have earlier suggested that power will accrue to assessors, as experts, who thereby gain access to the ears of both decision makers and the public at large, as well as gaining the legitimacy to define alternatives and make recommendations. *But who are these assessors and what internal values and external pressures will be oeprating on them to encourage and discourage the wise and evenhanded use of this power?*

The research needed to document who the assessors are and what values they tend to hold has not, to my knowledge, been done yet. Thus, much of what follows should be treated as hypothesis. The question is important enough, however, that some speculation is in order. Limited observation would suggest that two of the major defining characteristics of those who get involved in technology assessment are that they are "intellectuals" and "problem solvers." As such they could be expected to embrace most of the major value clusters of these two groups. For our purposes here the important "intellectual" values are truth, openness, and a faith in, and desire for, rationality. The relevant "problem solver" values are a pragmatic outlook along with a tendency to be "goal" versus "process" oriented, i.e., a somewhat greater concern for getting the primary job done, than for considering secondary effects or implications of the path taken. In addition to these more specialized characteristics, assessors share a number of common values and perspectives with other groups in society. For example, there is probably a propensity to be status and achievement oriented and, in our society especially, this would to some degree be correlated with a desire for economic success (or at least comfort). As a group, assessors probably tend to be socially aware

The Politics of Technology Assessment

and holistic in outlook (this may even be a self-selection criterion for those interested in TA). However, their holism and awareness is, to some degree, limited and influenced by a common experiential base which generally includes whiteness, maleness, college, and the requisite middle class background. Along with this would be a tendency toward a progressive, if not liberal, value set. One final self-selecting characteristic would be a special interest in, and appreciation of, technique and the technological aspects of a problem or activity. All of these characteristics are generalizations and simplifications, and thus distortions which must be treated with care and healthy skepticism. Hopefully, however, they are accurate enough to be instructive.[29]

Accepting these as the general characteristics of those typically involved in doing assessments, what can be said about the characteristics of the environment in which they will be operating? One of the most obvious characteristics is that it is an environment in which millions, if not billions, of dollars will be riding on the outcomes of decisions in which TA studies may have significant influence. From the assessor's point of view there are significant amounts of money riding on turning out studies of a nature which will facilitate getting additional contracts. *Along with the potential for cooptation, if not outright corruption, by the large economic stakes is the potential for cooptation by the enticements of power:* that is, the subtle and not so subtle psychic and material inducements and rewards that come from being close to the centers of power and wealth in society—the conferences in lavish settings, the invitations to testify before congressional committees, the handshakes and dinner table talks with important officials and dignitaries. *The potential rewards for going along and not rocking the boat are as substantial as the risks involved in being the bearer of bad news.* In addition, risk avoidance and self-maintenance are two of the primary characteristics of the bureaucratic and competitive environments in which much assessment work will be carried out. This potential for cooptation is exacerbated by the fact that at least some of the people involved in the technology assessment must have some expertise in the particular technology and thus have an obvious stake (both economic and psychological) in its perpetuation and protection.[30]

Perhaps the types of concerns raised above would be less important if technology assessments were more "scientific" and "technical" in nature. That is, if they could be repeated and verified in a more definitive way by a community of peers. However, this is not the case for a number of reasons. Full scale technology assessments have typically been too expensive and time consuming to be repeated for the sake of methodological development and verification. Apparently the view has been that the marginal returns for such an effort would not be as great as the returns from a TA in a totally different area needing assessment. In view of the many methodological problems inherent in TA this may be a reasonable assumption. And, while this may indeed be a poor "scientific" strategy, it may also be a necessary and wise "political" strategy. Tests of repeatability, at this point, might actually raise more questions than they would resolve and thus slow the growth of TA as a "legitimate" activity. Part of the explanation for this problem and for the concern about the potential for cooptation is inherent in the nature of the problems studied by TA. Much of technology assessment, and certainly the most controversial aspects, deals with social-political rather than technical problems. And one of the defining characteristics of social-political problems is that there are no clear-cut "right" answers. *Answers to social-political problems are typically ambiguous, value laden, subjective, multifaceted, dynamic, and untestable.*[31] *As such they leave room for considerable conscious and/or unconscious distortion, cooptation, and rationalization.* It is recognition of this potential which has already generated demands for public participation in TA studies as well as "public interest" groups whose major function is to review and evaluate technology assessment studies.[32]

The opening question of this section asked what internal values and external pressures will be operating on technology assessors to encourage and discourage the wise and evenhanded use of their power. The question is a loaded one since one person's evenhanded wisdom is another's biased foolishness. Nonetheless the preceding description of values and external pressures does suggest a system of countervailing forces which could serve to keep assessors within the general bounds of evenhandedness and, perhaps, wisdom.

As suggested earlier, a number of operative forces offer the potential for corruption, cooptation, and rationalization. These would be factors such as the high economic stakes; the enticements of power, status, and achievement; and self-censorship due to risk avoidance and self-maintenance. Still further threats are posed by the potential biases due to a limited experiential base, a technical orientation, a progressive-liberal value set, and a tendency to be "goal" rather than "process" oriented. Fortunately each of these "negative" factors is balanced to some degree by a second set of values and external pressures. *Foremost among these are the values of truth, openness, and rationality. These values have been at the heart of the movement to develop TA, and if they are subverted, that movement will be in great danger.* TA could quickly come to be regarded with the same skepticism as cost/benefit analysis in which almost anything can be shown to be worthwhile by the careful selection of analyst, variables, and discount rates. To some degree this skepticism is already operative and is resulting in mechanisms for peer and public review and evaluation. To date these have taken forms such as interest group oversight committees, peer reviews of early drafts with the inclusion of their critiques in the final reports, large conferences involving all interested parties, interaction with key decision makers, and the wide distribution of final reports. While a major objective of these procedures has been to maximize utilization of TA results, they should also reinforce the values of truth, openness and rationality, and the participation-oriented ones, in particular, should help to counteract any tendency to be goal-oriented at the expense of open and participatory processes.

In concluding this section it is instructive to shift away from detailed analysis to a more general level of argument. *The role of the assessor, or more generally the role of the expert, in the political arena depends for its power on a unique combination of the power of knowledge and the traditional forms of political power.*[33] *It is a relatively new role which the constitution did not foresee and which traditional institutions are still learning to cope with effectively.*[34] If we are to remain a democracy and not a technocracy we shall have to learn to integrate our experts into our democratic institutions. And if we are to solve our problems we shall have to integrate our scientific and technical knowledge into our political decision making. The evolving controls on technology assessment, as described above, would suggest that one of the best hopes for a successful integration is to maintain and strengthen the norms of truth, openness, and free public review and debate which have been major foundations for the success of both science and democracy.

Learning to Cope in a Political Environment

It should be clear by now, however, that for serious issues of importance to powerful interests, TA may be politicized to the point that platitudes such as truth, openness and free debate will not insure acceptance and utilization of TA results. *Successful integration of the scientific/technical perspective and the political perspective will require mutual accommodation and learning on the part of both communities.* And there are no guarantees that such accommodation and learning will occur.

From the political perspective there are ample reasons why powerful interests should be wary of TA gaining too much legitimacy and power. The holistic, planful, systematic approach of TA runs counter to many characteristics of current policy-making processes such as short time horizons, nonholistic disjointed incrementalism, pluralistic power centers, and bureaucratic politics.[35] One need only be reminded of the demise of Program, Planning, Budgeting Systems (PPBS) to realize that those interest groups that have operated effectively under these characteristics can be expected to resist changes from the status quo. And even those with less to lose from change may be reluctant to alter a decision-making process which works to some degree, which they understand, and which they are comfortable with. Where conflicts do exist between these characteristics and TA, assessors must treat them as constraints which must be lived within or consciously removed, but certainly not ignored. The remainder of this section discusses some of these problems, and opportunities, imposed on TA by its political environment and suggests some ways in which TA might effectively cope with these challenges.

Building a Constituency for TA Results

One such constraint is that *the results of TA studies must be relevant and timely with respect to policy decisions whose time has come.* To be too early or too late with relevant information may leave the study on the proverbial shelf gathering dust. Similarly, to be timely but to have emphasized issues which lack salience to decision makers will do little except hurt the image of TA. On the other hand, TA studies, and particularly OTA, must avoid being too responsive in the sense that they adopt too short a time frame or too narrow a view of what is relevant. Of course, TA will not be powerless with respect to timing and salience if assessors are willing and able to use effectively what power they have.

For example, to the degree that TA serves as a societal early warning mechanism for dangers and opportunities posed by technology, it is likely to be early with its results in the sense that it will lack a ready-made constituency for the output of its study. New problems are likely to fall within jurisdictional cracks which no one is responsive to or responsible for. If no groups have yet coalesced around an issue and no politicians have yet thought it worth championing, the odds are that those negatively affected are not very powerful or organized and/or there are no significant negative effects immediately visible. Similarly, if the impacts or opportunities are so far down the line, or so remote and hidden from view, it will be difficult for the imperfect speculations and "doomsday ravings" of a TA study to attract much enduring attention. Thus, this type of early warning TA study may prove difficult to pull off effectively. However, this function is important enough that efforts should be made to push back the political constraints. This type of TA may have to place a heavy emphasis on education, spending much of its efforts in dealing with the relevant

The Politics of Technology Assessment

publics as it attempts to build a constituency for its results. For those from a scientific and technical background this may prove to be difficult and uncomfortable work. It may also prove to be frustrating, as the first studies to jump into a jurisdictional vacuum may have a high probability of collecting dust.

Constraints of Coalition Building and Consensus

A different type of requirement placed on TA by the political environment is that *recommendations or policy alternatives must allow some degree of flexibility and room for maneuver.* Responsive action on the part of an individual policy maker or agency will require not just personal commitment, but also consensus on the part of a "winning coalition" with interests in the issue. Building and maintaining that coalition is likely to require some degree of maneuvering room to adjust for conflicting interests and ideologies. To the degree that an assessment cannot, in good conscience, leave policy makers with a significant degree of flexibility, its utilization and effectiveness will have to depend on making an exceptionally strong and well-communicated technical case. When the stakes are high, *the greater the degree of ambiguity and speculation in the results of a TA, the greater will be the potential for de-emphasizing its import and rationalizing inaction or self-serving behavior.* Consider, for examples, the typical self-serving responses to early warnings of smoking hazards, pollution, ozone depletion, or raditaion standards.

Problems of Value Articulation and Conflict

The political characteristics of coalition and consensus have other important implications for TA. TA, as suggested earlier, is a value-sensitive enterprise. This is logically apparent in the very term "assessment," i.e., to make a valuation. It is also historically apparent in the goal clarification steps of the "rational, systematic" policy processes and methods which have preceded and continue to aid and complement the TA process.[36] This value sensitivity when tied to a scientific tradition of explicitly stating underlying assumptions and propositions may lead some TA's to methodologies which employ or derive explicit statements of goals or value implications. Such methodologies have much to be said for them. They may, for example, be very useful in coalition building, by stressing the relation between the recommendation of the TA study and those specific values which are especially important to groups already in, or desired for, a coalition of supporters.[37]

However, in highly politicized situations the demands of coalition building may suggest that values and goals be left unsaid or ambiguous. "Politics makes strange bedfellows" when agreement on means is possible but agreement on ends is not. The need for action may lead to tacit agreement to ignore underlying conflict among coalition members. For example, consider the unique coalition of conservatives and liberals who have backed the concept of a negative income tax. TA must develop a political sensitivity such that the scientist's urge for truth and openness does not lead to needless conflict which only reduces the maneuvering room of relevant decision makers.

On the other hand that same political sensitivity can be used to determine those situations in which TA might be used as a means for generating conflict or for conflict resolution. Each of these situations presents its own risks and difficulties for an effective TA. But given that TA, by its very nature, is likely to generate "bad news" for someone, conflict situations will be unavoidable. This is compounded by the fact that it seems to be in the nature of the times, and in the nature of the problems studied by TA, that results are likely to be challenged and distrusted. An important consideration for TA is that *as conflict increases, there is likely to be a decrease in the degree of trust and legitimacy granted to those assessments attempting to take an evenhanded, neutral approach.* It is, of course, in conflict situations when trust and legitimacy become all the more important.

Assessors should be aware that when adequate trust and/or authority can be established there may be a unique opportunity for conflict resolution by taking on the role of arbitrator or ombudsman. For example, TA could provide a communication mechanism for involved parties and thereby start to build up mutual understanding. This might be done in the form of conferences, oversight committees, published surveys or special reports wherein common interests and nonzero sum alternatives could be pointed out. In other situations TA might take a highly intervention-oriented approach and recommend to some overarching authority that a particular decision network or pattern of responsibility be set up. Such an approach, if blessed with sufficient support to bring it off, could pre-empt the normal working of special interest politics and create an organizational framework designed to be in the "public interest."

TA as a Source of Unwanted Turbulence

Of course, when embroiled in the midst of a conflict situation, TA's legitimacy to speak for the "public interest" is likely to be challenged.

The Politics of Technology Assessment

Many involved actors will not accept TA as an "evenhanded" arbitrator or ombudsman, but will see it instead as one more actor threatening to tread on and influence the turn and autonomy of their organization. And from their own unique perspective, they may be right. *In its search for indirect, unintended effects and unexamined alternatives, TA is likely to create considerable turbulence for bureaucratic organizations whose preferences would tend more towards self-maintenance, risk avoidance, and closed organizational boundaries which block inputs from and accountability to, 'outside' stakeholder groups.*[38]

From the perspective of an individual policy maker, agency, or corporation, an independent TA may represent a troublesome loss of control over its environment. For example, it may mean that a negative evaluation of an individual's or agency's performance cannot be easily suppressed as in the case of in-house studies. Or it may mean that public information about one's behavior or intentions will open the door for control or counteraction.

Increased turbulence is also likely to flow from the increased depth and exposure of interest articulation which an independent TA is likely to produce. That is, many parties may not even know they are in conflict, or have tacitly agreed to ignore it, until a TA study shows it to be the case, and thus forces the issue. Additional turbulence may flow from the presentation of new alternatives or imagery previously suppressed by powerful interests.

Examples of this type of turbulence created by TA type activity are readily available in the fields of transportation and energy, e.g., the solar energy alternative over nuclear, or the mass transit alternative over private automobiles. Older, more "conserver" oriented, organizations may quickly learn to stay out of areas where TA studies and conflict might force them into a more turbulent or uncontrollable environment than they wish. While this in itself may reduce some forms of organizational externalities, it may also lead to less risk taking in areas where it is needed.

Deception, Influence and Cooptation of TA Studies

Many organizations or stakeholders, however, will not react passively to the potential negative implications of a TA type study. The study when viewed as a threat may receive biased or selective information, or no information at all from involved parties, e.g. "that information is classified" or "a competitive secret." *In conflict situations where one party is bound to be hurt by what the TA study reports, the project and its personnel may well be attacked as biased, out of date, misinformed, inept, etc.*

Thus, in those very situations where data may be hardest to come by, it may be doubly important that information sources be checked, verified, and documented wherever possible (in contrast to the understandable temptation, when data is scarce, to accept gratefully, and use, whatever data one can get). This may even include protecting oneself by sharing data (for purposes of confirmation or rejection) with organizations which did not reciprocate.

Assessors should learn to expect that stakeholders in conflict situations will attempt to use and influence TA studies to their own ends. As suggested earlier, this may be through financial support, cooptation, selected leaks of information, etc. Some TA's may be supported in attempts to give "progressive" legitimacy to what are really "reactionary" stalling tactics. (Boffey, in *The Brain Bank of America*, beautifully details these problems as experienced by the National Academy of Sciences.[39]) Others will be supported as mechanisms for indirectly subverting opponents too powerful to be attacked frontally. In other situations advocates of particular issues or interest groups will use information from TA studies (in and out of context), or carry out their own advocacy TA, in attempts to win on their issues.

Advocacy vs. "Evenhanded Analysis": Separate but Equal

This latter perspective, i.e., the use of TA by advocates, brings home the point that there are alternative models of TA. The two models most widely practiced and discussed in the literature are those of the "advocate's tool" and the "search for desirable choices."[40] Advocacy TA's are likely to occur when the stakes are high and trust is low.[41] Since these are characteristics of most important issues, *we can expect that some form of advocacy TA will develop in parallel with the more commonweal, or "public interest," oriented TA's*, emphasized in this paper, and currently funded by agencies such as the National Science Foundation and Office of Technology Assessment.

The distinction between the two models is important, for their differing perspectives have implications for the strategy and tactics of both analysis and effective utilization of results.[42] Advocacy, in its many forms and with its partisan perspective, is likely to emphasize narrow interests, limited scope, affective language, judicial proceedings, and high media exposure. Its intent is to influence those in authority and win its conflict. While many TA participants with scientific and technical backgrounds may be uncomfortable with an advocacy approach, it is important to realize that advocacy is an important

The Politics of Technology Assessment

mechanism for articulating values and opening up alternatives within the democratic process.

In contrast, "the search for desirable choices" model has a focus more oriented to containing conflict while achieving commonweal goals. [43,44] It is likely to emphasize multiple interests, wide-ranging scope, more neutral language, logic-based persuasion, compromise, and a somewhat lower media profile. The type of holistic information produced by this model of TA is a unique and important element in informed decision making about complex technosocial systems. As suggested earlier, *part of TA's long run success will depend on maintaining the distinction between the two models in both the minds of those doing and using TA studies.*

This section began with the suggestion that mutual accommodation and learning would be required on the part of both the political and scientific/technical communities. However, the forms of accommodation and learning emphasized here have tended to be accommodations by TA to the demands and style of the political sector. This should serve to reinforce the point that while TA has some power of its own, and may develop more, it is still a new and relatively weak entrant in the political arena. As such it must struggle with the dilemma of attempting to walk the fine line between an appropriate and useful degree of accommodation and self-defeating cooptation by special interests. Not until the TA movement has reached a more accomplished and mature stage, both intellectually and politically, can we expect to see a pattern of mutual accommodation and learning which is truly mutual.

Conclusions

Technology assessment, as a source of expert knowledge, must come to grips with how it shall use and control the special form of power that knowledge provides. Those involved in technology assessment must be consciously aware of their role, the values they hold, and how these may influence their outputs and their relationships with relevant others. TA, as a movement, is a worldwide phenomenon which will have to accommodate itself to many different political frameworks. In this country, if it is to be supportive of our democratic ethos, then it shall have to learn to use citizen participation and stakeholder inputs constructively and without compromising itself. It must aid in creating an informed citizenry as well as informed technical and political elites. It must learn to speak to diverse audiences, to use the press effectively, and to create vivid nontechnical imagery of alternative futures.

Mark R. Berg

Technology assessments will be carried out in a dynamic political and bureaucratic environment which acts and reacts in response to a constellation of events and forces, some of which will be the existence of TA studies. To be useful in this environment technology assessment will have to be policy oriented without needlessly reducing the flexibility and room for maneuver needed by policy decision makers. Doing this will require interaction with, and learning to cope with, politicians and bureaucrats who will often have very different time frames and motives than those of technology assessors. Within this political context TA must be prepared to have its results, techniques, motives, and personnel impugned by "political" opponents who carry the brunt of "bad news" generated by particular TA studies. Those involved in preparing assessments will increasingly have to be prepared to defend and justify their analyses both in the courts and in public debate.

Technology assessment's legitimacy comes from "being right" while its ultimate effectiveness comes from "being useful" and "winning." Those doing TA's must come to grips with the often uncomfortable fact that just "being right" may not be enough. Winning may require that assessors learn to use their "power of expertise" as a political force capable of mobilizing its own constituency as it institutionalizes and broadens its base of support. *In the long run, a meaningful and effective TA capability will be dependent on the successful integration and balancing of these dual requirements of the political and scientific/technical worlds.*

Notes

1. Ida R. Hoos, *Systems Analysis in Public Policy* (Berkeley: Univ. of California Press, 1972). To get a feel for the controversial nature of the issue see, also, Stephen M. Pollock, review of *Systems Analysis* in *Public Policy in Science*, 178, no. 4062 (November 17, 1972), 739-740.
2. For further examples see I. L. Horowitz, ed., *The Use and Abuse of Social Science*, Transaction Books, 1971; and R. F. Miles, ed., *Systems Concepts*, John Wiley & Sons, 1973.
3. Nathan Caplan, "Social Science Knowledge Utilization by Federal Executives," Center for Research on Utilization of Scientific Knowledge, University of Michigan, 1974.
4. Joseph F. Coates, "Technology Assessment: The Benefits... the Costs... the Consequences," *The Futurist*, 5, no. 6 (December 1971), 225-231.
5. Anthony Wedgwood Benn, "Technical Power and People," *Bulletin of the Atomic Scientists*, December 1971, pp. 23-26.
6. Guy Benveniste, *The Politics of Expertise* (The Glendessary Press, 1972).
7. Graham T. Allison, *Essence of Decision* (Little, Brown & Co., 1971).
8. Thomas S. Kuhn, *The Structure of Scientific Revolutions*, 2d ed. *(Univ. of Chicago Press, 1970).*

The Politics of Technology Assessment

9. Harold Lasswell, *Politics: Who Gets What, When, How* (Whittlesey, 1936). For insightful perspective on the inevitable linkage between politics and TA, see R. G. Kasper, J. M. Logsdon and E. R. Mottor, *Implementing Technology Assessments* (Program of Policy Studies in Science and Technology, George Washington Univ., 1974). See also, Hugh Folk, "The Role of Technology Assessment in Public Policy, in A. Teich, ed., *Technology and Man's Future* (St. Martins Press, 1972). pp. 246-254.

10. Benn, op. cit.

11. Robert L. Heilbroner, *The Making of Economic Society* (Prentice-Hall, 1962).

12. Benveniste, op. cit., pp. 30-34.

13. Donald N. Michael, *On Learning to Plan and Planning to Learn* (Jossey-Bass, 1973), pp. 118-119.

14. Donald A. Schon, *Beyond the Stable State* (Random House, 1971), pp. 123-128.

15. Benveniste, op. cit., pp. 119-135.

16. Eugene S. Schwartz, *Overskill* (Ballantine, 1971), pp. 59-73. (The author is not a "technological fix advocate," but elaborates on the impact of "residue problems.")

17. Donald. N. Michael, "What Are We Searching For," Address to the Science, Technology and Future Societies Seminar, Univ. of Michigan, Jan. 21, 1974.

18. Anthony Downs, *Inside Bureaucracy* (The Rand Corp., 1966).

19. Aaron Wildavsky, "The Politics of the Budgetary Process (Little, Brown & Co., 1964).

20. Francis E. Rourke, *Bureaucracy, Politics, and Public Policy* (Little, Brown & Co., 1969).

21. Philip Boffey, *The Brain Bank of America* (McGraw-Hill, 1975).

22. Folk, op. cit.

23. Benveniste, op cit., pp. 39-60.

24. Steven Ebbin, "Jamaica Bay/Kennedy Airport - Anatomy of Technological Assessment," *Technology Assessment*, 1, no. 1 (1972), 23-40.

25. K. Chen, K. E. Lagler, M. R. Berg, et al., *Growth Policy* (University of Michigan Press, 1974), pp. 193-207; and also, Donald N. Michael, *On Learning to Plan and Planning to Learn*, op. cit., especially Chaps. 11-16.

26. Harold P. Green, "The Adversary Process in Technology Assessment," in Raphael G. Kasper, ed., *Technology Assessment* (Praeger, 1972), pp. 49-69.

27. Michael B. Teitz, "Toward a Responsive Planning Methodology," in David R. Godschalk, ed., *Planning in America: Learning from Turbulence* (American Institute of Planners, 1974), pp. 86-110.

28. This potential political constraint was pointed out by Professor Paul Ray, Univ. of Michigan, in a most helpful critique of an earlier draft of this paper.

29. These speculations are largely corroborated by a survey of 184 participants in the World Futures Society's First General Assembly. The survey showed these futurists to be young (average in the thirties), male (86%) college graduates (92%), liberal (75% ranked themselves as left of center), and politically active (60% had taken active roles in political campaigns). Mark Wynn, "Who are the Futurists," *The Futurist*, 6, no. 2 (April 1972), 73-77.

30. Boffey, op cit., pp. 54-56.

31. Horst W. J. Rittel, and Melvin M. Weber, "Dilemmas in a General Theory of Planning," Working Paper No. 194, Univ. of California, Berkeley, Nov. 1972.
32. Public interest TA group forms, *TA Update*, 1, no. 1 (May 1974).
33. Benveniste, op. cit., pp. 3–21.
34. Vincent Ostrom, *The Intellectual Crisis in American Public Administration* (Univ. of Alabama Press, 1973).
35. These and other important characteristics of the policy process have been described in theory and practice by many political scientists during the past decades. For example, see Charles E. Lindblom, *The Policy Making Process* (Prentice-Hall; 1968); D. Braybrooke and C. E. Lindblom, *A Strategy of Decision*, (The Free Press, 1963); G. T. Allison, op. cit.; and A. Wildavsky, op. cit. For an insightful analysis of the implications of these characteristics for systematic analysis see C. Schultz, *The Politics and Economics of Public Spending* (Brookings, 1968).
36. K. Chen, K. F. Lagler, M. R. Berg, et al., op. cit., pp. 165–207.
37. M. R. Berg, K. Chen, and G. J. Zissis, "A Value-Oriented Policy Generation Methodology for Technology Assessment, Project Decision," University of Michigan, May 1975: See also C. Schultz, op. cit.
38. Donald N. Michael, *On Learning to Plan and Planning to Learn*, op. cit.
39. Boffey, op. cit.
40. Coates, op. cit.
41. William A. Gamson, *Power and Discontent* (The Dorsey Press, 1968).
42. For an insightful analysis of this question see R. G. Kasper, J. M. Logsdon and E. G. Mottur, op. cit., Chap. 15.
43. Gamson, op. cit.
44. Geoffrey Vickers, *Making Institutions Work* (Halsted Press, 1973), Chap. 9.

25. Science and the Formation of Policy in a Democracy

DUNCAN MACRAE, JR.

Today, science and democracy symbolise alternative modes of guiding society. Until recently, they were widely believed to support one another; science enlightened the public and provided it with new alternatives, while democracy allowed science to flourish. But in recent years, the values of science and democracy have increasingly

Reprinted with permission from *Minerva*, vol. XI, no. 2, April 1973, pp. 228–242.

Science and the Formation of Policy in a Democracy

appeared to conflict, especially in their application to public choices. In so far as public choices depend on expert information, science requires that this information be judged by experts rather than the electorate. Democracy, however, requires that the electorate have the ultimate power. Those who value democracy, or fear its erosion, sometimes see scientists as an elite serving special interests, or see applied science as simply unplanned and uncontrolled.

Science versus Democracy: Science in Policy

The conflict between expertness and popular rule is an ancient one, but in the case of science it assumes a distinctive form. Unlike other types of expertness which are constituted by wide experience and reflection or wisdom, science is specialised, and requires prolonged training which is not generally accessible. Its expertness is most fully realised in narrow fields where specialists can check one another. The specialists in each field constitute relatively closed groups, the boundaries of whose fields of knowledge do not always correspond to the requirements of practical problems with which policy must deal. The specialists in one area of policy are seldom the same persons as the specialists in another, and may not easily recognise one another's expertness. Because of the specialisation of science, it is very difficult for the generally informed citizen to hold his own against the expert.[1] In addition, the scientist's expertness in policy matters is expected to relate to means rather than ends.

Science and democracy may be in conflict in three areas of policy. The first area is "policy for science"; this chiefly affects scientists themselves, in their research, teaching, and professional practice, but does not generate major conflicts in society at large. Although other groups—taxpayers, students, patients—may contest the claims of scientists to determine what research should be done, these contests affect relatively limited interests in society. The second area and a politically more important one is of "science in policy";[2] there expert advice concerns policies which affect many persons other than scientists. In the third area, the standards and data of science may be drawn upon for the evaluation of the democratic process itself.

The values expressed by scientists in discussing "policy for science" illustrate a more general problem which is also related to the other two areas. Scientists have often justified support for their research on the ground that science fulfils one of a plurality of values, or standards of judgement, each of which governs a separate sphere of activity; science is thus seen as one among many cultural pursuits. A similar pluralism governs the criteria of excellence of the separate disciplines. Yet when policy choices require the reconciliation of these diverse standards of judgement, the pluralism provides no basis for doing so. The synthesis of these standards in more general criteria of decision, or ethical theories, may be necessary if such policy choices are to be guided intelligently.

The politically most significant area in which science and democracy can conflict is that of "science in policy"—i.e., in public policy affecting major aspects of society other than science. Democratic governments make use of science in many types of policies. They concern themselves with it both as direct sponsors—in nuclear armament, transportation, building, health services, welfare programmes—and as regulators of the private sector—controlling foods, drugs, the environment, and the effects of economic changes. And in so far as expert knowledge of political processes exists, governments may themselves be reformed by its guidance.

These applications of science are sometimes distinguished from science itself when the latter is restricted to basic research. But basic scientific research has come to be associated in the public mind with its applications. For this reason I shall define science broadly here, so as to include "science in policy."

Two aspects of science in policy have led to concern on the part of those who wish to preserve and improve democracy: first, the *complexity* of the problems which require political decisions, increased as it is by science and its applications; second, the *inequality* of knowledge between scientists and other citizens. In the face of these problems, we wish to reconcile the claims of science and democracy.

Among the proposals that have been made to reconcile these two values, some especially stress the problem of complexity, and seek to simplify the problems themselves. Thus it is conceivable that if technological developments could be slowed, ordinary citizens might easily learn to grasp the political problems which confront them. The unintended consequences of military technology and the garrison state,[3] and the development of devices for the civilian market with "externalities" not taken account of by the market, have drawn criticism on this score. If we could imagine a slowing of development of weapons technology, perhaps policies for both international arms control and civilian control of the military could be subject to broader popular discussion. Similarly, the external costs imposed by the use of automobiles, including traffic congestion and atmospheric pollution, might be better understood by the public if the increase in use of automobiles were less rapid.

Science and the Formation of Policy in a Democracy

Alternatively, systems of decision-making might be broken down into separate units or their internal decision-processes altered, in order to make the problems of government less complex and therefore more easily intelligible to the layman. Federalism and decentralisation to regions or urban neighbourhoods have been proposed as means to reduce the size of decision-making units. Technical functions such as health services have been included in these proposals. Under some conditions, however, the government of smaller units is less democratic or less just than that of larger ones.[4]

Changes in internal decision-making processes, which provide more direct feedback systems, have been proposed for local government and government organisations. These have included participatory democracy and co-determination—as in German industry. Such changes are open to criticism, however, from several points of view. Dahrendorf has criticised co-determination on the ground that it undermines party conflict at the national level.[5] But another problem of local feedback systems in a complex society derives from the possibility of "political externalities"—analogous to the economic effects that escape the calculations of the market. These occur in the foreign policies of democracies, but also for decision-making units smaller than the nation, such as organisations, communities, or states. The creation of separate government units in well-to-do suburbs has intensified urban problems for city residents.

Other proposals for reform stress the problem of inequality of knowledge. A conventional remedy for this problem is to raise the effective competence of major decision-makers by the use of technically trained staff. Thus scientifically trained persons are employed in advisory or staff roles in organisations. In addition, consultants or semi-expert advisers may be employed temporarily; but when academic scientists are consulted on policy, their advice is not normally subjected to the same scrutiny as are their contributions to scientific or scholarly journals. The employment of experts in either staff or consultant roles is normally associated with the distinction between means and ends. Experts are expected to be competent in choosing means; but the ends of organisations are to be chosen—in principle—by political representation or the market.

Science and the Informed Citizen

In this "engineering model," the employment of the applied scientist in an organisation in a role similar to that of an engineer provides for rapid communication of new technical knowledge to centres of decision, in so far as he masters new developments in his subject-matter and the organisation makes use of them. But in spite of its effectiveness in the use of scientific knowledge, this places decision-making in the hands of the leaders of the employing organisations. These organisations may in turn be regulated or controlled by the political system or the market; but this channel of transmission of scientific information does not serve particularly to enlighten the consumer, the citizen, or even the political elites (such as legislators or leaders of voluntary associations) who are concerned with the general policies of their government. In a society which considers itself democratic, and which wishes to control the possible biases of experts toward special interests or values, general scientific education must extend more widely.

Some of the defects of public discussion which is insufficiently informed on scientific matters are illustrated by the battery additive case in the United States. A product known as AD-X2 had been marketed in the late 1940s with the claim that it would improve the performance of automobile batteries. The National Bureau of Standards, on the basis of chemical tests and previous experience with battery additives, concluded that the product was worthless and published this conclusion in 1950-51. The manufacturer, Mr. Jess M. Ritchie, was threatened with removal of mailing privileges, but fought back publicly with support from congressional committees and the press. In 1953, the new Republican secretary of commerce sided with Mr. Ritchie on the grounds that the product had proved its worth in the market, and nearly succeeded in dismissing the director of the Bureau of Standards, the decision being reversed only after strong pressure from the scientific community.[6]

Perhaps a competent scientific adviser to the secretary would have induced him to resolve the matter on scientific rather than commercial grounds; but the slowness of Congress, journalists and the informed non-scientific public to perceive the secretary's erroneous basis of judgement illustrates the lack of scientific information in the possession of those who make public decisions. This is not to say that informed lay opinion will always support scientists, or that it should; a scientific community might show consensus and intense feeling and yet be in error, especially on a matter of policy which transcends its realm of expertness. But the possibility of resolving science-related issues on scientific rather than political grounds deserves wider recognition.

One possible solution to this problem lies in greater stress on sci-

Science and the Formation of Policy in a Democracy

ence in general education. This has been advocated as partly fulfilling the same function as classical education in an earlier epoch, and providing part of a general intellectual training for informed citizens and prospective leaders. But if this scientific education is given only to the young, it will become largely obsolete by the time they attain influence in society. Education in science must therefore be supplemented by the continuous transmission of information and the scientific education of adults.

One might imagine that elected officials could be given some scientific training. But they have little time for this training after election; and before election the group destined for such office cannot be readily identified. If the politicians are drawn from a generally educated public, without a selective bias toward special interests, then perhaps they could bring a moderate degree of scientific information to public office. As a result of this type of education, for example, the pool of scientific knowledge available to a legislative body in constituting its science-related committees would be increased; and the members of these committees could then appoint members of their staff and evaluate the testimony of witnesses with greater discrimination.

Recommendations in matters of policy typically require a broader range of knowledge than particular basic scientific disciplines afford. Interdisciplinary training, general education and particular combinations of specialities in applied fields are all necessary. Scientists who are to work on environmental problems must possess some knowledge of chemistry, biology, engineering, medicine, economics, and politics; and those conconcerned with urban housing should be able to combine the economics and technology of housing with urban politics and sociology.

The "well-informed citizen" must have a central role in the democratic control of technical decisions. Alfred Schutz distinguished the expert, with clear and distinct knowledge of a limited field; the man in the street, who knows recipes which are adequate for practical purposes; and the well-informed citizen, "who considers himself perfectly qualified to decide who *is* a competent expert and even to make up his mind after having listened to opposing expert opinions." The expert lacks this last competence because he is "at home only in a system of imposed relevances—imposed, that is, by the problems pre-established in this field."[7] This was Schutz's way of indicating the limitations of the closed communication systems of specialised sciences, which have relatively little contact with each other.

Duncan MacRae, Jr.

An argument parallel to Schutz's was made by Edward Shils. He noted that:

> One of the major features of American intellectual life in the present century is the attrition of the "educated public." The emptiness and the specialization of education are its major causes.... The critical and solicitous care for the whole in the proximate future must be the charge which American intellectuals take on themselves, if American society is not to become the victim of the parochial preoccupations of specialized technological experts. If academic social scientists and free-lance publicists or amateur social scientists are not critics on behalf of the whole society over the reasonably foreseeable future, no one else will be.

He saw a weak tradition of this sort of thought, but considered that "what it lacks above all are organs through which its rhetoric can be formed by expression and its judgment sharpened by dialogue."[8] I agree; but I believe that this dialogue should be incorporated into the system of professional scientific societies, scientific careers, and academic departments which lead their members to pursue certain topics more than others, and which protect the autonomy of their discourse against inexpert and irrelevant claims. In this way reasoned criticism among experts, a valuable aspect of science, could also contribute to the general education of citizens.

Representative Science

In trying to cope with the conflicts between science and democracy, we considered first the possibility of simplifying the problems themselves: then the application of the "engineering model" of policy advice by supplying expert advisers to decision-making organisations; and third, education of informed citizens, drawing as much as possible on the diverse resources of the university.

But there is still another way to cope with this conflict. The risk of technocracy lies in the possibility of uncontrolled power held by an elite and devoted to special values or interests rather than to the general welfare. We have sought the education of the citizen in order to control technology; but we can also consider reducing the biases of technologists as a group, through their selection and training. If those who apply science to policy bring to that task values and judgements beyond their strictly scientific competence, they might possibly improve the resulting policies, but might also make them worse.

In the application of science to policy, scientists might recommend policies which derive not only from their expertise, but also from values they bring with them from their membership in a particular

Science and the Formation of Policy in a Democracy

society, from their affiliation with particular groups in that society, or from attitudes which are traditional in their scientific discipline. Their national allegiances might blind them to the welfare of citizens of other countries. Their membership in higher social strata might render them insufficiently sympathetic to the problems of the poor in their own society. Their membership of a particular discipline may lead them to stress ultimate values which coincide with the central concepts of that discipline.[9] And certain general features of the norms and attitudes of science may also introduce bias into the formation of policy: an optimistic rationalism, a disregard for consultation of those affected by policy, a stress on specialised rather than more general views. All these might on occasion be undesirable.

For reasons such as these, inequalities in knowledge can conflict with the general welfare, especially if experts bring to policy decisions extrinsic values or interests which differ from the dictates of an ethic in which "the general welfare" is defined. And even though we cannot expect complete consensus on such an ethic, some of the special values mentioned above might well conflict with *any* such consistent general ethic. Conceivably the difficulties resulting from these extrinsic valuations could be reduced through the education of the scientist. Prospective scientists might be trained like Plato's guardians, with special attention to their character. They might be required to work periodically with other social strata as a form of continuing education, as in contemporary China. Or they might be trained to a strict neutrality in matters of valuation, so as to fulfil that ideal of applied science which entails dealing only with means and not with ends.

If these extrinsic values cannot be eliminated, an alternative might be found in "representative science."[10] This would require that each discipline should contain an adequate representation of various groups interested in the policies to which its knowledge is relevant. Thus a panel of scientific advisors would include distinguished members who are, for example, Negroes, women, or Republicans.

Such a proposal for representation of social groups, however, is difficult to formulate in general terms. The very bases of representation are matters of political controversy, and the groups granted representation at one time might later exclude other groups. Nor can we be sure how a person's origin will affect the values he carries with him into positions of influence; some will remain attached to their groups of origin, others reject them. And if we discard the notion of group representation and substitute that of representative political attachments or opinions, we are implicitly rejecting the possibility that some orientations might be more soundly or rationally based than others. It is important to retain a stress on reason and evidence, as contrasted with purely political or particularistic criteria of decision, in order to preserve the values which derive from general norms of objectivity and truthfulness with regard to matters of fact. This emphasis is important if rational ethical discourse is to be encouraged; the possibility of clear and consistent derivation of policy recommendations from precisely formulated ethical axioms or "hypotheses" might also be jeopardised by political criteria of decision about the scientific matter involved in a problem of policy. Although political judgements are central to the processes of decision regarding policy, there are serious risks if they are substituted for scientific or ethical judgements.

This stress on the values which scientists bring to the formation of policy, even if it should be feasible, implicitly favours expertise over democracy; it simply applies a concept analogous to "expertness" to the possession of proper values of character. Like the "engineering model" of policy advice, it accepts differences in status or power and tries to render their consequences less harmful or their processes of decision less ignorant. If we took this approach, we should in effect be redefining democracy as an aristocracy of the educated or a democracy of experts.

Redefining Democracy

Examination of the scientist's role in democratic policy formation thus leads us to ask more precisely what democracy entails. We are led to examine democratic institutions as means to the attainment of other ends beyond democracy itself—ends to which both science and democracy may be means, and in terms of which the conflicting claims of science and democracy may be reconciled.

The problem of redefining democracy is raised by the proposition that there are inescapable limitations of competence among some segments of the public which prevent their participating with equal competence in policy formation. The weakest part of this proposition, of course, is that some existing disparities might be removable rather than inescapable. Participation might have educative effects, and "freezing" a disparity of participation might prevent the levelling of given disparities.[11] Secondly, we must recall that the question of competence in "elite" and "mass" is not restricted to scientific knowledge but extends to other sorts of knowledge as well.

Schumpeter[12] argued that democracy consists in the competitors for positions of leadership presenting themselves to the popular vote; the voters could not be expected to understand all the complexities of policy or control it in detail. And several survey studies of voters' knowledge and attitudes have led to doubt about whether the voters' competence is sufficient to meet the standard of rationality which might have prevailed in the time of John Stuart Mill.[13]

Given these characteristics of the ordinary citizen, an escape from the inequality of knowledge is, so some writers claim, provided by pluralism. An essential ingredient of the pluralist solution is a set of elites within a variety of inter-connected associations or groups. These elites are expected to refine, reconcile, and channel those groups' political demands.[14] This proposal is closely akin to the "engineering model," except that it works through private associations rather than government.

A related modification of the notion of popular rule is the widespread assertion that democracy resides more in the carrying out of certain procedures than in any ascertainable power relations. Thus, popular elections with a universal franchise are seen as a criterion of democracy, as long as the public's choices are not overtly constrained and certain freedoms are preserved. But if the public's choices are in fact constrained, even in its own interest, by nominating procedures or by persuasion of public and private leaders, how are we to judge this? It is easiest simply to call the resulting system "democracy" and thus avoid the problem. However, the harder choices involve discrimination among various forms of government in which the public has more and less independent influence, and in which it needs to exercise a careful judgment as to which are preferable.

The Values of Democracy and Reasoned Normative Discourse

Our efforts to resolve the conflicts between science and democracy in questions involving "science in policy" have led us repeatedly to problems of ethics or of normative discourse. A discussion of "science and policy" is necessarily drawn toward a critical discussion of potentially conflicting evaluative standards. In considering policies which affect society at large, however, we cannot simply take the goals of science as a source of definition of social goals. The application of science to social policy must be seen as a means to the better or more effective accomplishment of ends external to its own activity. "Science for science's sake" is irrelevant in such a setting.

When we considered the role of the informed citizen, we touched on social criticism and criteria of relevance which might involve reasoned normative discourse. The extrinsic valuations which scientists make when they give advice on matters of policy might also be made more disciplined and explicit if they were shaped in reasoned discourse which was part of the scientist's training.

Thus science may also conflict with democracy in a third area. Both the findings of science—natural or social—and its procedures of discussion may reveal or illuminate conflicts between values. The findings of science may indicate conditions under which the practice of democracy is difficult, and may themselves undermine confidence in the workability of democratic institutions. As regards biological differences among individuals, the possibility that scientific research might lead to the discovery of systematic differences in innate capacity for political participation has been resisted. Research alleging to discover racial differences in ability has been sharply criticised, with respect to both the rigour of its methods and the motives of its authors.[15] In the analysis of electoral behaviour, there have been findings which seemed to challenge the view of a rational and participant electorate and stimulated debate as to whether an ideal model of democracy is realisable.

Not only the findings, but also the procedures of discussion of science may have bearing on the value-systems of democracy. There is a latent conflict between the values of clarity, consistency, and generality implicit in scientific discourse, and our actual practices in evaluative discussion, including the justification of science and democracy themselves. Much discussion of these questions falls short of the standards of clarity and consistency which scientists would require of their own theories and hypotheses.

Thus a still more significant problem concerns the possible subjection of the ideals of democracy themselves to penetrating logical analysis, such as science prescribes for its own theories. Much of our justification of democratic politics is patterned unthinkingly after economic justifications of market mechanisms. But in market behaviour itself, we are increasingly aware that the gratification of consumers' economic "demands" has led in advanced societies to the phenomenon of a "harried leisure class,"[16] as well as of a debt-ridden middle and working class. We see in this phenomenon not only a chaos of demands and "needs," but also a creation and manipulation of demands by means of advertising as well as by the rivalry of a competitive and mobile society. We ask then, "what *should* the consumers demand?" This question leads to a re-examination of the desirable directions for education and public communication.

The same questions can be asked about the economic notion of satisfaction of existing preferences through democratic politics. As television propaganda increasingly dominates election compaigns,

Science and the Formation of Policy in a Democracy

mobilising some remote segment of our attitudes in support of or opposition to a candidate (or his "image") we wonder whether popular rule indeed involves an aspect of ourselves which is worthy to rule. We ask, then, how the shaping and learning of citizens' attitudes and information *should* take place. We cannot, of course, advocate the "freedom" which would correspond to uncaused or random attitudes; but knowing that the public's political information and attitudes can be shaped to some degree by education, mass communications, and campaigns, we must ask through what processes they *ought* to be shaped. In asking this question, we become aware of the intrinsic conservatism of a procedure of choice which selects these processes "democratically"; by slighting the role of education and leadership, it holds public choice close to an existing state of opinion within an existing group. Democratic processes of choice render it especially difficult to create a social system with a new and perhaps superior system of values and preferences. They do not prevent change; the conservatism of democracy may produce not a static society, but change in an unknown and unanticipated direction, like unintended oscillation in a feedback system. The effects of this conservatism can still be felt even if the "democracy" of market and political system is produced by the actions of an elite which manipulates images to appeal to existing popular tastes or prejudices. In either case, they reduce the likelihood that a democratic society will reform its own basic values.

We are thus led to inquire about the rationale of democracy. Is it to be justified as a system which optimally combines individual values, as in welfare economics? Or is it a system providing for maximal consultation of citizens, so that policies will be supported and conflict minimised? Or could it conceivably correct the values of its citizens and lead them to prefer other policies, other goods, and other styles of life than those they had thought they wanted?

These questions lead us also to ask which of an individual's existing preferences *ought* to be combined in a system of social aggregation and which ought not. They lead, then, to a politically and morally oriented critique of welfare economics, making use of its technical sophistication but transcending the professional bias of economists that preferences are to be taken as given.

The conservative bias of democracy is also reflected in another fundamental set of evaluative choices: those concerning the criteria of *membership,* or the possible alteration in the composition of deciding groups. Who shall be "the people," whose rule is democracy? This question encompasses the alteration of the geographical boundaries of the political community and its subdivisions: policies for immigration and banishment; policies for the granting and deprivation of citizenship to residents; policies governing birth and (in extreme cases) death; and policies for the alteration of the genetic characteristics of the citizenry.

The criteria of democracy—of rule by the people—sharply limit the decisions which can be taken on such matters. A selfish version of democracy, justified on the principles of the market, seems least fit for them. It will tend to perpetuate the characteristics of the existing majority, whether desirable or undesirable. Officeholders may seek security by increasing the homogeneity of their constituencies. Migration, if governed in this way by politics, may increase local homogeneity and neglect the interests of outsiders. Perhaps there is also a more moral and enlightened type of democracy which can be expected to consider the interests of outsiders, but its theory will have to be elaborated on other grounds.

The greatest problem which science poses for democracy, therefore, is that of the critical, logical examination of the cluster of values which define democracy. What do they entail? Where do they conflict? To what other ends may they be considered means? The very discussion of these questions risks being undemocratic in at least one sense, if it is conducted by persons whose values are not representative of those of the general public. Yet if we are to examine whether a political system functions properly, or whether changes are needed, we must first be clear about what proper functioning *is*.

But this is also a problem for science. In the marriage of rationalism and empiricism which gave birth to modern science, the empirical component is not the only important one for questions of public choice. Ethical and evaluative questions can be discussed rationally, and philosphers are doing so. The rational conduct of evaluative discussion—dealing with problems which affect and involve science as deeply as those we have considered—is different from scientific analysis but the two activities should not be isolated from each other. To separate evaluative questions from science is perhaps to strengthen pure science, but also to weaken applied science by making it totally dependent for its guidance on unreflective standards and modes of evaluation. The values which guide the application of science are thus deprived of the rational component which is so essential to science's own internal functioning.

Scientists—or academic and professional groups which include scientists—still need to incorporate centrally into their own evaluative discourse the rationality which they value in the conduct of science itself. They may attempt to do so by an appeal to mankind at large,[17] but if they have the support of the university and of academic disciplines in the effort, they may find the task less difficult. If rational evaluative discourse is more closely tied to education in science, the foundations of the advice which experts give on problems of policy may be sounder, and they may more easily connect means with ends. But if it is also tied to general education, informed citizens may be better able to assess the advice of experts.

Conducting evaluative discourse in accordance with the pattern of scientific discourse which pays particular attention to clarity, consistency, and generality, we may narrow the range of possible ethical systems.[18] A consensus which is reached rationally may lead to more consistent and practicable political decisions; and this same limitation of admissible ethical foundations may also constrain the diversity of inconsistent individual ethical systems, such as we apply in the multiple roles of everyday life.

"Policy Analysis": A New Discipline?

The types of discourse and education we have advocated, if they are to be carried on, must be incorporated into a social structure which provides suitable rewards in a career within a defined sphere of expertness. Interdisciplinary programmes typically lack these rewards, but one conceivable remedy would be to provide them through a new discipline better adapted to the application of science to policy. It is conceivable that normative discourse, interdisciplinary communication, expert criticism of policy, and the undergraduate education of the informed citizen may all be combined with these rewards in the developing discipline of "policy analysis."

Graduate schools for training in "policy analysis," overlapping existing disciplines but defining a new domain of expertness, may provide a basis for the necessary careers and their corresponding rewards. A number of such schools have been organised in the United States in the past few years; some British universities have done the same. Their curricula share many common elements, including decision-criteria, such as cost-benefit analysis; statistical decision-theory; social indicators, evaluation-research and the measurement of desired outcomes; construction of models (e.g., economic models) linking policy parameters to outcomes; organisational politics; and case studies and internship to connect general principles to concrete situations. Joint degree programmes between "policy analysis" and professional training, including law, medicine, and engineering, are also being developed in certain American universities.

The members of the teaching staff of these schools may come to plan careers in the new discipline of "policy analysis," with its own professional societies, journals, occupational market and standards of competence. In addition, their journals might provide opportunities for reasoned advice and criticism related to public policies—advice which was itself public and subject to the review of expert colleagues. Universities might deliberately take this kind of activity into account in making appointments just as they take achievements in research into account. Such a system of communication within a defined sphere of competence might provide expert criticism of individual contributions, internal rewards within the university and a national market for competent persons.

It is important, moreover, to open these courses of study to undergraduates. Extension of the new discipline to undergraduate teaching would also help to broaden its scope beyond the engineering model and give a central place in it to reasoned normative analysis.

Persons trained in graduate schools of "policy analysis" would expect to go largely into non-academic employment and to conform somewhat to the "engineering model," supplying technical advice and skill to large organisations or for otherwise given ends. In this respect, graduate training in the analysis and assessment of policy would resemble training in modern engineering, medical or law schools, or the training of students for conventional professional roles in the medieval universities.[19] But undergraduate training would be designed to educate informed citizens, and not simply those expecting employment in large organisations or in professional practice. The informed citizen, himself concerned with the ends of policy, must arrive at his own system of values by rational and scientific analysis rather than by accepting it from an employer or profession. A central feature of training in "policy analysis" must therefore be reasoned, normative discourse. The adviser who works for a large organisation may have to suppress his own system of values; but having formulated his ethics clearly, he may have chosen better among employers in the first place. Thus for both undergraduates and graduate students in "policy analysis," the conventional aspect of the "engineering model" would be modified to include reasoned discussion of the ends of policy and critical evaluation of the goals of large organisations.

In the formation of informed citizens or the educated public, there is a particular opportunity for the social sciences. Their very difficulty in detaching themselves from the vocabulary and the perspective of

Science and the Formation of Policy in a Democracy

the citizen may also allow them to communicate more easily with him than can the natural scientist. The institutional arrangements within which this discourse takes place can accommodate the central and professional discourse of the social sciences as well as of "policy analysis." For the basic natural sciences, however, normative discourse may come no closer to their professional domain of expertness than undergraduate training. The central discourse of natural science is so self-contained and so fully divested of extrinsic evaluations that the introduction of normative discourse into the natural science disciplines seems unthinkable. Social science, however, contains such evaluations in its central discourse and has not achieved the elegant paradigms of natural science; parts of its discourse might therefore be explicitly normative. This normative discourse, if pursued actively, might contribute to public discussion and policy choice. Max Weber recognised the possibility of such critical discourse;[20] but he did not encourage its development through the necessary definition of disciplines and the proposal of a type of professional career devoted to its practice.

Other parts of the university, and experience outside the university, may also contribute to the development of informed citizens. Mid-career training of civil servants in "reverse sabbaticals" can educate both the civil servants and the students with whom they associate. Journalists who have made use of specialised knowledge of science in their work—including social science—may be intermediaries in this process, serving to educate the public themselves. For all these sorts of activities, a unified notion of training in "policy analysis" may be useful.

We have considered the motivation of university teachers to teach "policy analysis" and to engage in the reasoned public criticism of policies. The motivation of the informed citizens generally, however, has yet to be assured. The dissemination of technical information to citizens has often been a goal of undergraduate education; but we cannot be sure whether citizens thus trained will be motivated to inform themselves and participate in the making of public decisions. Persons who are paid to do so will do so. Scientists who are aroused by social problems related to their expert fields will spend some time on these problems—but often in a spirit of "doing good" together with colleagues, rather than in the spirit of rigorous criticism which would guide their professional work. A few experts may work on the staffs of voluntary organisations, whether public health specialists working for labour unions or economists for Ralph Nader.

One particular benefit of "policy analysis" might be a result of the infusion of expertise into the legislative function of "oversight of administration." In the United States, the criticism of executive actions is a function of Congress; but the reviews performed by its committees, or by the General Accounting Office, have been centred largely on whether the terms of the law were faithfully carried out. These terms might well be extended, however, to the expert study of whether the *purposes* of legislation were attained. Increasing attention is being devoted to evaluative studies by segments of the federal bureaucracy; but independent studies by research workers attached to Congress might provide a more independent criticism, less closely tied to the interests of the departments and agencies carrying out the policies. Either committee staffs themselves, the Office of Technology Assessment, or a social accounting office analysing the contribution of legislation and administration to changes in social indicators, might fulfil this function. Congress itself might thus increase the technical component of its reviews of the performance of administrative bodies, by asking whether organisations charged with producing health or education, or alleviating poverty, really do so. Such questions may be more important than whether every penny of the budget has been spent according to the provisions of the law. And, in this context social science deserves to be included in any assessment of "science and policy." Its potential role is already illustrated by economics.

The development of these capacities on the part of the legislature requires not only extensive technical staffs for political parties; it requires an increased appreciation of technical problems on the part of legislators themselves. The general competence required to supervise technical staffs must therefore be more widespread among potential political leaders through their training in university and in their professions, and must be rewarded by recognition and advancement. If public issues come to be debated more in terms of the measurable consequences of policies, there may be an increase in the demand for technical competence in assessment of these consequences. Journalists, leaders of voluntary associations, public spokesmen for policy-making organisations and committees appointed by professional associations to study problems of public policy, may all be informed for this debate by technical training in the analysis of policy provided by the universities.

These proposals for a new discipline and new legislative functions, even if open to criticism, should illustrate a more general problem which deserves attention. The systematic discussion of goals and the effort to design social institutions which will give effect to such discussion, are a major task involving social science and philosophy in the

Science and the Formation of Policy in a Democracy

guidance of "science in policy." At the heart of this problem is the design and evaluation of educational and governmental institutions. The value of expertness which science embodies, and its relations with the values of democracy, impel us to the analysis of this problem.

Notes

1. A. Schutz, "The Well-Informed Citizen: An Essay on the Social Distribution of Knowledge," *Social Research*, 13, no. 4 (December 1946), 463–472.
2. H. Brooks, "The Scientific Adviser," in Robert Gilpin, and Christopher Wright, eds., *Scientists and National Policy-Making* (New York: Columbia University Press, 1964), p. 76.
3. H. D. Lasswell, "The Garrison State and Specialist on Violence," *American Journal of Sociology*, 46, no. 4 (January 1941), 455–468.
4. Grant McConnell, *Private Power and American Democracy* (New York: Knopf, 1966).
5. Ralf Dahrendorf, *Society and Democracy in Germany* (Garden City, N.Y.: Doubleday, 1967), Chapter 2; M. Olson, "The Principle of Fiscal Equivalence," *American Economic Review*, 49, no. 2 (May 1969), 479–487.
6. See Samuel A. Lawrence, "The Battery Additive Controversy," *The Inter-University Case Program* (University of Alabama Press), no. 68, 1962.
7. Schutz, op. cit., pp. 466, 474.
8. Edward Shils, "The Intellectuals and the Future," *Bulletin of the Atomic Scientists*, 23, no. 8 (October, 1967), 13–14 (reprinted in *"The Intellectuals and the Powers and other Essays"* [Chicago: University of Chicago Press, 1972], pp. 225, 227).
9. See D. MacRae, Jr., "A Dilemma of Sociology: Science Versus Policy," *American Sociologist*, 6, Supplementary issue (June 1971), pp. 2–7; "Normative Assumptions in the Study of Public Choice," *Public Choice*, 16 (Fall, 1973), 27–41.
10. By analogy with J. Donald Kingsley, *Representative Bureaucracy* (Yellow Springs, Ohio: Antioch, 1944). See also Eliot Friedson, *Profession of Medicine* (New York: Dodd, Mead, 1970), pp. 372–375.
11. Peter Bachrach, *The Theory of Democratic Elitism* (Boston: Little, Brown, 1967), p. 45.
12. Joseph A. Schumpeter, *Capitalism, Socialism, and Democracy* (New York: Harper & Brothers, 1950), Chap. 22.
13. John Stuart Mill, *Utilitarianism, Liberty, and Representative Government* (London: Dent, 1910), pp. 218–227; Bernard R. Berelson, Paul F. Lazarsfeld, and William N. McPhee, *Voting* (Chicago: University of Chicago Press, 1954), Chap. 14; H. Daudt, *Floating Voters and the Floating Vote* (Leiden: Stenfert Kroese, 1961); P. E. Converse, "The Nature of Belief Systems in Mass Publics," in David E. Apter, ed., *Ideology and Discontent* (New York: Free Press, 1964), Chap. 6; V. O. Key, Jr., *The Responsible Electorate* (Cambridge, Mass.: Belknap Press, 1966).
14. William Kornhauser, *The Politics of Mass Society* (Glencoe, Ill.: The Free Press, 1959), Chaps. 1–4.
15. See D. MacRae, Jr., letter to the editor, *Science*, 143, no. 3604 (1964), 307–308.

Duncan MacRae, Jr.

16. Staffan B. Linder, *The Harried Leisure Class* (New York: Columbia University Press, 1970).

17. Gerald Feinberg, *The Prometheus Project* (Garden City, N.Y.: Doubleday, 1969).

18. D. MacRae, Jr., "Scientific Communication, Ethical Argument, and Public Policy" *American Political Science Review,* 66, no. 1 (March, 1971), 38–50.

19. Joseph Ben-David, *The Scientist's Role in Society* (Englewood Cliffs, N.J.: Prentice-Hall, 1971), pp. 46–50.

20. Max Weber, *The Methodology of the Social Sciences* (Glencoe, Ill.: The Free Press, 1949), p. 54.

Annotated Bibliography

Society

Bell, Daniel. "Notes on the Post-Industrial Society." *The Public Interest*, 6 (Winter 1966), 24–35. The notion of a qualitative change in our industrial technology with major implications for society.

Bevan, William. "Science in the Penultimate Age." *American Scientist*, 65 (1977), 538–546. Scientific and technological contributions are indispensable to deal with increasingly global-scale problems; scientists and technologists need also contribute to new social institutions able to cope with a new world.

Branscomb, Lewis. "Taming Technology." *Science*, March 12, 1971, pp. 972–977. Emphasizes the complex and long-range implications of technological developments that must be considered in its management.

Braudel, Fernand. The Mediterranean and the Mediterranean World in the Age of Philip Second, trans. Sian Reynolds, 2 vols., rev. 2d ed. New York: Harper & Row, 1976. Illustrates the different time frames by which to gauge change.

Commoner, Barry. "The Closing Circle," in Commoner, *The Closing Circle: Nature, Man and Technology*. New York: Knopf, 1971, pp. 293–300. Technology's effects on the environment call for political action to address basic human issues.

Ferkiss, Victor. "Technology and the End of Liberalism," in Ferkiss, *The Future of Technological Civilization*. New York: Braziller, 1974, pp. 52–60. Liberalism (premised on individualism, competition, and unlimited growth) is an unworkable philosophy for a technological society.

Gendron, Bernard. *Technology and the Human Condition*. New York: St. Martin's, 1977.

Gordon, Theodore J. "The Feedback between Technology and Values," in K. Baier and N. Rescher, eds., *Values and the Future*. New York: The Free Press, 1969, pp. 148–158. Values change and technology is a factor.

Gross, Bertram. "Technology in an Era of Social Revolution." *Public Administration Review*, (May/June 1971), 272–297. The most profound implications for our political institutions follow from the technological changes that qualitatively restructure our lives.

Hardin, Garrett. "The Tragedy of the Commons." *Science*, Dec. 13, 1968, pp. 1243–1248. People tend to overuse and destroy commonly owned re-

Annotated Bibliography

sources because there are no incentives for self-regulation and conservation.

Keniston, Kenneth. "The Human Toll of Technological Society," in Keniston, *The Uncommitted*. New York: Delta Books, 1960, pp. 420–424. Finds the price of technological progress too high in human costs; asks us to set our goals that technology may serve man rather than undermine our values.

Lawless, Edward. *Technology and Social Shock*. New Brunswick, N.J.: Rutgers University Press, 1977. Examination of 45 cases (plus 55 synopses) of technological development that misfired in some socially important way, with due consideration of the policy formulation processes involved.

Lear, John. "Where Is Society Going? The Search for Landmarks." *Saturday Review*, April 15, 1972, pp. 34–39. Introduction to the notion of social indicators and background for OMB's role.

Marcuse, Herbert. "Technological Rationality and the Logic of Domination," in Marcuse, *One-Dimensional Man*. Boston: Beacon Press, 1964. One of the prominent philosophers concerned with technology considers its impacts on mankind.

Toffler, Alvin. "The 800th Lifetime," in Toffler, *Future Shock*. New York: Random House, 1970, pp. 9–18 (Bantam edition); also "The Strategy of Social Futurism," pp. 446–487. Technology and accelerated change; the need for social, institutional "anticipation."

Weinberg, Alvin M. "Social Institutions and Nuclear Energy." *Science*, July 7, 1972, pp. 27–34. Discusses alternative institutional arrangements needed to manage and safeguard nuclear waste including intergenerational issues.

White, Lynn Jr. "The Historical Roots of Our Ecologic Crisis." *Science*, March 10, 1967, pp. 1203–1207. Western civilization's bent toward technology related to our cultural/religious heritage.

Winner, Langdon. *Autonomous Technology*. Cambridge, Mass.: M.I.T. Press, 1977. Call for social guidance of technological change.

The Polity

Benn, Anthony Wedgwood. "Technical Power and People: The Impact of Technology on the Structure of Government," *Bulletin of the Atomic Scientists*, 27 (December 1971), 23–26. Science and technology are sources of power; as such, their control is of critical importance to a democracy.

Chase, Edward T. "Politics and Technology." *Yale Review*, 52 (1963), 321–339. Technology induces political stresses of far-reaching significance.

Coates, Joseph. "Technological Change and Future Growth: Issues and Opportunities." *Technological Forecasting and Social Change*, 11 (1977), 49–74. Foresees major technological changes over the course of the next three decades requiring several roles of government: guiding interplay of basic market forces (e.g., risk management), meeting information needs to inform public policy choices, reform of regulatory structures, and new R & D directions.

La Porte, Todd. "The Context of Technology Assessment: A Changing Perspective for Public Organization." *Public Administration Review*, 31 (1971), 63–73. Technological solutions to social problems may threaten

Annotated Bibliography

primary social values and institutional independence (by tightening interdependences among government, industry, and academia).

Lowi, Theodore. *Poliscide*. New York: Macmillan, 1976. Community impacts of the quest for an R & D facility.

MacRae, Duncan Jr. "Technical Communities and Political Choice." *Minerva*, 14 (Summer 1976), 169-190. Proposes development of "technical communities" as providers of expertise on applications of science, in place of scientific communities; addresses issues concerning the use of expertise in the policy process.

Nelkin, Dorothy. "Scientific Knowledge, Public Policy, and Democracy." *Knowledge*, 1 (September 1979), 106-122. Scientific knowledge, a commodity, profoundly affects the distribution of political power in democratic societies.

Nelson, Richard. "Intellectualizing about the Moon-Ghetto Metaphor: A Study of the Current Malaise of Rational Analysis of Social Problems." *Policy Sciences*, 5 (1974), 375-414. Contrasts the potential for technological solutions between the defense/space and social sectors.

Press, Frank. "An Agenda for Technology and Policy." *Technology Review*, 80 (January 1978), 51-55. Exploration of some of the scientific-technological issues confronting us, and call for enlightened public involvement in assessing the complex socio-technological relationships.

Price, Don. "The Diffusion of Sovereignty," in Price, *The Scientific Estate*. New York: Oxford University Press, 1965, pp. 57-81. Technical information exerts considerable influence on congressional committees and an effectively decentralized executive branch.

Sayre, Wallace S. and Bruce L. R. Smith. *Government, Technology and Social Problems* (monograph). Institute for the Study of Science in Human Affairs, Columbia University, 1969. Governmental structure depicted as a chief obstacle to the use of technology to solve social problems.

Schooler, Dean Jr. "The Future of Science and Policy Making," in Schooler, *Science, Scientists, and Public Policy*. New York: Free Press, 1971.

The Economy

Binswanger, Hans, and Vernon Ruttan. *Induced Innovation: Technology, Institutions and Development*. Baltimore: Johns Hopkins Univ. Press, 1978. Examines the role of economic motivation in prompting innovation.

Brooks, Harvey. "Models for Science Planning." *Public Administration Review*, May/June 1971, pp. 364-374. Contrasts different models of the R & D system with intriguing implications.

Carter, C. F. "Government and Technology." *Nature*, May 15, 1965, pp. 652-654. Staunch statement in favor of governmental promotion of technological development in the U.K.

Chandler, Alfred Jr. *The Visible Hand: The Managerial Revolution in American Business*. Cambridge, Mass.: Belknap Press, 1977. Technology has led to emergence of large multinational corporations that don't follow the invisible hand, requiring governmental action.

David, Edward Jr. "Science Futures: The Industrial Connection." *Science*,

Annotated Bibliography

March 2, 1979, pp. 837–840. Urges closer linkage of academic and industrial communities in interdisciplinary R & D.

Eads, George, and Richard Nelson. "Governmental Support of Advanced Civilian Technology: Power Reactors and Supersonic Transport." *Public Policy,* 19 (1971): 405–427. Case studies of two technological developments.

Galbraith, John Kenneth. "The Imperatives of Technology," in Galbraith, *The New Industrial State.* Boston: Houghton Mifflin, 1971, pp. 11–20. Most consequences of technology derive from its unremitting pressure toward specialization of skills, organizations, and products.

Heilbroner, Robert. "Economic Problems of a Postindustrial Society." *Dissent,* 20 (1973), 163–176. The technology-induced transition to "postindustrial" society will disrupt the environment and the capitalist system, but offer a world in which we can make our own future.

Holloman, J. Herbert. "Government and the Innovation Process," *Technology Review,* 81 (May 1979), 30–41. Considers policy actions for the federal government to boost innovation; contrasts strategies followed in different industrialized countries.

Holloman, J. Herbert. "The U.S. Patent System." *Scientific American,* 216 (June 1967), 19–27. Patents play an important role in the innovation process; revision in the law is needed.

Johnson, Leland J. "Government Regulation and Technological Advance," in *Rand: 25th Year.* Santa Monica, Calif.: Rand Corporation, 1973, pp. 125–136. Regulatory strategies merit careful consideration to avoid inhibition of technological advance.

Kelly, Patrick, Melvin Kranzberg, Federick Rossini, Norman Baker, Frederick Tarpley, Jr., and Morris Mitzner. *Technological Innovation: A Critical Review of Current Knowledge.* San Francisco, Calif.: The San Francisco Press, 1978. Denotes the complexities in the process and institutionalization of technological innovation.

Lawson, C. W. "The Hidden Hand and the Control of Technology." *Technology and Society,* 7 (1972), 137–139. How effective is the market in developing socially desirable technology? (Comments on High Gibbons, "A Note on the Control of Technology," *Technology and Society,* 7 (1972), 64–66.

Mansfield, Edwin. "Contribution of R & D to Economic Growth in the United States." *Science,* February 4, 1972, pp. 477–486. R & D appears to contribute to productivity gains, and we may be underinvesting in certain civilian R & D sectors.

National Science Foundation, Office of National R & D Assessment. *Technological Innovation and Federal Government Policy.* Washington, D.C.: National Science Foundation, publ. no. NSF76-9, January 1976. An overview of the issues.

Research Management, 20 (January 1977). Special issue on national science and technology policy. Barriers to innovation can be countered by new federal policies.

Schumacher, E. F. "Buddhist Economics," in Schumacher, *Small Is Beautiful: Economics as if People Mattered.* New York: Harper and Row, 1975, pp. 53–62. Western economics leads to disregard for the environment and wanton use of resources; an alternate economics aims to maximize human well-being with the minimum, not the maximum, of consumption.

Utterback, James. "Innovation in Industry and the Diffusion of Technology."

Science, February 15, 1974, pp. 620-626. Consideration of innovation as a process, and the relationship thereto of science and federal actions.

Wiesner, Jerome. "Has the U.S. Lost Its Initiative in Technological Innovation?" *Technology Review,* 78 (July/August 1976), 54-60. Present policies jeopardize our ability to find and use needed new technology.

World Affairs

Brodie, Bernard. "The Impact of Technological Change on the International System: Reflections on Prediction." *Journal of International Affairs,* 25, no. 2 (1972), pp. 209-223.

Brzezinski, Zbigniew. *Between Two Ages: America's Role in the Technetronic Era.* New York: Viking, 1970. Portrays America as the font of technology and technology as a vital force in moving mankind toward large-scale cooperation.

Foster, William C. "Technological Peace." *Impact of Science on Society,* 22 (July/September 1972), 235-242. Arms control rests on technological characteristics.

Gustavson, M. R. "Evolving Strategic Arms and the Technologist." *Science,* December 5, 1975, pp. 955-958. New technology essentially dictates national defense policy.

Nelson, Richard. "World Leadership, the 'Technological Gap' and National Science Policy." *Minerva,* 9 (1971), 386-399. Counters Boretsky's concerns by asserting that the technological gap between American and European technology is more than 100 years old; our policy should not be based on maintaining comprehensive technological leadership.

North, Robert C., and Nazli Choucri. "Population, Technology, and Resources in the Future International System." *Journal of International Affairs,* 25, no. 2 (1971), 224-237.

Pavitt, Keith. "Technology, International Competition, and Economic Growth: Some Lessons and Perspectives," *World Politics,* 25 (1973), 183-204. Small and medium-sized industrial powers deal with modern technology differently than the two superpowers, and, quite possibly, more effectively.

Rathjens, George. "The Dynamics of the Arms Race." *Scientific American,* 220 (1969), 15-25. A Pandora's box of technological possibilities confounds negotiations and decisions concerning missiles and nuclear warheads.

Schultze, Charles. "Balancing Military and Civilian Programs," in Schultze, *National Priorities: Military, Economic, and Social.* Washington, D.C.: Public Affairs Press, 1969, pp. 30-53. Tradeoffs highlight the R & D process.

Seaborg, Glenn. "Science, Technology, and Development: A New World Outlook," *Science,* July 6, 1973, pp. 13-19. An upbeat statement of the need for science and technology on an international front.

Skolnikoff, Eugene B. "The Changing Ingredients of Foreign Policy," in Skolnikoff, *Science, Technology and American Foreign Policy.* Cambridge, Mass: M.I.T. Press, 1967, pp. 3-19; also, "New Imperatives," pp. 301-316. Foreign policy and technology entwined.

Wade, Nicholas. "Third World: Science and Technology Contribute Feebly to

Annotated Bibliography

Development." *Science*, September 5, 1975, pp. 770-776. Explores roles of developed vis-à-vis undeveloped countries in regard to technology.

Yarmolinsky, Adam. "Military Sponsorship of Science and Research," in Yarmolinsky, *The Military Establishment*. New York: Harper and Row, 1971, pp. 283-301.

The Federal Executive

Benveniste, Guy. "The Prince and the Pundit," in Benveniste, *The Politics of Expertise*. Berkeley, Calif.: Glendessary Press, 1972, pp. 3-21. Experts and planners have a difficult, but vital, role to play.

Burger, Edward Jr. "Effective Institutions for Presidential Science Advice." *Technological Forecasting and Social Change*, 9 (1976), 337-347. Provision of scientific information to inform policy-makers is difficult to sustain against political influence; thoughtful exploration of past advisory mechanisms and ongoing needs.

Cronin, Thomas E., and Norman C. Thomas. "Federal Advisory Processes: Advice and Discontent." *Science*, February 26, 1971, p. 771.

Katz, James E. *Presidential Politics and Science Policy*. New York: Praeger, 1978. Traces federal science and technology policy from World War II onward; interprets presidential power as a decisive force on it.

Killian, James, George Kistiakowsky, Jerome Wiesner, Donald Hornig, Lee DuBridge, and Edward David, Jr. "Science Advice for the White House." *Technology Review*, 76 (January 1974), 8-19. Personal insights of the first six presidential science advisers on their role and the influence of politics on the relationship between the president and his science advisers.

Kohlmeier, Louis Jr. "The Regulators and the Regulated," in Kohlmeier, *The Regulators*. New York: Harper and Row, 1969, pp. 69-82. A glimpse into the realities of interaction among lobbyists, federal regulatory agencies, and the Congress.

Lyons, Gene. "The President and His Experts, *Annals of the American Academy of Political and Social Sciences*, 394 (1971), 36-45. Loyalty to an expert's profession may compete with loyalty to the president.

McElroy, William. "The Global Age: Roles of Basic and Applied Research." *Science*, April 15, 1977, pp. 267-270. Notes role for lead institutions and long-term support.

Peters, B. G. "Insiders and Outsiders—Politics of Pressure Group Influence on Bureaucracy." *Administration and Society*, 9 (1977), 191-218. Useful commentary on the institutional relationships, often addressing technologies.

Primack, Joel, and Frank Von Hippel. "The Uses and Limitations of Science Advisors," in Primack and Von Hippel, *Advice and Dissent: Scientists in the Political Arena*. New York: Basic Books, 1974, pp. 38-48. Weaknesses and strengths of the advisory system illustrated with respect to uncovering the implications of DDT.

Roback, Herbert. "Do We Need a Department of Science and Technology?" *Science*, July 4, 1969, pp. 36-43. Consideration of the possibility provides a framework for discussing important features of the executive's functioning with respect to technical matters.

Seidman, Harold. "Concluding Observations," in Seidman, *Politics, Position, and Power: The Dynamics of Federal Organization.* New York: Oxford University Press, 1970, pp. 271-286. Poses enduring questions concerning the merits of executive branch reorganizations from the inside perspective of the Budget Bureau (now OMB).

Skolnikoff, Eugene, and Harvey Brooks. "Science Advice in the White House? Continuation of a Debate." *Science,* January 10, 1975, pp. 35-44. Consideration of the basic premises for and roles of scientists and engineers providing advice to the president.

U.S. Congress, Senate Committee on Commerce, Science, and Transportation, "A Legislative History of the National Science and Technology Policy, Organization, and Priorities Act of 1976," Washington, D.C.: U.S. Government Printing Office, 1977. Details the deliberations and issues raised in reinstitution of a presidential science adviser.

Congress

Cahn, Ann, and Joel Primack. "Technological Foresight for Congress." *Technology Review,* 75 (March/April, 1973), 39-48. Perspective on the creation of OTA—legislative history and expectations for the new office.

Decker, Craig. "A Preliminary Assessment of the Congressional Office of Technology Assessment." *Journal of the International Society for Technology Assessment,* 1 (1975), 5-26. Sees inherent difficulty in the match between a disjointed incrementalist Congress and neatly rational assessment efforts.

Huddle, Frank. "Political Adaptation to a Technology Surfeited Society." *Denver Law Journal,* 47 (1970), 629-643. Congressional efforts to improve its ability to cope with technology that has a momentum of its own.

Mosher, Charles A. "Needs and Trends in Congressional Decision-Making." *Science,* October 13, 1972, p. 134.

Shick, Allen. "The Supply and Demand for Analysis on Capitol Hill." *Policy Analysis,* 2 (1976), 215-234. Relationship between analysts and their congressional clients.

U.S. Congress, House Commission on Information and Facilities. "The Office of Technology Assessment: A Study of Its Organizational Effectiveness," 1976; also, House Subcommittee on Science, Research, and Technology, "Review of the Office of Technology Assessment and Its Organic Act," 1978. Washington, D.C.: U.S. Government Printing Office. Congressional reviews of their creation—OTA.

U.S. Congress, Legislative Reference Service, *Technical Information for Congress.* Prepared for House Committee on Science and Astronautics. Washington, D.C.: U.S. Government Printing Office, 1969 (revised, 1971). Still useful collection of cases illustrating congressional decision processes concerning technology; introduction poses the questions of concern to Congress.

The Courts

Curlin, James. "Saving Us from Ourselves: The Interaction of Law and Science-Technology, and Comments by Ernest Jones, Joseph Coates, and

Annotated Bibliography

Philip Bereano." *Denver Law Journal,* 47 (1970), 651-673. Exploration of the role of the law in controlling the social costs of technology.

Green Harold P. *The National Environmental Policy Act in the Courts.* Washington, D.C.: The Conservation Foundation, May 1972.

Green Harold P. "The Role of Law and Lawyers in Technology Assessment." *Technology Assessment,* 1, no. 1 (1972), 47-52.

Johnson, Ralph. "The Role of the Courts as Environmental Protectors." *Washington Public Policy Notes.* Seattle: University of Washington, Institute of Governmental Research, 1973. Notes the liberalization of standing, restriction of government's sovereign immunity, and judicial review of substantive (as well as procedural) concerns as enhancing judicial control over executive branch actions.

Katz, Milton. "Decision-making in the Production of Power," *Scientific American,* (September 1971), 191-200. Discusses technology assessment, internalization of "external" costs, and the use of taxation, regulation, and private civil actions at law as ways to balance intended and side effects of technological developments.

Tribe, Laurence. *Channeling Technology Through Law.* Chicago: Bracton Press, 1973. In-depth examination of the roles of the law vis-à-vis technology.

Tribe, Laurence. "Legal Framework for the Assessment and Control of Technology," *Minerva,* 9 (1971), 243-255. Contrasts three models of legal control of technology—issuing specific directives, modifying market incentives, and changing decision-making structures.

State and Local Government

Bickner, Robert. "Science at the Service of Government: California Tries to Exploit an Unnatural Resource." *Policy Sciences,* 3 (1972), 183-199. A case discussion of the use of four different sources of scientific and systems expertise.

Elazar, Daniel J. "The Resurgence of Federalism." *State Government,* 43, no. 3 (Summer, 1970), 166.

Hayes, Frederick, and John Rasmussen. *Centers for Innovation in the Cities and States.* San Francisco: San Francisco Press, 1972.

Maher, Theodore. "Power to the States: Mobilizing Public Technology." *State Government,* 45 (Spring 1972), 124-134. Technology at the state level needs to be user-focused to yield visible benefits; various arrangements can assist state executive branches and legislatures.

Maher, Theodore J. "Public Technology for State Government." *State Government,* Summer, 1971, p. 142.

Meyerson, Martin. "Environmental Quality Assessment in State Government," *State Government,* Summer, 1972, p. 192.

Sapolsky, Harvey. "Science Advice for State and Local Government." *Science,* April 19, 1968, pp. 280-284. Explores this still uncertain issue.

Citizen Participation and Preferences

Arnstein, Sherry. "A Working Model for Public Participation." *Public Administration Review,* 35 (1975), 70-73. Describes the experience of formal citizen participation in a technology assessment.

Annotated Bibliography

Arnstein, Sherry. "A Ladder of Citizen Participation." *Journal of the American Institute of Planners,* 35 (July 1969), 216-224. An eight-rung ladder characterizes citizen participation efforts on the critical dimension of citizen power—ranging from manipulated nonparticipation to citizen control.

Benveniste, Guy. "The Prince and the Pundit," in *The Politics of Expertise.* Berkeley, Calif.: Glendessary Press, 1972. Experts and planners have a difficult, but vital role to play.

Clark, Terry. *Community Power and Policy Outputs.* Beverly Hills, Calif.: Sage, 1973. Good review and bibliography.

Gofman, John. "Nuclear Power and Ecocide: An Adversary View of New Technology." *Bulletin of the Atomic Scientists,* September 1971, pp. 28-32. An appealing case presenting the need for challenge to institutional support for a technology.

Kennard, Byron. "Some Methods and Criteria for Public Participation in Technology Assessment," *Journal of the International Society for Technology Assessment,* 1 (September 1975), 43-46. An advocate of participatory processes considers the technology assessment process.

LaPorte, Todd and Daniel Metlay. "Technology Observed: Attitudes of a Wary Public." *Science,* April 11, 1975, pp. 121-127. Survey shows public support for science, but deep concerns about technology.

Mazur, Allan. "Opposition to Technological Innovation." *Minerva,* 13 (Spring 1975), pp. 58-81. Comparison between public opposition to fluoridation and nuclear power leads to generalizations about the generation of local oppositional activity.

Michael, Donald. "Democratic Participation and Technological Planning," in Alan Westin ed., *Information Technology in a Democracy.* Cambridge, Mass.: Harvard University Press, 1971, pp. 291-300. Technology poses dilemmas for effective, participatory planning. (This volume notes challenges posed by communication technology for political processes.)

White, Gilbert. "Formation and Role of Public Attitudes," in Henry Jarrett, ed. *Environmental Quality in a Growing Economy.* Baltimore: Johns Hopkins Press, for Resources for the Future, 1966, pp. 105-127. Public attitudes regarding technological developments and their likely effects are important, difficult to formulate, and hard to measure.

Winner, Langdon. "On Criticizing Technology." *Public Policy,* 20 (Winter 1972), 36-59. Pleads for radical, participatory questioning of technological developments; such actions as technology assessment are not bold enough.

Public Choice

Amara, Roy C. "Toward a Framework for National Goals and Policy Research." *Policy Sciences,* 3 (1972) 59-69.

Baram, Michael. "Social Control of Science and Technology." *Science,* May 7, 1971, pp. 535-539. Framework relates various forms of policy determination to the stages of technological innovation on which they bear; for example, court actions and citizen efforts usually enter too late in the process in many respects.

Brooks, Harvey, and Raymond Bowers. "The Assessment of Technology." *Scientific American,* 222 (February 1970), 3-10. Classic statement on the purpose of technology assessment and ways to institutionalize it, based on a pivotal National Academy of Sciences report.

Annotated Bibliography

Campbell, Donald. "Reforms as Experiments." *American Psychologist*, 24 (1969), 409-429. Proposes carefully designed experiments to assess the effects of innovations in a societal context.

Casper, Barry. "Technology Policy and Democracy," *Science*, October 1, 1976, pp. 29-35. Questions the notion of trying to dissociate the technical and political facets of an issue in a science court; suggests instead that an adversarial format be used to place all the issues before the public and the policy makers; forget the scientific judges.

Cellarius, Richard, and John Platt. "Councils of Urgent Studies," *Science*, August 25, 1972, pp. 670-676. Call for organization of scientific efforts to focus on research on solutions to listed problem areas.

Ezra, Arthur. "Technology Utilization: Incentives and Solar Energy." *Science*, February 28, 1975, pp. 707-713. Application of solar energy to homes explored in terms of the technological delivery systems involved.

Green, Harold. "The Adversary Process in Technology Assessment." *Technology and Society*, 5 (1970), 163-167. Value choices inherent in technology assessment should not be decided by an elite body of professionals; suggests we need to institutionalize a "devil's advocate" function to bring out reasons against particular developments.

Holloman, J. Herbert. "Technology in the United States: The Options Before Us." *Technology Review*, July/August 1972, pp. 32-42.

Kantrowitz, Arthur. "Controlling Technology Democratically." *American Scientist*, 63 (September-October 1975), 505-509. Proposal for a science court.

Porter, Alan, Frederick Rossini, Stanley Carpenter, and A. T. Roper. "Critiques of TA/EIA," in Porter, Rossini, Carpenter, and Roper, *A Guidebook for Technology Assessment and Impact Analysis*. New York: North-Holland, 1980, pp. 455-470. Pro's and con's of technology assessment as a strategy for public management of technological development.

Rescher, Nicholas. *Scientific Progress*. Pittsburgh: University of Pittsburgh Press, 1978. Argues for a deceleration of science.

Tribe, Laurence. "Technology Assessment and the Fourth Discontinuity: The Limits of Instrumental Rationality." *Southern California Law Review*, 46 (1973), 617-660. Harmonious relations between modern societies and their technologies require that choices be seen as part of an integration process, not as means to given ends.

Wenk, Edward Jr., and Thomas Kuehn. "Interinstitutional Networks in Technological Delivery Systems," in J. Haberer, ed., *Science and Technology Policy*. Lexington, Mass.: Lexington Books, pp. 153-175. The technology delivery system provides a very workable framework for consideration of technology policy.

Index

Advanced Research Projects Agency, 209, 297
Advisory system. *See* Scientific advisory system
Alternate media movement, 442-444
American Association for Advancement of Science, 251-262, 259-260, 317
Apollo program, 36, 304, 309
Appropriate technology, 17, 44-45
Arnold, Gen. H. H., 195
Arrow, Kenneth, 62
Atomic Energy Commission, 111, 192, 196, 297

Bacon, Francis, 133
Bacon, Roger, 133
Baram, Michael S., 451-474
Bateson, Gregory, 439, 445
Baumol, William, 404
Bazelon, David L., 356-365
Berg, Mark R., 474-496
Biomedical research, 278-280
Blair, J. F., 406
Boretsky, Michael, 161-188
Bowles, Edward, 194
Branscomb, Lewis, 339
Brookings Institution, 276
Brooks, Harvey, 35-57, 77, 138, 330, 339
Brown, Harold, 209-211, 335, 338
Buckley, Oliver E., 199
Bundy, McGeorge, 208
Bush, Vannevar, 192, 198

Carey, William D., 276-277
Carroll, James D., 416-434
Carson, Rachel, 269, 426
Carter, Jimmy, 340
Casper, Barry M., 315-345
Center for Naval Analysis, 195, 264
Center for Science in the Public Interest, 441

Citizen participation:
 feedback mechanisms, 438-440
 litigation as means of, 420-423
 mechanism for, 25-26
 public understanding of science, 95, 423, 434-448
 role of, 15, 25, 95
 social movement of, 438-444
 use of media, 442-444
Civil liability, 348-350
Clean Air Act, 467
Climate modification, 228-231
Coates, Joseph, 337, 424
Commoner, Barry, 426, 453
Communist party, science doctrine of 124-125
Compton, Karl T., 191
Conant, James B., 192
Congress, 21, 110
 authorization process, 301-302, 308-311
 committee procedures, 311-312
 investigations and reviews of, 302-305
 politics and personalities, 305-308, 337-344
 relations with executive branch, 320, 331-337, 439
 research budgeting in, 297-314
 science advice and expertise, 311-314
 science fellowship program, 315-327
 science and technology policy, 266, 319
 See also Office of Technology Assessment; Technology Assessment Advisory Council; Legislative Reference Service
Consumer protection, 451-452
Consumer sovereignty, 41, 47
Cost-internalization, 350-353
Council of Economic Advisers, 65, 279
Council on Environmental Quality, 256, 462-463

525

Courts. *See* Judiciary
Culliton, Barbara, 274-295
Culture:
  intellectual trends, 65-68
  scientization of, 86-88

Daddario, Emilio Q., 328, 330, 334, 337, 339, 341, 423-425
David, Edward, 208
Decision-making, 105, 458-460
  command and control, 114-123
  consensus building in, 489
  public vs. scientists in, 357-360
  role of courts, 361
  under uncertainty, 363-365
  See also Policymaking; Technology assessment; Science and Technology Policy
De Jouvenel, B., 416
Democracy:
  redefinition of, 504-505
  versus science, 497-499
  values of, 505, 509
Departments of federal government. *See* U.S. government
Descartes, 133
Dewey, John, 11, 64
Domestic Policy Council, 256
Drell, Sidney, 341
DuBridge, Lee, 208, 268

Economic concepts:
  balance-of-trade, 168-173, 218-219
  capital investment, 173-174
  corporate power, 41-42
  effect of R & D, 174-180
  innovation and, 148-160
  labor productivity 39, 46, 221-223
  productivity and growth, 38, 163-168, 171
  technological products, 169-171
Eddington, 95
Ehrlich, Paul, 86, 453
Einstein, Albert, 11, 112
Eisenhower, Dwight D., 201-202, 209, 387
Ellul, Jacques, 87, 89, 263, 420
Energy crisis, 14, 46-47, 139-142
Environment:
  alteration of, 227-234
  crisis, 14, 18
  impacts of technology on, 43
  pollution and externalities, 49-50
  quality of, 453-454
Etzioni, A., 416
Expertise. *See* Technical and scientific expertise; Professional ethics

Federal Council for Science and Technology, 207
Federal executive branch, 20, 110, 267, 275, 302, 313, 340, 439
  science and technology in, 251-262
  See also U.S. government
Federal government departments. *See* U.S. government
Federalism, 368, 372, 399, 411, 499
Federal Water Pollution Control Act, 467
Fogarty, John, 306
Foster, John S., Jr., 210
Freedom, and principle of responsibility, 108-114
Free enterprise, 150
Futurism, 61, 77

Galbraith, John K., 153, 263
Gendron, Bernard, 11
Glenn, John, 338
Golden, William T., 199
Government:
  relations with industry, 149-152
  science and technology issues, 19-26
  technology of, 156-160
  See also Federal executive branch; Congress; Judiciary; State and local government; Citizen participation; Public choice; U.S. government
Great Society, 276, 298
Greb, G. Allen, 190-215
Green Revolution, 142-143, 240
Gross, Bertram, 438

Handler, Philip, 360
Henderson, Hazel, 434-448
Heraclitus, 65, 68, 70, 77
Humphrey, Hubert, 335

ICBM, 115, 199-201, 298
Industrial innovation:
  federal policies toward, 16, 152-155
  government-industry relations, 149-152
  See also Innovation
Industrialization, 60, 71, 86, 161, 220-225
Industrial Revolution, 60, 161
Innovation, 35, 64, 148-160
  cross-cultural transfers of, 54
  definitions of, 37, 67
  disincentives for public, 405
  effect of government policy on, 157
  goals of, 50-54
  by institutions, 400-405, 408
  organizational and financial, 71
  social and market, 35

Index

in state and local government, 397–414
technological evolution and, 40
Institute for Defense Analysis, 195–196, 264, 272
Institutions:
complexity of, 86, 191, 243, 419, 498
information and change in, 440–442
innovativeness of, 400–405, 408
international, 226–246
technological trends in, 69
International technology transfers, 180–186, 215–225

James, William, 58
Johnson, Lyndon B., 208, 276, 279, 330, 376, 427
Joint Research and Development Board, 192–193
Judiciary:
civil-liability trends, 348–350
cost-internalizing by, 350–353
role in risk regulation, 356–365
social value of litigation, 353–355
technical information for, 23
technological ethic in, 347–350

Kapp, K. W., 440
Keller, K. T., 200–201
Kemeny Commission, 36
Keniston, Kenneth, 416
Kennedy, Edward, 318–319, 323, 328–329, 334–336, 340
Kennedy, John F., 207–208, 276, 279
Khrushchev, N., 124–126
Killian, James R., Jr., 195, 202, 205, 207, 209, 379
Kistiakowsky, George, 204, 207, 209
Knowledge. *See* Social knowledge
Kuhn, Thomas, 475

Lambright, W. H., 398
Lane, Robert E., 76
Legislative Reference Service, 264, 313
See also Congress
Liability doctrine, 347–356
See also Risk assessment; Judiciary
Lipset, S. M., 76
Local government. *See* State and local government

MacRae, Duncan, Jr., 496–514
Magnuson, Warren, 291
Malone, T. F., 228
Mansfield, Mike, 291
Marcuse, Herbert, 419
Marxism, 58, 60, 67, 81, 127
Maslow, Abraham, 435, 446

Mason, Edwards, 216
Materialism, 60, 69, 81, 92
Materialization of values, 419
McCormack, Mike, 326
McNamara, Robert, 191, 210, 276, 281
Mesthene, Emmanuel, 11, 57–80, 416
Meynaud, Jean, 93
Military-industrial complex, 14–16
Military R & D, 190–214
Millikan, Clark, 199, 204
Mills, Wilbur D., 308
Mishan, E. J., 92, 439
Mosher, Charles, 329, 333
Muskie, Edmund, 423
Myrdal, Gunnar, 215–225

Nader, Ralph, 86, 269, 426, 441, 445, 451, 511
National Academy of Engineering, 259, 264, 424
National Academy of Sciences, 259, 261–264, 313, 360, 408, 424–425, 430, 452, 492
National Environmental Policy Act, 23, 447, 458, 462–470, 481
National Institute of Health, 278–280, 289, 292–294, 300, 306–308
National Science Board, 259
National Science Foundation, 251, 259, 261, 277–278, 284, 297, 301, 313, 376, 379, 380, 398, 403, 463, 479, 481, 492
appropriations for, 301–308, 323
Natural Resources Defense Council, 362
Nichols, Kenneth D., 200
Nixon, Richard, 208, 268, 275, 279, 281, 288, 300, 398, 428
Nordhaus, William, 165

Oceans policy, 234–236
Office of Science and Technology, 279, 298–299, 304, 313
Office of Technology Assessment, 316, 320, 412, 447, 465–468, 479–481, 488, 492
history of, 327–345
rationale for, 329
See also Congress
Olsen, Mancur, 446
O'Neill, Paul H., 286, 289, 293–294
Operations analysis, 100, 104
Oppenheimer, J. Robert, 111
Organizations. *See* Institutions

Participation. *See* Citizen participation
Participatory technology, 416–431
definition of, 417

Patent policy, 156–157
Perl, Martin L., 262–274
Peterson, Russell W., 328, 345
Policy analysis, 509–513
  See also Technology policy analysis
Policy-making, 66, 74–76, 101–102, 251–255
  See also Decision-making; Technology assessment: Science and technology policy
Political choice, 418
Political consensus, 76–77
Political system, 94
Political trends, 74–78, 91–93
  in science and technology, 138–146
Politics:
  definition of polity, 74
  fundamental task of, 78, 106
  ideals of, 91
  motivations of, 134
  spectrum of science to, 122, 134
Pollution, 230–233
Population:
  birth control, 225, 241–242
  international implications, 240–242
  world growth of, 231
Post-industrial consumer, 436–437
Powell, John Wesley, 132
President's Science Advisory Committee, 191, 206–209, 263, 267, 330, 338
Press, Frank, 340
Price, Don K., 12, 95–131, 368
Productivity. *See* Economic concepts
Professional ethics, 112–113, 117, 123
Professions:
  four estates of, 105–108
  limits to competence, 100–104
  role in science policy, 97–100
Program planning and budgeting, 65, 284
Public choice:
  issues of, 26–30
  public opinion and, 154, 418
  reforms in process of, 462–470
  science policy and, 496–514
  spectrum of, 12, 95–131
  technology and, 11–12
  technology assessments in, 476–496
Public goods, 69, 93
Public interest groups, 426–427, 441–444
Public technology, 398
Public understanding of science, 95, 417, 423, 434–448, 499–502
  See also Citizen participation; Democracy

Ramo, Simon, 89–91, 204
Rand Corporation, 194–195, 264, 407

Randor, Michael, 406
Representative science, 502–504
Research and development:
  budget decisions for, 15, 255–271, 299–304
  definition of, 38
  economic effects of, 372–373
  federal support of, 50–52
  management of, 256–258
  state expenditures for, 369, 374–377
Research and Development Board, 193–194, 198–199, 202
Revelle, Roger, 132–146
Risk assessment, 356–365
Roback, Herbert, 297–315
Rockefeller, David, 461, 467
Roessner, J. David, 397–414
Roszak, Theodore, 80–93

Sapolsky, Harvey M., 367–397
Satellites, 236–239
Sawhill, John C., 281, 286
Schlesinger, James R., 281, 340–341
Schon, Donald, 148–160
Schultze, Charles L., 276, 279–281
Schutz, Alfred, 501
Science:
  control of, 96
  fear of, 110
  motivations of, 122, 134
  policy for, 497–499
  role in democracy, 496–513
  social responsibility of, 132
Science and technology policy:
  advocacy of, 258–260
  challenges in field, 29, 103
  dilemma of, 17
  national, 143–146, 260–261, 314
  politics of, 15, 95–97, 112, 496–513
  science in policy, 497–498, 513
  See also Office of Science and Technology; State and local government; Decision-making; Technology assessment
Science Policy Research Division, 264, 313
Scientific advisory system:
  evaluation of, 265–266
  history of, 251–274, 367–369
  limits of, 136–138, 357, 382–389
  role of, 102, 190, 369
  skeptical view of, 287–294
  in state government, 379–391
  See also President's Science Advisory Committee, Office of Science and Technology
Scientific community, 96, 271–272, 322–323

Index

Scientific establishment, 264-265, 269-271, 322-323
Skolnikoff, Eugene B., 226-246
Social adaptability, 445-447
Social alienation, 416
Social change, 60-78
Social choice, 63, 418.
  See also Public choice
Social feedback, 438-440
Social goods, 69, 93
Social indicators, 66
Social knowledge:
  abstractions, 100
  applications of, 133
  consequences of, 11, 67, 87
  definition of, 64, 75
  intellectual trends, 65-68
  power of, 128
  systematic, 97
Social trends, 68-74
Social management of technology:
  checks and balances in, 118-120
  goals of, 50-55, 138, 158-159
  shortfalls of, 13
Social sciences, political acceptance of, 135
Social trends, 68-74
Social values. *See* Values
Socio-technical systems, 35, 156-160
  common characteristics of, 138
  conceptual framework for, 455-462
  decision-making in, 458-460
  definition of, 149
Space policy, 236-239, 309
  See also Apollo program; Sputnik
Sputnik, 206, 208, 297-298, 372, 387
Standardization, 45
State and local government:
  innovation in, 397-411, 404-408
  public technology, 408-411
  science advice for, 379-391
  science policy in, 367-379
  weaknesses in science resources, 24-25
State science policy, 367-397
  functions of, 390
  limits to, 387-389
State Technical Services Act, 373, 379
Strategic planning, 254-255
Systems:
  analysis, 89
  approach, 89-90, 474-476
  feedback in, 438-440
  See also Socio-technical systems; Scientific advisory system

Teague, Olin, 337-338
Technical and scientific expertise:
  limits of, 100, 103, 111
  political elites, 499-502
  power of, 484-487
  technocratic elites, 83, 480-482
  See also Professional ethics
Technocracy, 91-93, 502
  See also Technical and scientific expertise
Technological alienation, 416
Technological anaesthesia, 420
Technological arms race, 190
Technological change:
  history of, 13, 50-51, 217-218
  institutional responses to, 14, 58, 61
  legal liability for, 347-356
  legitimization of, 420
  process of, 58-62, 68, 73, 77, 162
  evolution of, 40, 103
  trends of, 41-50, 64-78, 161-187, 218, 226
Technological choice, 58-59
Technological coalescence, 80, 489
Technological determinism, 37
Technological dislocations, 154-155
Technological evolution, 40
Technological fix, 15, 20, 478
Technological innovation. *See* Innovation
Technological liability, 354, 347-356
Technological responsibility, 355
Technological values, 62-64
Technology:
  current state of, 162-173
  definition of, 11-12, 35, 87
  ends and means of, 11, 19, 28, 118, 259
  of government, 156-160
  political-economic view of, 161-173
  social effects of, 37, 68, 114
  substitution, 46
  transfer, 54, 64, 180-186, 215-225
Technology and society:
  human needs, 50-55
  issues of, 13-19
  technological determinism, 37
  trends, 68-74, 161-188
  social effects, 57, 60, 68, 347
  See also Culture; Institutions
Technology assessment:
  conceptual framework for, 455-462
  constituency for, 488-489
  definitions of, 423, 475, 509-510
  merger with public policy, 474-476
  methodology for, 482-484
  politics of, 474-494
  risk assessment, 51, 356-365
  See also Decision-making; Technology policy analysis; Systems
Technology Assessment Advisory Council, 333, 338
Technology forecasting, 77, 137, 244

Index

Technology policy analysis:
  applications of, 28
  approaches to, 26
  definition of, 27, 509–510
  methods of, 20, 66, 74, 87, 100–104, 254–255, 369, 482–484
  need for, 25, 158, 184, 254, 261, 509–513
  See also Technology assessment
Technology transfer 54, 64, 180–186, 215–225
Technostructure, 263
Tribe, Laurence H., 347–356
Truman, Harry, 196, 199–200

Urban-industrialism, 80–82
U.S. government:
  Department of Defense, 264, 297–298, 309–311, 419
  Department of Energy, 20, 321–322
  Department of Health, Education, and Welfare, 306, 402
  Department of Transportation, 398, 411, 462–463
  Environmental Protection Agency, 458, 467
  Federal Communications Commission, 468
  Food and Drug Administration, 458, 461
  General Accounting Office, 303, 333, 338, 403, 512
  House Committee on Science and Astronautics, 313, 330, 397
  Law Enforcement Assistance Administration, 402, 410
  National Aeronautics and Space Administration, 206, 278, 308–311, 322, 383, 402–403
  National Bureau of Standards, 20
  National Oceanic and Atmospheric Administration, 463
  National Security Council, 256
  Nuclear Regulatory Commission, 358–359, 362–363
  Occupational Health and Safety Administration, 358, 452
  Office of Management and Budget, 255–257, 261, 274–295, 299
  Office of Science and Technology Policy, 279, 298–299, 304, 313, 398
  Senate Energy and Natural Resources Committee, 321
  See also Congress; Federal executive branch
USSR, 115, 124–125, 191, 196

Value analysis, 74
Value articulation, 489–490
Value clusters, 63
Values, 58, 62–64, 73–75, 81, 97, 419
  of democracy, 505–509
  information and change of, 440–442
  materialization of, 419
Vogt, Evon, 68, 78
Von Braun, Wernher, 201
Von Karman, Theodore, 195, 196, 204

Walsh, John, 274–295
Weather modification, 228–230
Weinberger, Caspar, 281, 285
Welfare economics, 507
Wenk, Edward Jr., 335, 339
White, Lynn, 59
Wiesner, Jerome, 204, 208, 210, 330, 335, 341, 343
World Food and Agricultural Organization, 240
World Population. *See* Population

Yin, Robert, 405, 410
York, Herbert F., 190–215, 209

Ziman, John M., 36, 87–88

Science, Technology, and National Policy

Designed by R. E. Rosenbaum.
Composed by The Composing Room of Michigan, Inc.
in 10 point Baskerville, 2 points leaded
with display lines in Optima Bold.
Printed and bound by Vail-Ballou Press, Inc.

Library of Congress Cataloging in Publication Data
Main entry under title:

Science, technology, and national policy.

Bibliography: p.
Includes index.
1. Technology and state—United States—Addresses, essays, lectures. 2. Science and state—United States—Addresses, essays, lectures. I. Kuehn, Thomas J., 1948-   II. Porter, Alan L.
T21.S35   338.4'76'0973   80-66900
ISBN 0-8014-1343-5
ISBN 0-8014-9876-7 (pbk.)